Ecology of Aquatic Management

Chris Frid

and

Mike Dobson

Prentice
Hall

An imprint of **Pearson Education**

London · New York · Toronto · Sydney · Tokyo · Singapore · Hong Kong · Cape Town
New Delhi · Madrid · Paris · Amsterdam · Munich · Milan · Stockholm

Pearson Education Limited
Edinburgh Gate
Harlow
Essex CM20 2JE

and Associated Companies throughout the world

Visit us on the World Wide Web at:
www.pearsoneduc.com

ISBN 0 130 86610 5

British Library Cataloguing-in-Publication Data
A catalogue record for this book is available from the British Library

Library of Congress Cataloging-in-Publication Data
Frid, Chris.
    Ecology of aquatic management / Chris Frid, Mike Dobson.
      p. cm.
    Includes bibliographical references (p.).
    ISBN 0-13-086610-5 (pbk.)
    1. Fishery resources. 2. Aquatic resources. 3. Ecosystem management. I. Dobson,
Mike. II. Title.

SH327.5 .F74 2002
333.95−dc21   2001058091

10 9 8 7 6 5 4 3 2 1
06 05 04 03 02

Typeset in 9.5/13pt Concorde by 35
Produced by Pearson Education Asia Pte Ltd.,
Printed in Singapore

Ecology of Aquatic Management

We work with leading authors to develop the
strongest educational materials in biology,
bringing cutting-edge thinking and best learning
practice to a global market.

Under a range of well-known imprints, including
Prentice Hall, we craft high-quality print and
electronic publications which help readers to understand
and apply their content, whether studying or at work.

To find out more about the complete range of our
publishing, please visit us on the World Wide Web at:
www.pearsoneduc.com

Contents

Preface

Humans have been exploiting aquatic environments for as long as it has been possible to recognise man as a species. However, growth in human populations, greater urbanisation, industrial development and the need for greater supplies of protein have all contributed to a massive increase in the ways and extent to which we exploit the aquatic environment. The degradation of environments by pollution initially led to calls for controls and regulation; similarly, declines in fish stocks led to attempts to provide management regimes based on biological science. With the continuing development of pressures on aquatic environments, there is a growing need to provide a means of ensuring that future developments are sustainable and that the biological richness of the planet is protected. The United Nations conference on the environment in Rio in 1992 provides a strong codifying lead in this area and has established the issues of biological diversity (both the richness and functioning of ecosystems) and sustainability in international treaties.

Against this background, this volume considers current exploitation practices, but from an **ecological** perspective. The ecological principles that constrain the levels of exploitation possible and the systems' responses to exploitation are developed and placed in the wider ecological context. The application of these principles to formulate sustainable management regimes is described and case studies presented. While the role of ecology in the management of aquatic resources is emphasised, the influence of economics and politics on the management of ecological processes is also discussed. However, ecological science is only part of the decision making process and readers are directed to Huxham and Sumner (2000) for a fuller consideration of the role that science plays in the socio-political process. In this work the emphasis is upon general themes and processes; it is not a catalogue providing extensive lists of species exploited or the types of waste disposed of. However, readers will be directed to sources of such information as appropriate.

While recognising the inherent links between the various processes operating in the aquatic environment, for reasons of clarity we have divided the volume into four parts.

Part 1, Water as a Resource, covers water itself as a resource, and the ecological impacts of its exploitation by humanity. Water is an essential biological requirement for life and as such is used to support human life by providing drinking water and in agriculture to support our food supply. However, water also has a series of unique physical properties that make it culturally valuable. Using water in these two ways generally has only incidental effects on water quality, although these can often be quite major. Each of these uses can result in changes to the physical environment, both in terms of volume of water and structure of aquatic habitats, as attempts are made to ensure an adequate supply.

Part 2, Aquatic Organisms, considers aquatic organisms themselves as exploitable components of the environment. A detailed understanding of the ecological processes by which organisms increase their numbers is required in determining the levels of harvesting which a population can support and how the harvest should be controlled to ensure that it is sustainable. The biological basis of harvesting is that populations produce more offspring than are required to replace themselves. Some of this excess is lost to natural mortality, predation and disease, but the remainder are available for harvest. Understanding the population ecology is necessary to determine the intensity of harvesting that will lead to least damage to the exploited population. Most harvesting of aquatic organisms is carried out on wild populations. However aquaculture, in which the exploited species are raised and closely managed, is a very long established technique for several freshwater fish species, and is becoming more common for both marine and freshwater species as wild stocks suffer from over-exploitation.

Part 3, The Aquatic Environment, considers the aquatic environment itself as a resource. Historically waste disposal and the extraction of mineral resources were the principle means by which the aquatic environment was exploited. Each potentially has an impact on both water quality and the physical environment. Therefore they can be major influences on aquatic systems. More recently a third type of exploitation has become important – the aquatic environment as a recreational resource. During the 20th century recreation developed as one of the most important human activities and now influences many shallow water aquatic systems. Its influences are less easy to establish than for waste disposal and extraction activities. In part this is an issue of scale but also of the different recreational uses having different effects.

Part 4, Synthesis, attempts to bring together the various impacts covered and to consider requirements for effective management strategies. These need to consider the ecological requirements of a system whilst, at the same time, appreciating that human exploitation is a necessary component of the modern ecology of aquatic systems. We introduce the various issues associated with integrating ecological science into the environmental management framework in this section.

The structure of this book maps on to the various types of use to which aquatic resources are put. However, it must be emphasised that there are overlaps between these. Shipping often results in pollution which, although incidental to the use of water as a transport medium, has the same ecological consequences as deliberate waste disposal. Another example is recreational taking of aquatic wildlife, shooting wildfowl for sport and bait collection for angling are technically types of harvesting, subject to the same constraints as other 'fisheries'. Culturally they are better regarded as recreational activities and this has major implications for the way that they are managed. The importance of considering management and sustainability holistically, rather than with respect to a single use of a water body, is therefore a key feature that is emphasised.

While this book has been written primarily to meet the needs of undergraduate and masters students taking courses in aquatic biology, aquatic ecology, resource management, environmental science, environmental management, water resource management, and introductory courses in fisheries and aquaculture, we hope that it will also be useful to a wider audience. We assume a basic familiarity with aquatic ecology and ecological principles and would recommend that those wishing to develop such an understanding read the companion volume, *Ecology of Aquatic Systems* (Dobson & Frid, 1998).

While we hope some readers will read the text from end to end, we recognise that many will seek to 'dip into it' and for this reason we have laid the text out with plenty of markers. We also use text boxes to allow detailed examination of case studies and issues not central to the text.

Given the breadth of the material we have set out to cover we cannot claim expertise in all areas. We therefore wish to acknowledge all those who have inspired us and whose expertise we have drawn on, either formally through reference to their work or informally in discussions over the years. In particular we would like to thank Bob Clark, Robin Clark, Stewart Evans, Steve Hall, Simon Greenstreet, Christina Lye, Peter Olive, Nick Polunin, Stuart Rogers, the Freshwater Ecology Group at Manchester Metropolitan University and the Marine Ecology Research Group at the Dove Marine Laboratory (Department of Marine Sciences & Coastal Management, University of Newcastle). For editorial assistance, thanks are due to Alex Seabrook, Pauline Gillett and Karen Mclaren at Pearson Education and to Carol Weiss for secretarial support and checking the references! Finally, special thanks to Susan and Deirdre for their support and understanding and to Hannah and Lawrence for letting their dad out to play!

Acknowledgements

We are grateful to the following for permission to reproduce copyright material:

Table 1.1 from G.J. Young, J.C.I. Dooge and J.C. Rodda (1994), *Global Water Resource Issues*, Cambridge University Press; Figure 1.1 from M. Dobson and C. Frid (1998), *Ecology of Aquatic Systems*, Pearson Education Ltd.; Figure 1.2 from www.metoffice.com/research/hadleycentre/models/modeldat/HadCM3_IS92a_map_P_ann19601990_20702100.gif, Met Office; Figure 1.4(b) from R.T. Paine (1980), Food webs: linkage, interaction strength and community infrastructure, *Journal of Animal Ecology*, **49**, 667–685, Blackwell Science Ltd.; Table 1.2 from W. Dall, B.J. Hill, P.C. Rothlisberg and D.J. Sharples (1990), The biology of the Penaeidae, *Advances in Marine Biology*, **27**, 1–484, Academic Press Ltd.; Figure 2.2(a) from M.F. Collier, R.H. Webb and E.D. Andrews (1997), *Experimental Flooding in Grand Canyon*, copyright © 1997 by Scientific American, Inc., all rights reserved; Figure 2.2(b) from P.J. Boon (1993), Distribution, abundance and development of Trichoptera larvae in the River North Tyne following the commencement of hydroelectric power generation, *Regulated Rivers: Research and Management*, **8**, 211–224, © John Wiley & Sons Limited, reproduced with permission; Figure 2.3 from J.G. Tundisi and M. Straskraba (eds) (1999), *Theoretical Reservoir Ecology and its Applications*, Backhuys Publishers; Figures 2.6 and 2.7 from A.J.M. Smits, P.H. Nienhuis and R.S.E.W. Leuven (eds) (2000), *New Approaches to River Management*, Backhuys Publishers; Figure 2.8 from T.M.L. Wigley and P.D. Jones (1987), England and Wales precipitation: a discussion of recent changes in variability and an update to 1985, *Journal of Climatology*, **7**, 231–246, Royal Meteorological Society; Figure 2.9 from J.A.A. Jones (1997), *Global Hydrology: Processes, Resources and Environmental Management*, Pearson Education Ltd.; Table 2.3 from N.C. Eno, R.A. Clark and W.G. Sanderson (eds) (1997), *Non-native Marine Species in British Waters: A Review and Directory,* Joint Nature Conservation Committee; Box 2.2 Figure from American Studies Department, University of Virginia; Box 2.3 Figures from E.L. Mills *et al.*

(1993), Exotic species in the Great Lakes: a history of biotic crises and anthropogenic introductions, *Journal of Great Lakes Research*, **19**, 1–54, reproduced by permission from International Association for Great Lakes Research; Figure 3.6(b) from M.S. Bartlett and R.W. Hiorns (eds) (1973), *Mathematical Theory of the Dynamics of Biological Populations*, Academic Press; Figure 3.12 from J.A. Gulland (ed.) (1977), *Fish Population Dynamics*, © John Wiley & Sons Limited, reproduced with permission; Figure 3.17 from K.R. Allen (1980), *Conservation and Management of Whales*, reprinted by permission from Washington Sea Grant Program and University of Washington; Figure 4.4 from J.C. Castilla and L.R. Duran (1985), Human exclusion from the rocky intertidal zone of Central Chile, *Oikos*, **45**, 391–399, © 1985 Munksgaard International Publishers Ltd., Copenhagen, Denmark; Box 5.1 Figure 2 from T.J. Pitcher and P.J.B. Hart (1983), *Fisheries Ecology*, © Kluwer Academic Publishers and with kind permission from P.B. Hart, originally from D. Glen (1974), Unilever salmon farm in Scottish sea loch, *Fish Farming International*, **1**, 12–23, with permission from Heighway Publications; Box 5.2 Table from M.T. Khalil and H.A. Hussein (1997), Use of waste water for aquaculture: an experimental field study at a sewage-treatment plant, Egypt, *Aquaculture Research*, **28**, 859–865, Blackwell Science Ltd.; Table 5.1 from J.F. de L.G. Solbe (1982), Fish farm effluents: a United Kingdom survey, *Report of the EIFAC Workshop on Fish Farm Effluents*, Food and Agriculture Organization of the United Nations; Figure 5.4 from C. Folke and N. Kautsky (1992), Aquaculture with its environment: prospects for sustainability, *Ocean and Coastal Management*, **17**, 5–24, copyright 1992 with permission from Elsevier Science; Box 5.3 Figure 2 from J.R. Brown, R.J. Gowen and D.S. McLusky (1987), The effect of salmon fishing on the benthos of a Scottish sea loch, *Journal of Experimental Marine Biology and Ecology*, **109**, 39–51, © 1987 with permission from Excerpta Medica Inc.; Table 5.2 from H. Rosenthal, D. Weston, R. Gowen and E. Black (1987), *Report of the Ad Hoc Study Group on Environmental Impact of Mariculture*, The International Council for the Exploration of the Sea; Figure 6.2 from H. Güttinger and W. Stumm (1992), An analysis of the Rhine pollution caused by the Sandoz chemical accident, 1986, *Interdisciplinary Science Reviews*, **17**, 127–135, IOM Communications Ltd.; Box 6.1 Figure 1 from M. Ruivo (ed.) (1972), *Marine Pollution and Sea Life*, Fishing News Books, Blackwell Science Ltd.; Figure 6.4(a) from M.L. Miserendino and L.A. Pizzolón (2000), Macroinvertebrates of a fluvial system in Patagonia: altitudinal zonation and functional structure, *Archiv für Hydrobiologie*, **150**, 55–83, E. Schweizerbart'sche Verlagsbuchhandlung; Table 6.1 from E.R. Gundlach and M.O. Hayes (1978), Vulnerability of coastal environments to oil spill impacts, *Marine Technology Society Journal*, **12**, 18–27, © Marine Technology Society; Figure 6.9 from W. Salomons, B.L. Bayne, E.K. Duursma and U. Förstner (eds) (1989), *Pollution of the North Sea: An Assessment*, © Springer-Verlag; Figure 7.2 from A.G Brown and T.A. Quine (eds) (1999), *Fluvial Processes and Environmental Change*, © John Wiley &

Sons Limited, reproduced with permission; Figure 7.4 from J. Andrews and D. Kinsman (1990), *Gravel Pit Restoration for Wildlife: A Practical Manual*, The Royal Society for the Protection of Birds; Figure 7.7 from P.F. Kingston (1987), Field effects of platform discharges on benthic macrofauna, *Philosophical Transactions of the Royal Society, London, Series B*, **316**, 545–565, The Royal Society and P.F. Kingston; Figure 8.6 from H.B. Cott (1961), Scientific results of an inquiry into the ecology and economic status of the Nile crocodile, *Transactions of the Zoological Society of London*, **29**, 211–356, The Zoological Society of London; Figure 8.7 from G. Howells and T.R.K. Dalziel (eds) (1992), *Restoring Acid Waters: Loch Fleet 1984–1990*, p. 295, Fig. 14.2, with kind permission from Kluwer Academic Publishers and R.W. Battarbee; Figures 8.8 and 8.9 from M.S.C. Havard and E.C. Tindal (1994), The impacts of bait digging on the polychaete fauna of the Swale estuary, Kent, UK, *Polychaete Research*, **16**, 32–36, School of Biosciences, Cardiff University; Figure 8.10 from U.K. Pollingher, T. Zohary and T. Fishbein (1998), Algal flora in the Hula Valley – past and present, *Israel Journal of Plant Sciences*, **46**, 155–168, Laser Pages Publishing Ltd.; Figures 9.4–9.6 from K.R. Clarke and R.M. Warwick (1994), *Change in Marine Communities: An Approach to Statistical Analysis and Interpretation*, PRIMER-E Ltd. and Plymouth Marine Laboratory.

In some instances we have been unable to trace the owners of copyright material, and we would appreciate any information that would enable us to do so.

Chapter 1 How do Humans Impact Aquatic Systems?

1.1 Introduction

Water is, of course, necessary to sustain human life. More than this, however, its control and large scale exploitation are essential to maintain modern human societies, while its living systems offer a rich variety of exploitable components. For these reasons, aquatic systems are exploited throughout the world, so that very little of its water is not influenced by human activity.

The world's water resources are huge. The Earth's surface and atmosphere contain around 1.4×10^9 km^3 of water, although this is very unevenly distributed (Table 1.1). The vast majority of water is in the oceans, whereas exploitation is concentrated on freshwater systems, a tiny

Table 1.1 *Distribution of the world's water (from Young et al., 1994)*

Category	Total volume (km³ × 10³)	% of total	% of fresh water	Replacement period
Oceans	1 338 000	96.5		2650 years
Groundwater	33 930	2.46	30.1	1400 years
Soil moisture	16.5	0.001	0.05	1 year
Glaciers	24 064.1	1.74	68.7	9700 years*
Permafrost	300	0.022	0.86	10 000 years
Lakes	176.4	0.013		17 years
Freshwater	91	0.007	0.26	
Saline	85.4	0.006		
Marshes	11.47	0.0008	0.03	5 years
Rivers	2.12	0.0002	0.006	16 days
Biological water	1.12	0.0001	0.003	
Atmospheric water	12.9	0.001	0.04	8 days
Total water	1 385 984.61	100		
Total freshwater	35 029.21	2.53	100	

*Estimated turnover time for Greenland ice sheet; other areas of permanent ice have more rapid turnover, although typically greater than 1000 years.

Figure 1.1 *The hydrological cycle. Although movement is mainly one way (anticlockwise in this case), there are mechanisms which act in the reverse direction, of which the most important are shown here as dashed lines. Major storage compartments are boxed. P = precipitation (rain and snow); E = evaporation; T = transpiration (from Dobson & Frid, 1998).*

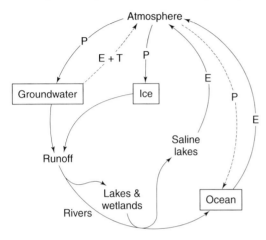

fraction of the total. The Earth's water systems are, however, interconnected into a dynamic hydrological cycle (Figure 1.1), the constant mobility of water having three major effects with respect to exploitation: it ensures that supplies are replenished, it provides kinetic energy, and it acts to disperse and dilute any contaminants added to a water body.

Human activity is now impacting water resources worldwide. The nature of these impacts ranges from changes in water quality to alteration of water quantity and localised disruption of the water cycle. Impacts are being felt by living organisms and the integrity of ecological systems is being threatened. We are now in a situation where water, or at least human use of water and its products, needs to be carefully managed. A key goal of such management is to ensure that human societies may continue to exploit water resources indefinitely and this inevitably means that ecological systems must be protected or restored. Achieving this requires a detailed understanding of the ecological processes that occur in aquatic systems and how human activity is impacting upon them.

1.2 Definitions

Various terms are key to this subject and appear many times in the text. It is therefore important to provide some definitions.

Exploitation

Exploitation may be defined as the use by humans of water or its products for their own benefit. Therefore, in any circumstance in which a component is extracted from a water body, whether it is sediment, living organisms, or even the water itself, it is being exploited. To this may be added processes whereby a physical or chemical property of the water is used, without removing anything from the water body. This includes taking advantage of the mobility of liquid water to disperse a contaminant added, or exploiting the ability of water to support low density objects to provide a means of transport.

Management

Management of aquatic systems refers to mechanisms whereby their exploitation is controlled or regulated. Normally it is applied in the sense of control to reduce or reverse detrimental effects on the supply or quality of the resource being exploited, or to reduce the negative effects of exploitation on ecological systems within the exploited water.

Resources

Resources are the components of a water body that are exploited. The term is normally applied to components that can diminish through over-use, such as stocks of fish. Resources can also refer to requirements for maintenance of a healthy ecosystem, whose availability or quality may be impaired by human exploitation. If the processes that create the resource still operate, then it can be described as renewable, as that component removed by humans can be replaced. If the supply is finite and not replaceable, or its creation requires a time scale well beyond the scope of human activity (e.g. fossil fuels), then it is termed non-renewable. The degree to which a resource is renewable is, of course, largely a function of the rate of human extraction relative to the rate of creation, so it is perhaps better to apply the terms renewable and non-renewable to the rate of exploitation rather than to the resource itself.

Sustainability

Sustainability refers to the management of resources in a way that does not deplete them and therefore ensures their continuation. The Convention on Biological Diversity (signed at the UN Earth Summit in Rio de Janeiro) defines sustainable use as 'the use of components of biological diversity in a way and at a rate that does not lead to the long-term decline of biological diversity, thereby maintaining its potential to meet needs and aspirations of present and future generations'. This essentially recognises that biological systems have the ability to replace themselves and provided that the type and level of use does not compromise this, then the use can be continued indefinitely – for the benefit of present and future generations. Sustainability, therefore, refers to a level of exploitation that does not exceed the capacity for self-renewal.

Pollution and contamination

Pollution is defined as the introduction of substances or energy into the environment, resulting in deleterious effects to humans, human activities or other living components of the environment. Contamination is the introduction, directly or indirectly, of substances or energy into the environment such that levels are altered from those that would have existed without human activity. The common perception of pollution as something released by human activity into the environment, therefore relates to contamination. It is also commonly assumed that this 'pollution' is a 'bad' thing. Under the definitions given above, pollution is a detrimental influence by definition, but contamination is not necessarily harmful.

1.3 What do aquatic systems provide?

There is a diverse range of ways in which water bodies are exploited, and the remainder of the book is divided into four parts to reflect this. The essential features of exploitation are, however, outlined here.

Water as a resource

Water itself is a resource, exploited in two fundamentally different ways. The first is water as an essential biological requirement for life, the second is its physical properties as a liquid which make it culturally valuable. Influences of these modes of exploitation upon water quality are generally incidental, although often major, but each can result in changes to the physical environment, both in terms of volume of water and structure of aquatic habitats, as attempts are made to ensure an adequate volume for the required purpose.

Aquatic organisms

Aquatic organisms are exploitable components of the environment. An effective understanding of harvesting, and particularly of its sustainability, requires a detailed understanding of the ecological processes by which organisms increase their numbers, and therefore the intensity of harvesting that will lead to least damage to the exploited population. Most harvesting is carried out on wild populations with minimal management, but aquaculture, in which the exploited species is raised and closely managed, either in enclosures or in the natural habitat, is a very long established technique for several freshwater fish species and is becoming more common as wild stocks suffer from over-exploitation.

The aquatic environment

The aquatic environment itself is a resource. Historically, aquatic environments have been exploited for two main purposes – waste disposal and extraction of mineral resources. Each has a direct impact on both water quality and the physical environment and, as such, strongly influences aquatic systems. More recently a third type of exploitation has become important – the aquatic environment as a recreational resource. Although always used to a small extent as an amenity, during the 20th century recreation developed as one of the most important human influences upon shallow water aquatic systems. Its influences are less easy to categorise than are those of traditional exploitation, different recreational uses having different effects, but each has some specific requirements, which are often at odds with those of the other uses.

Synthesis

The diversity of methods by which aquatic environments are exploited is enormous, but the effects of different uses are often very similar, so that common patterns can be identified. Furthermore, although the classification outlined above demonstrates natural differences in the types of use of aquatic resources, it must be emphasised that there are overlaps between these. For example, the use of water as a communication medium results in pollution which, although incidental to the use of the water, has the same ecological con-

sequences as deliberate waste disposal. As further examples, shooting wildfowl for sport and bait collection for angling are technically types of harvesting, subject to the same constraints as other 'fisheries', but culturally they are better regarded as recreational activities.

The effects of exploitation all point to a need for effective management strategies, which consider the ecological requirements of a system whilst, at the same time, appreciating that human exploitation is a necessary component of the modern ecology of aquatic systems. The importance of considering management and sustainability holistically, rather than with respect to a single use of a water body, is therefore a key requirement for success.

1.4 The need for understanding

Overlaid upon direct use of water bodies, for purposes such as extraction, transport or harvesting, are more profound changes, global in extent. These processes impact aquatic systems in ways that are often difficult to predict, but effective management needs an awareness of their influences.

1.4a Human population

The human population is rising and, although the rate of increase is declining, even the best estimates predict several decades and a couple of billion extra people before it levels off. Therefore, the demand for water and its products is going to rise. Management for sustainability must take into account predicted increase in requirement for aquatic resources, rather than simply trying to satisfy today's demand.

1.4b Climate change

There is increasing evidence that human impacts are changing the world's climate. This does not simply result in warmer temperatures worldwide,

Figure 1.2 *Predicted changes in precipitation as a result of human impacts. The map is based on the Hadley Centre's HadCM3 coupled atmosphere–ocean circulation model, and compares average annual precipitation from 1960 to 1990 with predicted annual precipitation from 2070 to 2100 (based on information provided by the Met Office).*

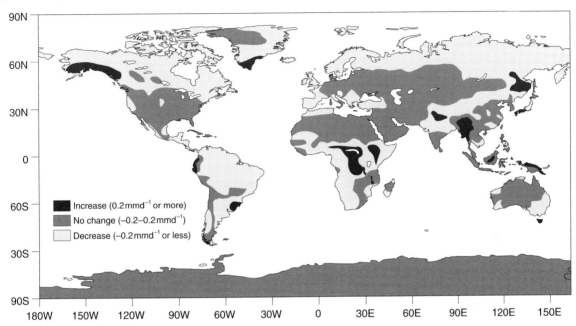

but will cause major changes in weather patterns and climate generally. For example, total rainfall may not change, but its distribution is likely to alter (Figure 1.2). This has important implications for fresh waters, all of which are fed ultimately by precipitation, mainly in relatively small headwater regions of river systems. If rainfall declines, river discharge will be reduced, and so the amount of water available downstream will decline. It is possible that these changes will occur rapidly, in the space of a few decades, which is too little time, even without political barriers, to modify human processes that require water.

Another widespread prediction of the effect of climate change is a rise in sea level as ice caps melt and the oceans undergo thermal expansion. Any change in sea level will have an immediate impact upon coastal and estuarine habitats, while low-lying freshwater wetlands will also be impacted. In areas heavily modified by human activity, such as the coastal regions of western Europe, a small rise in sea level without any associated management would probably have a positive effect on natural coastal environments, by creating large areas of new wetland on coastal plains. In practice, however, sea level rise is likely to initiate major management, including engineered sea defences. Most prognoses for sea level change predict a rise in water levels, but some suggest that there may be regional reductions. In the north Atlantic, for example, the Greenland ice sheet exerts gravitational pull on the ocean; if the ice sheet is reduced in size, this pull is reduced and sea levels around western Europe and eastern Canada may fall. Obviously, coastal management strategies are going to be determined by the direction that the sea moves but in either case, inadequate knowledge of ecological processes and interactions will compromise the effectiveness of management strategies implemented.

1.4c Loss of habitat

On a more local scale, but happening worldwide, is habitat loss. The edges of all types of aquatic system support often complex and diverse habitats, including a variety of wetland types. All of these are important hydrologically as water storage zones, and ecologically as permanent habitats for many species and temporary habitats for others which then play a key role in open water bodies. Human impacts affect these systems to a great extent, and the main trend is towards their destruction. Much human exploitation of aquatic systems requires a complex infrastructure, including modifying the form of the water body itself. Channelising a river to improve navigability or drainage reduces the complexity found in a natural system, while construction of dams, barrages, sea defences and extraction schemes alters patterns of water movement, sediment transport and natural fluctuations in water level. Development of port facilities, salt pans and other industrial structures results in major loss of such habitats globally. Flood control and drainage also destroy many, severing rivers, estuaries and coastal seas from a major influence on their form and function. Flood control and drainage are outside the direct scope of this book, but can have serious influences on many types of exploitation. Most of these influences are expected to result in reduced diversity, of both habitats and species, but in some cases new environments are created in artificial water bodies such as reservoirs and canals.

1.4d Altering species' abundance and distribution

The most obvious way in which the abundance of a given species is reduced is through direct harvesting. However, harvesting can also lead to enhanced numbers, if the population is carefully managed, while aquaculture and recreational angling have led to the deliberate enhancement of numbers and expansion of the distribution of many economically important species. Indirectly, the populations of many species are altered, normally detrimentally, by pollution, habitat loss, excessive removal of key species in their food web and a host of other mechanisms.

Introductions occur in a variety of ways, but may be conveniently divided into deliberate translocation and accidental introduction. Deliberate translocations, particularly of fish, for harvesting and recreation are very common throughout the world. Aquatic organisms have also been translocated as pest control agents, including herbivorous fish for weed control and insectivorous fish to control mosquitoes (**Box 1.1**). In addition to this process, however, many species have been moved unintentionally, either directly in association with boats or indirectly following the creation of canals connecting formerly separate catchments. As a result of these activities, exotic species are now a major problem in aquatic ecosystems throughout the world.

1.4e Interactions

The influences of the impacts outlined above often interact. For example, sea level rise has an immediate influence upon coastal habitats. The direction of the influence is determined by the management response. Increasing sea level may induce more flood defence management, further destroying coastal habitats. Alternatively, it may precipitate a managed retreat, allowing salt marsh, itself a very effective flood control system, to develop. The latter requires abandonment of farmland but, in doing so, may create a new nursery habitat for commercially important sea fish. The potential problem of climate change altering rainfall patterns is exacerbated by increases in

Box 1.1 Exploiting fish for pest control

Fish are amongst the most widely exploited organisms on the planet, almost entirely for food (**Chapters 4, 5**), but also for recreational purposes (**Section 7.2**). However, some species have been exploited for their perceived ability to reduce numbers of pest organisms. Many freshwater fish species are insectivorous and some lake and wetland species are voracious consumers of mosquito larvae. Therefore, they have been used as biological control agents, in an attempt to reduce numbers of mosquitoes, and known consumers of mosquitoes have been translocated from their native habitat into new environments which apparently lack a suitable natural mosquito consumer. The success of these translocations has been mixed.

One of the most widely translocated species is the western mosquito fish (*Gambusia affinis*). This species is native to the lower Mississippi catchment and the coastal region of the Gulf of Mexico, west of Alabama in the USA and Mexico. In the eastern Gulf region and the southern Atlantic coast it is replaced by the closely related eastern mosquito fish (*G. holbrooki*), which has probably also been translocated, albeit to a lesser extent.

Gambusia has a reputation as an effective mosquito control agent, as a result of which it has been introduced indiscriminately. In the first decade of the 20th century it was transported to Hawaii, since when it has been carried to locations throughout North America. It was also taken to Italy, from which introductions further afield in tropical Africa and central Asia were

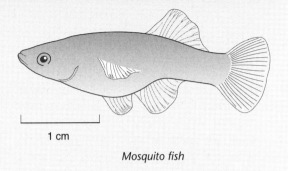

1 cm

Mosquito fish

carried out. Now it is established in localities throughout the tropical and warm temperate parts of the world, including localities as isolated as Easter Island, Fiji and the Cook Islands.

Gambusia is an effective mosquito control agent in its native range, but has had major negative influences in most environments into which it has been introduced (Rupp, 1996). It is an aggressive species that has eliminated native fish species through a combination of direct competition and predation. In some cases, it has eliminated more effective native mosquito predators, thus exacerbating the mosquito problem (Courtenay & Meffe, 1989). Furthermore, it has severely impacted non-target species, including frog tadpoles (Goodsell & Kats, 1999) and, by selectively predating zooplankton grazers, has precipitated algal blooms. Now, *Gambusia* control is a key conservation requirement in lakes and wetlands in many parts of the world.

populations, particularly in the developing world where there is a shortage of money to finance alternative sources of water.

When faced with interconnected problems such as these, only a detailed understanding of ecological interactions within aquatic systems will allow management decisions to be based upon accurate predictions of their direct and indirect effects. Hence the need to understand the ecology of aquatic systems which, in turn, requires an understanding of basic ecological principles.

1.5 Key ecological concepts

For the reasons outlined above, it is important to understand the mechanisms by which species interact within ecological communities. It is also important to appreciate the extent of our understanding of these processes, because management strategies can only be as good as the quality of data upon which they are based. Fundamentals of ecology and hydrology are introduced, where appropriate, in various places in the book, but several general features are relevant throughout and so are introduced here.

1.5a The community and the ecosystem

An ecological community is defined as the assemblage of interacting living organisms within a location or habitat. This simple description hides a long established problem with definition of the community: whether it is adequate to describe it simply as a list of co-occurring species or whether it is a dynamic unit, the interactions being an important component of the definition. However, effective descriptions of a given community normally incorporate interactions between species, such as feeding links (see **Section 1.5c**).

Normally, the community is defined in terms of an identifiable habitat, so that adjoining habitats, if clearly different, can be assumed to support different communities. Thus we can talk of a pond community and a grassland community in surrounding pasture, or a benthic community on the bed of a water body and a separate pelagic community in the water column above it. In each of these examples, there will be interactions between adjacent systems – the pelagic zone providing detritus as an energy source of the benthic system, or the grassland being the main habitat of pond predators, for example. However, in each case the interactions between organisms within a community are stronger, more abundant and more diverse than those with organisms in other communities. So, for example, although some invertebrates within the pond may be eaten by terrestrially based predators, or may even spend part of their life cycle in the terrestrial environment, while in the pond they interact mainly with other pond organisms.

The ecosystem may be defined as the biotic (living) components of the environment along with the abiotic (non-living) components with which

they interact. Therefore, the ecosystem concept expands the community to include features such as weather, geology and soil.

Community structure

There are currently two approaches to explaining community structure, although they are not mutually exclusive. One school of thought – the **assembly approach** – holds that real communities do not spontaneously burst into existence, rather they are built up gradual subject to certain constraints. The actually assembly process imposes structure on the community. The alternative view – the **dynamic stability approach** – proposes that, whatever the processes that produce communities, the end result will have to be a system that is stable, and therefore we need to understand the stability of food webs.

1.5b The niche

Every species occupies a niche within a community. Although difficult to define with any precision, a species' niche encompasses the sum of all of its interactions with other species and with components of the abiotic environment. Each type of interaction will have quantifiable dimensions, so a niche can be represented as an interacting series of dimensions. As with community structure, it is easiest to visualise the niche in terms of feeding relationships. Figures 1.3a and b show a single niche dimension for a hypothetical predator, in this case the size of prey consumed. Any individual will have an optimum prey size, neither too big for its mouth to cope with nor too small to give enough energy return for the effort of eating it. On either side of this optimum, it can consume larger and smaller prey, but with decreasing efficiency. The species as a whole will show a similar pattern – the size of individuals will be variable, but most will be close to the mean.

Figure 1.3 *(a) Representation of an individual's niche dimension on two axes. For any component of the niche, there will be an optimum value. For example, optimum food particle size will be determined by features such as the mouth size of the organism, handling efficiency declining with deviation from this size. (b) Representation of a population's niche dimension. Each individual will have its own optimum for a given resource, but the organisms within a population will cluster around a mean size, with relatively few appreciably larger or smaller than this. (c) Effect of intraspecific competition. The niche width increases as individuals that exploit the periphery of the niche dimension are favoured by reduced competition. (d) Effect of interspecific competition. The niche width of each species contracts away from the zone of overlap.*

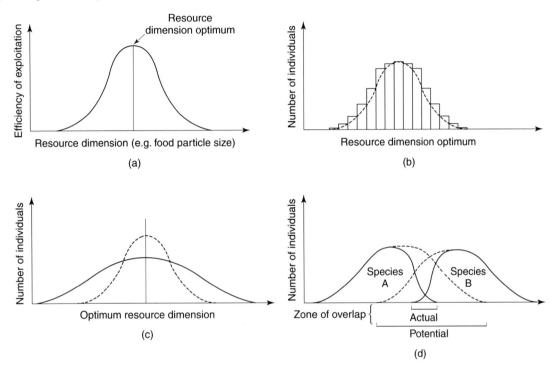

1.5c Representing the community

A community can be represented in a straightforward fashion as a list of co-occurring species, or with indications of relative densities, allowing diversity to be determined. It can also be described in terms of its physical structure, although this is useful only for systems in which vegetation provides such a structure. More useful is to incorporate the interactions between species.

Perhaps the single most important parameter structuring any community is the availability of energy. Energy is fixed in a biologically available form by primary producers and then passes through a series of trophic levels, a process considered in detail in **Sections 3.2** and **3.3**. The mechanism for the movement of energy between organisms is feeding interactions and these have, therefore, become the most common way in which community structure is represented.

Food chains and food webs

Food chains are the simple representation of the steps and transformations that organic material goes through from primary producer via

9

Figure 1.4 *Ways of representing feeding interactions. (a) Food chain. A simple linear representation of passage of energy between species, all of which are at different trophic levels. The trophic levels themselves are identified. (b) Food web, demonstrating more of the complexity of feeding interactions that is found in real systems. A rocky shore food web – Cape Flattery, Washington, USA (from Paine, 1980).*

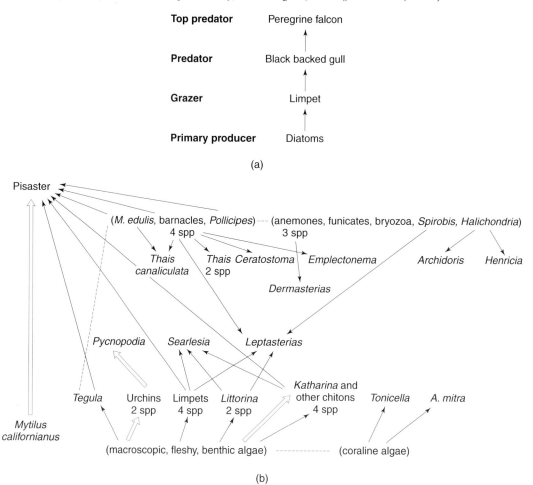

(a)

(b)

intermediate consumers to final consumer – the top predator (Figure 1.4a). Each component of a food chain can be classified as a basal species, either a primary producer or a consumer of detritus, and therefore with no prey species, a top predator, which no other organisms predate, and an intermediate species, having both predators and prey. In practice, food chains often rely on aggregations of species into groups which can be represented in this simple chainlike manner. Food chains are attractive representations of ecological

systems; they appear simple and elegant and allow ready comparisons of the number of trophic levels in a particular ecosystem and the efficiency of transfer of energy across the levels. They also allow easy differentiation of trophic levels (Figure 1.4a): stages through which energy passes from its input by primary producers or detritus.

Food webs (Figure 1.4b) are more realistic representations of the feeding interactions in an ecosystem than food chains (Paine, 1980), in that they recognise that each trophic level will

Figure 1.5 *Subset of a food web from an acid stream, demonstrating the effecting of resolution in determining food web patterns. Trophic species (combinations of species assumed to have similar niches) are shown in boxes. (a) A simplified version of the subset, similar to the level of resolution found in most published food webs. At higher trophic levels, most species are differentiated, but towards the base the number of 'trophic species' increases. (b) The same subset, incorporating what is known about actual species and their feeding interactions in this system. In this case, current evidence suggests that the trophic species is made up of true species that are, in most cases, functionally equivalent: they all feed on similar material and in turn are preyed upon by a similar predator assemblage. Note, however, that* Attheyella crassa *differs from the other microcrustacea in that it only has one predator. Note also that there are still three undifferentiated groups that may represent further trophic species rather than true species. S.fulig. =* Sialis fuliginosa *(alderfly larva); P.consp. =* Plectrocnemia conspersa *(caddis larva); D.spp. =* Diacyclops *spp.; P.fimb. =* Paracyclops fimbriatus; *E.serr. =* Eucyclops serrulatus; *A.vern. =* Acanthocyclops vernalis; *B.spp. =* Bryocamptus *spp.; A.cras. =* Attheyella crassa; *M.brev. =* Moraria brevipes; *A.quad. =* Alona quadrangularis: *A.rust. =* A. rustica; *OST =* Ostracoda *(from Lancaster & Robertson, 1995).*

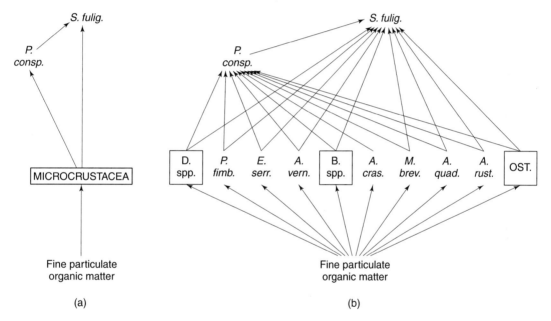

(a)   (b)

normally have many species, and few predators are confined to a single prey species. The food web therefore attempts to incorporate the true complexity of feeding interactions in natural systems.

An inevitable problem with attempting to create a food web for a complex natural community is that of resolution: large or clearly important species, including top predators, are discriminated at the species level, or even adults and juveniles may be distinguished. At lower trophic levels, however, a suite of species may be present whose feeding interactions we are unable to differentiate, or we may not know exactly how many species of certain taxonomic groups are actually present. In these situations, such organisms are generally lumped together in a food web as **trophic species** or **biospecies** – groups of true species which are assumed to feed on the same range of prey and in turn are preyed on by the same suite of predators – so that they are functionally equivalent (Figure 1.5).

Many food webs have been published, from both aquatic and terrestrial environments, and extensive analysis of webs across many environments has demonstrated many common features, suggesting that there are predictable 'rules' by which species assemblages are interconnected and therefore by which communities are constructed (Yodzis, 1989).

1.5d Rules in food web structure

Energy flow

Energy flow within a food web is one way of structuring the web. Very rarely are there cycles in which two species eat each other; when these do occur, it is normally as different stages in the life cycle which are better differentiated as biospecies. For example, many predatory invertebrates, such as beetle larvae, eat frog tadpoles; however, adult frogs also eat beetle larvae. Therefore, although two species are eating each other, the frog is best regarded as two biospecies, each with a different niche depending upon the stage in its development.

Proportions of species at different levels

The proportion of biospecies which are basal species, intermediate species and top predators is remarkably constant (Yodzis, 1989), such that:

$$B/S = 0.19$$
$$I/S = 0.53$$
$$T/S = 0.29$$

where S is the species richness (number of biospecies), B is the number of basal species, I the number of intermediate species and T the number of top predators (based on an analysis of 62 webs by Briand & Cohen, 1984).

From this it follows that the ratio of prey to predators is also a constant:

$$(B + I)/(I + T) = 0.89$$

Omnivory

Omnivory refers to consumption of food from two or more trophic levels. It is generally considered to be rare in food webs; in other words, a species will normally feed only on species in the trophic level immediately beneath it. Although apparently true for terrestrial food webs, omnivory is frequent in the aquatic food webs that have been determined to a high resolution (e.g. *Sialis fuliginosa* in Figure 1.5), suggesting that absence of omnivory is not a universal feature of food webs generally.

Links in the food web

A link is the line drawn on the web indicating a feeding interaction between biospecies. The maximum number of potential links is:

$$S^2 - S = S(S - 1)$$

The **connectance** of a food web can be calculated by comparing the actual number of feeding links (L) with the theoretical maximum:

$$C = L/S(S - 1)$$

thus giving a measure of the complexity of a web (Figure 1.6). The general pattern is for C to decrease as S increases, and for a scale relationship to exist such that:

$$L_{BI} = 0.27L \ (L_{BI} \text{ links from Basal to Intermediate etc.})$$
$$L_{BT} = 0.08L$$
$$L_{II} = 0.30L$$
$$L_{IT} = 0.35L \text{ (Briand \& Cohen, 1984)}$$

Compartmentalisation

Decreasing connectance with increasing complexity demonstrates that a large species assemblage divides itself trophically into a series of relatively discrete components known as compartments. The reason for this is that **linkage density** (mean number of links per species) remains relatively

Figure 1.6 *The relationship between potential links and connectance in a food chain.*

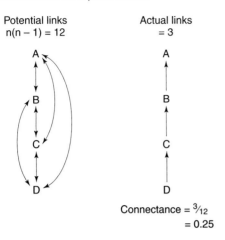

Potential links
$n(n - 1) = 12$

Actual links
$= 3$

A   A

B   B

C   C

D   D

Connectance $= {}^3\!/_{12}$
$= 0.25$

constant because, as more species are added, they naturally start to form mini-food webs within the main web. Within this, groups of species can be identified which are exploiting a similar food resource, often with a degree of overlap. Such species groups are called **guilds**.

Keystone species

A keystone species is a single species whose role is critical to the maintenance of the community, and whose removal will cause a major change in community structure. The best known examples of keystone species are marine, and their importance has been determined by examining the effects of their removal.

The American lobster (*Homarus americanus*), found in coastal waters off eastern Canada, may be a keystone in determining the structure of subtidal communities. Lobsters eat sea urchins (*Stronglocentrotus droebachiensis*) which, in turn, graze large seaweeds such as *Laminaria*, a major structural component of the habitat. Lobsters are heavily exploited off eastern Canada and a consequent decline in their populations has been blamed for a population explosion of sea urchins which, in turn, resulted in decimation of large

areas of seaweed. This alters the habitat completely; one of its consequences is that lobster recruitment declines, young lobsters requiring seaweed beds for shelter. Therefore, lobsters are unable to increase in numbers to take advantage of the abundance of sea urchin prey, and the community is permanently altered.

The role of the lobster as a keystone species is controversial, because there is contradictory evidence that sea urchins undergo natural cycles in abundance, determined by periodic disease outbreaks (Elner & Vadas, 1990). The starfish *Pisaster ochraceous*, found in the rocky intertidal of western North America provides a less equivocal example. The keystone role of this species was determined by experimental removal of individuals, following which the mussel *Mytilus californianus* increased in numbers to dominate the system, to the detriment of other attached species of invertebrates and algae (Paine, 1974). The natural system supports a wide diversity of basal invertebrate species, all of which are preyed upon by *Pisaster*. However, the starfish have a net beneficial effect on most of their prey species by keeping in check the highly competitive *Mytilus*. If the top predator is removed, *Mytilus* takes over and species richness declines.

Food chain length

While long chains, up to nine links, do exist in the literature, most are far shorter than would be expected by chance, and chains longer than five links are extremely uncommon.

There are two hypotheses to explain why food chains are so short. The **energetic hypothesis** proposes that the length of food chains is limited by inefficiency of energy transfer along the chain, while the **dynamic stability hypothesis** proposes that population fluctuations at a low level are transferred and magnified up the chain, making the upper levels unstable if their food supply fluctuates below the amount required to sustain viable populations.

Of these two hypotheses, the latter is probably the most realistic, and the reason top predators need great mobility. However Briand & Cohen (1987) found that mean food chain length neither correlated with environmental variability (as suggested by the dynamic stability approach) nor with primary production (as might have been expected under the energetic approach). There was, however, a strong link with the **dimensionality**, two-dimensional systems such as rocky shores and most terrestrial habitats having fewer links per chain than three-dimensional systems, such as the pelagic zone of lakes and the ocean.

1.5e Problems with food webs

Although straightforward in principle and intuitively easy to understand, food webs as representations of community structure have many problems. The problem with having to combine trophically similar species, thereby losing resolution, has been mentioned and is addressed in more detail below, but there are other problems with the standard presentation of food webs.

As the diversity of a community increases, the complexity of the food web increases very rapidly. As such the pictorial web also becomes more complex and so difficult to understand, while simplification to make a readily appreciated pictorial representation leads to divergence between reality and representation. Food webs are variable in time and space – importance of different prey species varies seasonally in many systems, while juveniles and adults tend to have different diets. Furthermore, few systems are closed, interactions occurring with other adjacent communities, whereas webs tend to downplay processes by which energy is imported or exported.

Food webs undoubtedly provide a useful visualisation of the trophic interactions in an ecosystem. While the representation of a food web containing hundreds of taxonomic species into a limited number of biospecies is recognised as an unavoidable simplification, it is much harder to incorporate factors such as the change in webs through time, other than by presenting different food webs for different seasons.

Effects of resolution on food web structure

The Ythan Estuary is a small, compact estuary on the east coast of Scotland that has been the subject of study for over 40 years. The food web for the estuary comprises 92 species – 1 mammal, 26 birds, 18 fish and 44 invertebrates (Hall & Raffaelli, 1991). This assemblage is supported by three resource groups – macroalgae, phytoplankton and detritus. In this taxonomically resolved web there were 409 observed (directly observed or by gut content analysis) links. In this web food chain lengths were longer and the degree of omnivory higher than in most published webs. Other properties (number of links, proportion of basal, intermediate or top species, number of predators and number of prey taxa) were similar to those of other published webs. Hall & Raffaelli (1991) concluded that the differences between the Ythan food web and others in the literature are a consequence of its being more finely resolved, in other words most of its species are true species with little aggregation into biospecies (see also Hall & Raffaelli, 1993). They also could find no evidence of the web being divided into a number of functional compartments (Raffaelli & Hall, 1992). This, along with the lack of evidence for keystone predators, supports the notion that, although strong functional links will lead to compartmentalisation in food webs (as suggested by Paine, 1980), it is not a ubiquitous feature of food webs.

Are food webs randomly constructed?

In recent years the development of a **null model** approach to ecology has led to the development of a random assembly approach for food webs. Kenny & Loehle (1991) constructed random food webs, with the connections between species randomly chosen from all possible connections, but with no consideration of realistic biological interactions. The connectance curves for these random webs fitted in the range of those of published webs, raising the possibility that food webs are in fact randomly connected! Such studies do not prove this, but serve to caution us that at present we cannot distinguish food web structures from random patterns, and therefore it is premature to start proposing models of interactions to produce the observed structures.

1.5f Non-feeding interactions

The simplification of real communities into simple food webs disregards many trophic links and completely ignores a host of other habitat interactions, including symbioses, parasitism and habitat structure forming species (Polis & Strong, 1996). Organisms are exposed to many sources of mortality, of which predation is only one; just because two species interact trophically does not necessarily mean that they are closely linked ecologically (Paine, 1988, 1992). The abundance of a population is a consequence of a combination of factors affecting the birth rate (e.g. food supply to the previous generation, environmental conditions), recruitment success, environmental conditions for the adults, disease and predation. In such circumstances it is likely that changes in one component of a food web will quickly be damped out as it spreads out through the linked 'species' (Strong, 1992).

An important structuring process in ecological communities is **competition**. If the food resource is limiting, the large number of individuals clustered close to the mean size will be in direct competition for food, while slightly smaller or larger individuals will suffer less competition and so be more successful. In this way, intraspecific competition (competition occurring between members of the same species) will tend to expand the species' niche dimension (Figure 1.3c). If two different

Figure 1.7 *If complete overlap between species occurs along one niche dimension, consideration of a second dimension often demonstrates lack of overlap. In this example, overlap is broad or complete for pairs of species on each dimension separately, but when both are considered together there is minimal overlap. If two dimensions still result in significant overlap, then further dimensions must be explored.*

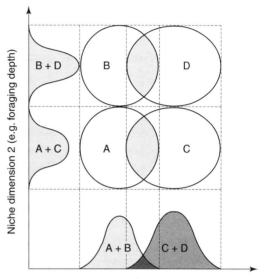

Niche dimension 1 (e.g. food particle size)

species are exploiting the same resource in the same community, then those individuals of each species in the area of overlap will suffer interspecific competition and be less successful than individuals outside the range of overlap. Interspecific competition therefore leads to species' niche dimensions contracting away from the range of overlap (Figure 1.3d).

Within each community, each species has a unique niche, so if there is apparent overlap in one dimension, other dimensions will show avoidance of overlap (Figure 1.7). When all niche dimensions are taken into consideration, the maximum overlap between any two species is around 50%. Any more than this, and competition will be so severe that the less competitive species will be driven to local extinction. In practice, interspecific

Figure 1.8 *Feeding and non-feeding interactions between three river pool species in North American rivers: sunfish (*Lepomis cyanellus), *salamander (*Ambystoma barbour) *and isopod (*Lirceus fontinalis). *Feeding interactions are shown by solid arrows, non-feeding interactions by dashed arrows; the thickness of the arrow is proportional to the importance of the interaction. + and – signify a positive and negative effect, respectively, on the recipient species. Although sunfish is a predator of isopods, it has an indirect positive effect through suppressing predation by the salamander (after Huang & Sih, 1991).*

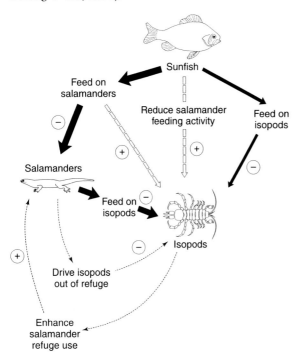

competition is rarely observed in natural systems, because the process of elimination of one species by another is normally rapid, but there are many examples of introduced species out-competing and eliminating natives.

Some interactions are indirect. Huang & Sih (1991) determined that an apparently straightforward food web in stream pools in North America hid some important non-feeding interactions. In this system, salamanders eat isopods, while sunfish eat both, thereby acting as omnivores (Figure 1.8). However, by feeding on salamanders, sunfish indirectly benefit isopods; the presence of sunfish makes salamanders less likely to leave predation refuges and therefore less able to forage. Sunfish, in turn, feed preferentially on salamanders, because they are larger. Therefore, although predatory upon isopods, sunfish have a net positive effect on their numbers through indirect processes (Figure 1.8).

1.6 Biological diversity

1.6a Definition

The term biological diversity, or biodiversity as it is often phrased, took on a special meaning after the UN Conference on the Environment held in Rio de Janeiro in 1992. At the conference the delegates signed the Convention on Biological Diversity (CBD) and in so doing agreed to implement measures that protected the natural biological diversity found on Earth. So what then is biological diversity and why does it need to be conserved?

The CBD defines biological diversity as 'the variability among living organisms from all sources including, inter alia, terrestrial, marine and other aquatic ecosystems and the ecological complexes of which they are part; this includes diversity within species, between species and of ecosystems'. This definition therefore recognises two components of biological diversity: the biological composition and the preservation of the ecological complexes of which they are part, that is to say ecological functionality.

1.6b Types of biological diversity

The diversity of life can be assessed at a number of hierarchical levels. The term biodiversity appeared in the conservation literature in the early

1980s; it was poorly defined but often referred to the diversity of species in a system. This is now generally regarded as the middle tier of the hierarchy. The simplest scheme involves three levels of biological diversity (Norse *et al.*, 1986): genetic, species and ecosystem.

The least visible level of biological diversity is that contained within the genetic material of the population. Most species consist of separate, to some extent isolated, populations. These tend to differ genetically, either due to local selection (for example, salmon being adapted to the conditions in their maternal river system) or due to genetic drift and the accumulation of random mutations. So genetic diversity includes at least two distinct levels: within population and between population. This genetic diversity is the basis for evolution; its loss, while not necessarily leading to the immediate loss of a species, may prevent it from adapting to meet future changes in the environment.

Species diversity is the most intuitive of the three levels of biological diversity. At its most straightforward, it is simply a count of the number of species in an area – the species richness (S). In addition there are many indices of diversity which purport to improve upon this simple measure; Magurran (1988) provides an excellent introduction to these.

There are many biological patterns to species richness, and therefore scales at which it can be measured, so it is important to clarify the differences in types of diversity.

- α-diversity (alpha diversity) is the diversity of species within a given community at a single locality, and is therefore often referred to as within-habitat diversity.

The following patterns relate to the ecosystem component of diversity.

- β-diversity (beta diversity) refers to the change in species diversity across habitats or ecosystems, and is often referred to as between-habitat diversity. Within a given locality it can incorporate the α-diversity of each identifiable

Table 1.2 *Species richness and endemism of Penaid shrimps in four marine regions*

Biogeographic region	Number of species	Number of endemic species
Indo-west Pacific	125	124
East Pacific	16	16
West Atlantic	21	18
East Atlantic	16	3

Source: Dall *et al.* (1990).

community into a total regional species richness. It can also be used to refer to the species pool for a given type of habitat. Several discrete components of the habitat may each contain a subset of the total number of species found in all the components combined. For example, a number of different lakes will have a total, the species pool, incorporating all those species that could potentially occur in each lake (the β-diversity), while the actual species in each lake make up its α-diversity.

- γ-diversity (gamma diversity) refers to diversity at large geographical scales. It incorporates patterns such as that of species richness for most taxa being higher in the tropics than at high latitudes. It also includes geographic clines within a given latitudinal range, including, for example, a common decline in marine species richness away from the Indomalayan region (Table 1.2).

1.7 Ecological functioning

1.7a Disrupting ecological systems

If human exploitation causes the abundance or distribution of even a single species to be altered, then it has had an ecological impact. Organisms within a single aquatic system are, however, linked in a complex series of interactions, so that an impact on one species will also affect others and, ultimately, the entire ecological community. For this reason, human impacts extend far beyond those species that are directly affected by human activity, such as harvestable species or those most sensitive to pollutants.

The importance of the entire ecosystem in maintaining sustainable numbers of an exploited species is clear: each species relies upon others as sources of food, habitat, predators of potential competitors and a whole host of other reasons. Its requirement for a suitable physico-chemical environment is also obvious, but here too, ecological processes are important as many environmental parameters are determined by living organisms and their interactions. Indeed, human life depends heavily upon ecological processes to maintain an appropriate environment and, as water of sufficient quantity and quality is an essential for human life, and living aquatic resources play a key role in providing resources, ecological processes operating within aquatic systems are of direct importance to ourselves.

1.7b Alternative stable states

Many systems exhibit alternative stable states – two or more different types of community organisation, the predominant state in any one locality being determined by the site's history. From a human perspective, there may be a more desirable and a less desirable state, human intervention switching the system from the former to the latter. The following examples illustrate this process. In each case, the community structure is shifted by a major human intervention from one type to another, the new structure being resistant to successional processes that would return it to its former state.

Coral reefs and ship groundings

When a ship runs aground on a coral reef it will physically impact the reef. When the ship is

Figure 1.9 *The site of a ship grounding on a coral reef in the Red Sea. Notice the unconsolidated rubble produced by the impact with the reef and the lack of new coral growth on this mobile material. (Photo by Susan Clark)*

removed – either under its own power, towed away or by sinking into deep water – the reef will be physically scarred and the reef material broken away by the vessel will remain as unconsolidated rubble (Figure 1.9). Recovery of the impacted material can proceed quite rapidly. If the loose material falls down the reef slope into deep water and a new hard reef face of solid carbonate rock is exposed, then coral recruits will settle and a new functioning reef system will develop at the site. If, however, a large amount of unconsolidated debris remains on site, this will prevent coral recruitment.

Corals are relatively slow growing and new coral recruits are very vulnerable while they are small. Even low levels of wave action will be enough to move the rubble; this will dislodge recruits that have settled on the rubble particles and abrade recruits on adjacent areas of reef. In this case the system may switch to an alternative state as the rubble is seasonally colonised by ephemeral algae which provide a dense cover during periods of stability and further exclude coral recruits. During the stormy season these algae are eroded and abraded, but they re-establish again as soon as the storm season is over. These changes

cascade through the remainder of the ecosystem as the reef fish are displaced and replaced by small algal grazers.

Lake eutrophication

Lakes subject to high nutrient loading will normally be dominated either by macrophytes or by algae. In both cases the primary producers benefit from the high nutrient levels and each type suppresses growth of the other. Macrophytes absorb extra nutrients and suppress wave action that would otherwise re-suspend bottom sediments and lead to shading, while high densities of algal cells contribute to water turbidity and the consequent shading suppresses potential vegetation growth. From a human perspective, the former is normally the most desirable, but the macrophyte system is easier to destroy and replace with the algal system than vice versa. If macrophytes are disturbed or killed by chemical weed killers, mechanical damage by boats or excessive grazing, the water will remain nutrient-rich but algae, with life cycles measurable in days, will be able to respond to the unused nutrients more rapidly than macrophytes, whose life cycles require several months. Therefore, the system will change to an algal dominated system, equally stable to the macrophyte system it replaces but much more difficult to remove.

1.7c Ecosystem services

Unlike harvestable resources, or the water itself, it is difficult to quantify the importance of ecological processes in economic terms. However, their importance must not be under-estimated. The indirect benefits that living organisms provide have been termed **ecosystem services**.

Ecosystem services

Humans continually exploit natural ecological processes, often unwittingly, to ensure that their environment remains healthy. The self-cleaning mechanisms in water bodies are widely understood. Riparian strips of natural or semi-natural vegetation effectively remove contaminants such as excess nutrients from groundwater, while wetlands clean polluted water, removing not only nutrients but organic matter and heavy metals. Ensuring clean water for the domestic supply is made easier by ensuring that the catchment in which the water is collected is as free as possible from development, including agriculture. On a more global scale, oceans in particular play an important role in nutrient cycling and in removing excess inputs such as carbon dioxide from the atmosphere. These processes are biologically mediated and, although it is sometimes possible to identify specific groups of organisms that are involved (for example, denitrifying bacteria removing excess nutrients in wetland soils; marine phytoplankton absorbing carbon dioxide and facilitating its incorporation into deep ocean sediments), the integrated nature of ecosystems means that detrimental influences on species not directly involved will almost certainly have knock-on effects that could influence the efficiency of the process.

We have little understanding at present of the importance of biodiversity in ecosystem processes – whether all the species are required or whether the system would function perfectly well with just a few key species – but the first hints that biodiversity in itself is a key component of a healthy ecosystem are being identified. For example, leaf litter decomposition is an important ecological process in upland streams, terrestrially derived leaf litter providing the main energy source for the entire river. Leaf litter breakdown is mediated by physical processes, invertebrate detritivores and microbial decomposers (Gessner *et al.*, 1999). It has long been appreciated that removing one of these components reduces the efficiency of detritus breakdown, and therefore the availability of energy elsewhere. However, Jonsson *et al.* (2001) have demonstrated that, even within one of the

biological components, the more species that are present, the more efficient the process is. They determined that processing rates were more rapid with increasing numbers of invertebrate detritivore species, species richness being more important than numbers of individuals. A challenge for ecological science is to determine the extent to which this conclusion, derived from a relatively small component of a small system, is analogous to the rest of the world. Each species has its own unique niche (**Section 1.5b**), and therefore a subtly different influence on the environment to other similar species, so optimal ecosystem functioning may be dependent upon optimal diversity. Ensuring a healthy environment for humans may, therefore, be more dependent upon high diversity than we currently appreciate.

When we understand the mechanism behind an ecological service, we are able to manipulate the environment to our benefit. An incomplete understanding of the process, or assumption that the service required is the role of one component of the natural system, can lead to problems.

1.8 Summary

- Water is necessary to sustain human life, while its exploitation for various purposes is essential to maintain modern human societies.

- Human use of water impacts upon ecological systems, whose health is often essential to ensure continuity of the components exploited. Therefore, a clear understanding of ecology is essential.

- Key ecological concepts include the community and the niche. Community descriptors such as food webs are valuable, but their limitations need to be appreciated.

- Food webs share many common features, suggesting the presence of natural 'assembly rules'.

- Non-feeding interactions such as competition are also important interactions within communities.

- Biological diversity is a key measurement of ecological systems. There are many biological patterns to species richness, and therefore scales at which it can be measured, to which the terms α-diversity, β-diversity and γ-diversity can be applied.

- Human exploitation has ecological impacts on aquatic systems. Any impact on one species will also affect others and, ultimately, the entire ecological community.

- The indirect benefits that living organisms provide have been termed ecosystem services. These are of equal importance to human society as direct benefits.

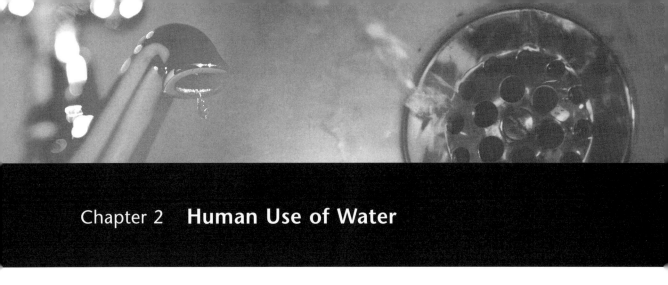

Chapter 2   Human Use of Water

2.1 Introduction

2.1a Uses for water

Water is, of course, an essential requirement for life, but in addition to this it has many properties that make it valuable to human society. Its physical properties are considered in this chapter, while those relating to its use as a cooling fluid and its solvent properties are dealt with in **Chapter 6** in relation to pollution.

Domestic supply

Humans need water for direct consumption in the form of drinking and cooking, and for sanitation. As affluence within a society increases, the amount of water used domestically also rises. The amount drunk per capita from a domestic supply may remain constant, or even decline if industrially manufactured drinks are bought. However, consumption for washing increases and novel uses may become important, including watering gardens or maintaining swimming pools.

Irrigation

Irrigation – the application of water to crops – accounts for 70% of human consumption of water. Around 12% of the world's cultivated land is irrigated, mainly in arid and semi-arid areas, but increasingly in naturally wet areas, such as western Europe. As populations increase, so the use of water for irrigation is likely to rise (Figure 2.1).

Figure 2.1 *(a) Area of land under irrigation through the 20th century. (b) Changes in volume of water used by human activity. The volume increased tenfold during the 20th century, with the majority used for irrigation purposes* (based on Young *et al.*, 1994).

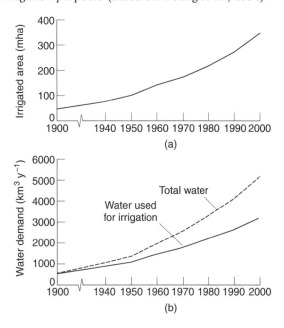

Power generation

Some of the earliest structural alterations to water bodies were to harness water power. In these early industrial plants, such as flour mills, the water power was applied directly to the machinery. More recently, water power has been harnessed to generate electricity, which is used by multiple consumers remote from its source, and the scale of power generation has therefore increased. As the demand for power and the development of electricity has increased, so power generation has become a more important use for water resources.

Power is generated in two main ways. In rivers, the kinetic energy released when water drops in elevation, as a consequence of gravity, is exploited. The higher the hydraulic head (the vertical distance that the water falls), the greater the power generation potential, so the best rivers are in steep valleys, which normally occur only at high altitudes. The same principle is exploited by tidal barrages, which use the difference in water level between high and low tides as the hydraulic head. The second means by which power may be generated is to capture the energy in wave action. Tides are essentially very long wavelength waves. The sea surface is continually disturbed by wind waves and swell and the difference in height between the crest of a wave and the trough represents a source of potential energy.

Industry

Many industrial processes, such as brewing and food processing, require large volumes of clean water to be incorporated into the manufacturing or associated cleaning process. Such developments require a reliable source of water, in the same way as domestic supplies, but often in much greater volumes. Whereas the earliest water supply reservoirs for urban centres were developed mainly for domestic water supply, therefore, some recent developments have been to guarantee water supply for industry.

The use of water as a cooling medium is of major importance in power generation using fossil and nuclear fuels, and in heavy industrial processes such as steel-making. The cooling process adds heat to the water, which is generally then returned to the water body from which it is extracted; this extra heat is best considered as a form of pollution, and is considered further in **Section 6.5**.

Transport

Use of water for transport, initially of people and their personal possessions (including domesticated animals) and then for commercial trade, is one of the oldest types of exploitation. For example, the first humans arrived in Australia around 60 000 BP, a feat that required crossing large stretches of water, even with the advantage of lowered sea levels at this time.

Water transport is still of crucial importance to modern human societies, with the vast majority of goods transported overseas being carried by boat rather than by air or pipeline. Navigable inland waters are also important, particularly in heavily populated parts of the world.

In the United Kingdom, canals were widely used for transporting industrial products before the development of railways; between 1750 and 1840, some 2800 km of canal were constructed. Almost all of these canals are now either abandoned or used solely for recreational boating. They are now, therefore, often very valuable as wildlife habitats, particularly those in urban areas that act as wildlife corridors. They can, however, also act as corridors for translocation of species between catchments (see **Section 2.7b**).

2.1b Infrastructure

Dams and reservoirs

Many of the uses for water described above can be enhanced by dam construction. A dam is a structure built across a river channel in order to retain a greater depth or more constant volume of water behind it than normally occurs. The most effective way to ensure water supplies in non-arid areas is to dam river valleys, allowing the river to fill up the resulting impounded area and create a reservoir. Such structures dampen seasonal fluctuations in flow by storing high discharge volumes during wet seasons, and they also raise the water level, increasing the hydraulic head of water and ensuring continuity of supply for power generation. If the dam is low, it may simply fill the river to its natural bankfull level, but normally dams create a reservoir, wider and deeper than the river channel and displaying features of a lake rather than a river.

Dams have been employed as water storage systems for thousands of years, but for most of human history they were very small structures in headwater streams, rarely creating a reservoir greater than a few square metres in area. They were limited in size by the absence of appropriate technology to build large structures. During the 19th century, earth dams were developed that could be built across wide valley mouths, up to 1 km across, and therefore store volumes of water equivalent in size to large lakes. During the 20th century, concrete technology allowed high dams to be built, so that narrow but high valleys could be effectively dammed. The number of high dams (15 m or higher) increased from around 5000 in 1950 to 40 000 in 2000, in addition to which there are approximately 800 000 smaller dams. Between them, these dams can store 20% of the Earth's renewable runoff, and around 60% of the 200 largest rivers arc at least moderately affected by impoundment or diversion.

The marine environment is relatively free from barrages, as their construction is only possible across narrow estuaries or other indented coastal features. Estuaries have been dammed to create freshwater lakes, for aquaculture (**Chapter 5**) or recreational water-based activities (**Chapter 8**). Tidal barrages are a relatively new development,

designed to exploit tidal movement to generate electricity. They allow the rising tide water to enter an impoundment, then release it once the water level outside has dropped, creating a hydraulic head.

Wave power

Commercial exploitation of wave energy has to date been limited, but a number of prototype schemes have been developed. Some of these are on-shore and use the increased height of waves in shallow water to produce a greater difference in height. The waves are generally channelled into a flume where they either directly drive a turbine, or the force of the air pushed in front of the wave is used to turn a turbine. Such on-shore schemes involve large-scale engineering in the coastal zone. The alternative is offshore wave energy plants. There have been designs involving the waves moving a buoyant piston in an enclosed tube, the movement of which forces air through a turbine, but the most widely known design is the 'Salter's duck'. This is a floating structure which rocks up and down as waves pass under it; the rocking motion pumps hydraulic fluid past a turbine. Large numbers of 'ducks' would be needed, moored together in strings, to generate significant quantities of energy. This offshore design would reduce the wave energy inshore of the 'ducks'. While this could have advantages in reducing the need for coastal defence works, it may also have major ecological effects on those systems which need the energy input from the waves to keep them clear of sediment (rocky reefs) or supplied with food.

Water transfer schemes

Water transfer schemes are mechanisms employed to carry water from areas of surplus runoff to areas where demand exceeds local supply. Water may be transferred from its source directly to the area where it is used or, in the case of **inter-catchment transfer** (ICT), from one river to another, the receiving river then transporting the water to an appropriate extraction point. If the source and sink rivers are geographically distinct from each other – they discharge separately into the sea or, if entirely landlocked, their waters never mix naturally – then this type of transfer is better referred to as an **inter-basin transfer** (IBT). In all cases, channels are created to transport water, although they may be either surface channels or pipelines. Normally, water transfer schemes include dams to facilitate storage of water on the source and/or sink rivers. Some examples are discussed in **Box 2.1**.

Channelisation, ports and canals

Travel by boat does not necessarily lead to alteration of the environment, although as boats became bigger, more complex port and harbour facilities were required, with consequent detrimental impacts on riparian habitats. Initially, human population and trading centres were situated mainly on navigable water bodies. In recent centuries, however, increasing size of vessels and development of population centres close to sources of industrially valuable deposits, such as coal and iron, led to the need for water transport remote from naturally navigable water bodies. As a consequence, rivers have been straightened and deepened to allow passage of vessels, while artificial canals have been dug where no suitable river exists.

Water extraction

Infrastructure for extraction is very localised, normally consisting of an artificial side channel or pipe into which water is diverted by gravity or by pumping. Large industrial plants are frequently situated adjacent to large water bodies, from which they extract cooling water (see **Section 6.5**). This passes through a series of grilles to trap objects that may block or damage cooling pipes, then is released at a discharge point.

Desalination of sea water is a method of obtaining fresh water in some areas of the arid tropics and subtropics. There are no effects of depletion, as water is extracted directly from the sea, and detrimental effects are limited to localised influences around the extraction point, although these can include locally significant quantities of pollutants associated with the desalination process (Resources Agency of California, 1997).

2.2 The river basin as an ecological unit

Human influences upon the physical environment of aquatic systems are confined almost exclusively to inland and coastal waters. Furthermore, issues such as water extraction, flow regulation and impeding flow are wholly or mainly pertinent to fresh waters. Therefore, appreciating the influences of human activities upon freshwater systems requires an understanding of the links that occur within river basins. These are explored more fully in Dobson & Frid (1998), but a summary of the main features is presented here.

Water on the Earth's land surface moves downhill. Apart from some localised runoff near the point of precipitation, this movement is within river channels, which converge into river basins. Almost all lakes and freshwater wetlands are intimately associated with river basins, often either as sources or sinks of water, but generally as features that simply slow down the passage of water to the sea. The one-way movement of water ensures that characteristics of upstream parts of a river and its catchment will influence water quality downstream.

A river basin is an integrated system. Longitudinal links between different parts of the river system occur, mainly in the form of upstream zones influencing downstream sections, but with some reverse interactions. Also of importance are lateral links between the river and its adjacent catchment.

Temperature and oxygen concentration

River systems rise at relatively high altitude and flow to lower altitudes. Therefore, unless a river is large enough to cross climatic zones and flows towards a polar region, its average temperature will inevitably rise with increasing distance from the source, reducing the water's capacity to hold oxygen in solution. As distance from the source increases, a river will normally be carrying an increasingly greater load of contaminants, both particulate and in solution, derived from the catchment. If high in organic matter content, this will increase the water's biochemical oxygen demand (BOD; see **Section 6.2a**).

Sediment transport

Rivers transport sediment eroded from their catchments. This is normally picked up in the upper reaches, in the erosion zone. It is then carried downstream to be deposited at lower altitudes. A river system may have a single major deposition zone, where it reaches the sea and loses its momentum in an estuary or delta; alternatively, it may have a series of deposition zones inland, including those from which sediment is continually deposited and re-eroded.

Biological links

Physical and chemical links are matched by biological links. Although primary production may be locally important, rivers depend heavily for their energy upon inputs of detritus from the surrounding catchment. This is especially true of rivers that are heavily shaded by riparian vegetation, which impedes light penetration but provides a direct input of leaf litter. Beyond the erosion zone, most rivers are too deep or turbid to support much primary production, and inputs of detritus from the immediate catchment are small relative to the volume of water. Instead, a large proportion of their energy derives from detritus carried downstream from the headwaters. In temperate regions, inputs of detritus to headwaters may be highly seasonal, coinciding mainly with autumn leaf fall, but a combination of physical attrition, microbial decomposition and invertebrate consumption breaks this detritus down and releases it into the water column over an extended period, providing a continuous energy supply for the river further downstream.

Lateral links

In the headwaters, lateral inputs from the catchment are the primary determinants of the nature of the river channel. Water derives mainly from the catchment, while the role of terrestrial vegetation as a supply of energy has been described above. Further downstream, however, the role of the catchment does not necessarily decline. Lateral inputs continue, although they are reduced relative to the total volume of water as the river picks up tributaries and increases in size. However, beyond the erosion zone many rivers flow across floodplains, into which their flood waters extend during periods of high discharge. This allows lateral interactions to occur; as the flood water spreads across a floodplain it deposits sediment and detritus, but water receding back into the river will take detritus with it. Therefore, flooding is a two-way process, allowing both river and catchment to exchange nutrients, energy and sediments.

Upstream movements

Sediment and detritus, carried passively by the flow, are only able to move downstream or laterally onto the floodplain. However, some living organisms move upstream, creating an important reverse link. Anadromous fish, such as salmonids, move from the sea into the headwaters of rivers in order to spawn (see **Section 5.7**, **Box 5.1**). Spawning salmon transport biomass and nutrients from the sea to headwaters. Pacific salmon of the genus *Oncorhynchus* undergo annual spawning runs, following which adults die *en masse*. Their decomposing carcasses and the eggs that they have laid provide important sources of nutrients and energy, both to stream ecosystems (Bilby *et al.*, 1996) and, by activity of terrestrial predators such as bears, surrounding riparian systems (Hilderbrand *et al.*, 1999). Atlantic salmon (*Salmo salar*) mortality is much lower than that of Pacific salmon and many return to the sea after spawning. Inevitably, however, some will die during the spawning run, contributing marine-derived nutrients to the river system (Elliott *et al.*, 1997).

Role of floods

Flooding is a natural feature of all rivers subject to variable flow. Headwater rivers, with relatively high inputs directly from the catchment, respond rapidly to precipitation events. Rainfall brings increased runoff from the catchment and discharge rises accordingly. Therefore floods are weather dependent and, from the perspective of river organisms, unpredictable (or stochastic). A flood event is a situation of stress, to which organisms need to be adapted if they are to persist. A common strategy is to exploit flood refugia – areas of the river channel where increases in flow are dampened, reducing the possibility of being washed downstream. Flood refugia occur in backwaters, downstream of obstacles such as large boulders or debris dams, in deep pools and in sediments beneath or beside the river channel. They are features contributing to the heterogeneity of the river channel structure and therefore the heterogeneity of water movements within it.

In mountainous areas, rivers are often tightly constrained within narrow channels and floods produce extreme conditions of high discharge and velocity, with powerful erosive forces. Where the valley widens, however, the river in flood will be able to burst its banks, spreading water, sediment and energy across a floodplain. Dissipating energy in this way reduces the erosive force of the water

so that, further downstream, water levels will rise but most of the destructive power of the flood is reduced.

Stochastic floods are important determinants of benthic community structure in headwater rivers. By causing frequent disturbance they act as a reset mechanism, creating disturbed habitats in which diverse species assemblages persist. A flood acts as a major disturbance event but the community normally recovers rapidly, demonstrating strong resilience to disturbance episodes. Indeed, such disturbance is probably necessary to allow many species to persist – without frequent floods, sedimentation, overgrowth with macrophytes or establishment of competitors may lead to elimination of flood-adapted species.

Away from the headwaters, flooding is determined by inputs from the catchment. If inputs are from multiple tributaries, the river is unlikely to flood in response to single weather events affecting a small part of the catchment, but instead shows seasonal changes in discharge. Large rivers draining mountain ranges with a clear snowmelt period, or a catchment with distinct wet and dry seasons, have predictable seasonality to their flooding (Figure 2.2). Flood waters rise slowly, as increased water is brought down from many tributaries and, rather than a stressful event, the flood is a feature to which organisms become adapted. Large rivers typically have seasonally flooded wetlands, and fish species in particular synchronise their life cycles to these, moving out of the river channel to spawn in the spatially heterogeneous and food-rich newly flooded area, their offspring gradually migrating to the main channel as flood waters slowly recede.

2.3 Effect of dams and barrages

Building a dam across a river creates a barrier to free movement along the channel. This has a variety of impacts upon channel processes, which can be divided into those having their greatest

Figure 2.2 *(a) Effect of a dam on river discharge patterns. Colorado River, Grand Canyon, USA. Prior to closure of the Glen Canyon Dam in 1963, the Colorado River was marked by major seasonal fluctuations in flow. Since closure, discharge has become approximately constant, with small daily fluctuations marking changes in demand for hydroelectric power. The discharge for spring and summer of 1996 is presented, illustrating the effect of the controlled flood which commenced in March of that year (from Collier et al., 1997). (b) Daily fluctuations in water level below a dam as a result of hydroelectric power generation, showing a constant daily rhythm during winter (above) and a more intermittent series of fluctuations during spring (below), when demand for power is lower. River North Tyne, Kielder Water, northern England (from Boon, 1993).*

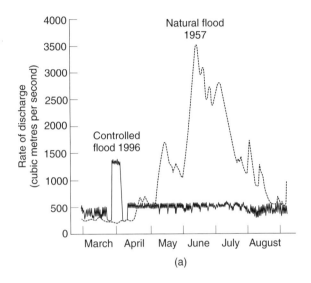

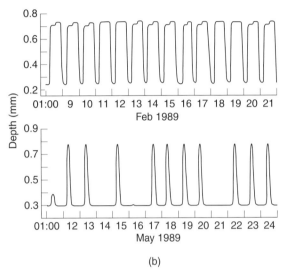

(a)

(b)

effect upstream of the dam and those operating downstream.

2.3a Upstream of the dam

The ideal place to situate a dam is in a position where the valley is relatively narrow and high, so that a large body of water can be contained by a relatively small construction. Such river valleys are generally confined to the erosion zone, where the river is torrential and subject to stochastic flow events. Therefore, the water body upstream of the dam undergoes a fundamental change in structure as it is converted into a less mobile lentic form. Changes in the physico-chemical environment, and therefore in community structure, wrought by dam construction may be divided into short-term and long-term influences.

Short-term temporal changes

Initial changes are very rapid, as the lotic system changes into a lentic environment, and as flooding converts a terrestrial habitat into an aquatic one. The rate of filling is determined by local conditions; dams constructed in high discharge areas for hydropower generation may fill over a few weeks or even days, whereas those in arid areas constructed to provide irrigation water may take many years to fill and, in some cases, may never reach their theoretical capacity. As the reservoir fills, however, terrestrial vegetation is killed and begins to decompose, while the river deposits organic matter into the newly lentic environment. A consequence of this is eutrophication, which may be very rapid and unpredictable. Phytoplankton, uncommon in the flowing river, are able to establish and rapidly take advantage of the high nutrient loads. Oxygen levels, however, decline and where large amounts of terrestrial vegetation have been inundated, deoxygenation of the reservoir depths is common. In tropical reservoirs, the water close to the base of the dam may rapidly become anoxic; the anoxic zone gradually creeps outwards, killing benthos in the original channel and forcing mobile species, such as fish, into the littoral zone and the mouths of tributaries (Agostinho *et al.*, 1999).

The timing of dam closure may have important consequences for faunal composition upstream of the dam. The Paraná River in Brazil supports several migratory fish species, for which the Sete Quedas Falls, separating the Upper Paraná River from the Middle Paraná River, were a barrier

to upstream movement. When Itaipu Reservoir was impounded in 1982, it flooded the waterfall, allowing free movement of fish upstream, but creating a barrier in the form of the dam further downstream. The dam was closed in November (late spring), trapping between it and the falls 13 species that migrate upstream in early spring, and allowing them access to the Upper Paraná River for the first time. Sixteen other species, which migrate upstream in early summer, were confined to the region downstream of the dam (Agostinho *et al.*, 1999). In this example, delaying dam closure by two months would have captured more migratory species and resulted in a very different fish fauna in the reservoir and the upper catchment of the river.

Figure 2.3 *Changing abundance of fish in river–reservoir system: the Itaipu Reservoir, Brazil. Frequency of abundance (%) is shown for three years – 1977, 1987 and 1997. The river was impounded in 1982 (from Agostinho* et al., *1999).*

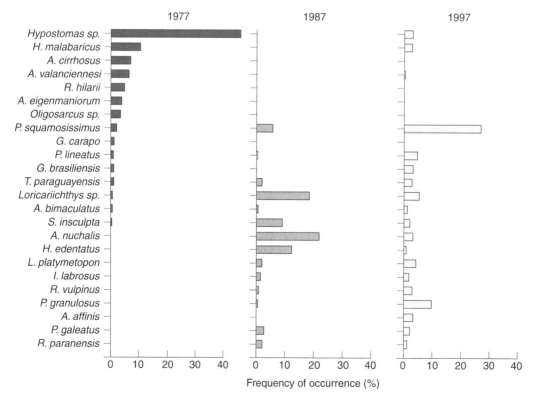

Long-term temporal changes

Initially the organisms inhabiting the reservoir are river species. As the inundated area expands, species that require lotic conditions are rapidly eliminated, but their replacement by lake species is determined by the proximity of suitable sources of colonists. Species previously present in low numbers may become dominant (Figure 2.3). Plankton, which are unable to persist in rapidly flowing rivers, will rapidly establish, particularly if there are natural lakes upstream of the dam. The community within the reservoir is, however, likely to undergo frequent changes in structure as new species continue to establish, frequently augmented by deliberate introductions into the newly created lake (**Box 4.3**). Fluctuations may be exacerbated by the disruptive effect of artificially controlled fluctuations in volume of the reservoir, whose effect is felt particularly by benthos in the shallower regions.

Dams trap around 98% of sediment carried by a river and, in so doing, gradually infill. Sediment is deposited unevenly across the bed of the reservoir; in small reservoirs it may be evenly sorted, with coarser sediments close to the mouth and progressively finer sediments, richer in organic matter, in increasing proximity to the dam. In large reservoirs, sedimentation may be centred around tributary mouths. Deposition alters benthic habitat structure and may bring with it large quantities of nutrients and organic matter, exacerbating eutrophication and deoxygenation problems.

Drawdown

Unlike natural lakes, water supply reservoirs undergo drawdown, major and often rapid reductions in volume in response to extraction (Figure 2.4). Drawdown limits the ability of aquatic macrophytes to establish in the littoral zone and therefore to stabilise sediment. It also influences benthic fauna. An established reservoir will normally support a benthic fauna typical of lakes, adapted to the soft sediments that aggregate on its bed. Drawdown exposes the benthic environment to the air, often for extended periods, and as organisms such as oligochaete worms, chironomid larvae and bivalve molluscs, which dominate these systems, are relatively immobile they can suffer high levels of mortality. Benthic invertebrate diversity and density is expected to decline in natural lakes from the littoral to the profundal zone, whereas in reservoirs the maximum species richness and biomass is found immediately below the level of maximum drawdown the previous season. Above this line, exposure to air and, in temperate regions, winter ice kills most individuals (Kaster & Jacobi, 1978).

Movement of organisms

Dams restrict longitudinal movements of animals, particularly migratory fish. Where salmon are important, this can be rectified by providing 'fish ladders', artificial channels that either slope gently or comprise a series of low steps no higher than the rapids over which the fish are able to jump. Another mechanism to maintain populations of migratory fish is to include salmon traps, which capture fish below the dam as they move upstream. They are then carried by road around the dam and released in the reservoir upstream.

2.3b Downstream of the dam

Flow regulation

When a dam is built across a river it dampens fluctuations in flow, whether seasonal or stochastic; indeed, this is the purpose of many dams, not to store water but to regulate flows. By storing water that would otherwise spill onto a floodplain, a barrage reduces the amount that reaches flood wetlands. Such wetlands are then threatened with drying, and with reduced fertility following severance of the sediment normally deposited by riverine floods (see **Box 2.2**).

River channel form is established over a long period of time. The structure of a channel is

Figure 2.4 *Examples of dams, illustrating two of the mechanisms by which they modify the aquatic environment. (a) Errwood Reservoir, northwest England. Although normally full, with little fluctuation in volume, extended dry periods lead to major draw-down, as extraction rate exceeds replenishment. This reduces the volume considerably and also exposes benthic environment to the air. (b) Llyn Brianne dam, Wales, showing the plume of cold deep water released into the outflow river. The discharge is forced into a spray in an attempt to re-aerate it. (Photos by M. Dobson)*

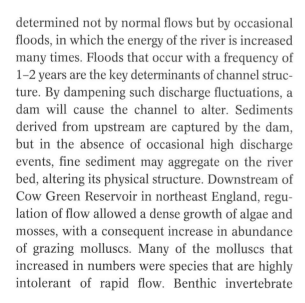

(a)

(b)

determined not by normal flows but by occasional floods, in which the energy of the river is increased many times. Floods that occur with a frequency of 1–2 years are the key determinants of channel structure. By dampening such discharge fluctuations, a dam will cause the channel to alter. Sediments derived from upstream are captured by the dam, but in the absence of occasional high discharge events, fine sediment may aggregate on the river bed, altering its physical structure. Downstream of Cow Green Reservoir in northeast England, regulation of flow allowed a dense growth of algae and mosses, with a consequent increase in abundance of grazing molluscs. Many of the molluscs that increased in numbers were species that are highly intolerant of rapid flow. Benthic invertebrate

biomass was also enhanced by a novel food source – large numbers of plankton washed out of the reservoir. However, although biomass and abundance of benthic invertebrates increased overall, several taxa typical of torrential streams disappeared and were replaced by species more typical of slow-flowing rivers. For example, most stonefly species were lost as a consequence of dam construction (Armitage, 1977).

It is not only the magnitude of flow that may be affected by a barrage, but also its variability (Figure 2.2). Dams normally serve to dampen fluctuations, by capturing and retaining unusually high discharges. In some situations, however, the opposite may occur: a dam constructed to supply hydroelectric power may release water in daily surges, corresponding to the peak demand periods for electricity (Figure 2.2). Some dams built for this purpose then require a second dam a short distance downstream to capture these surges and moderate the flow further downstream.

Sediment starvation

The middle reach of a river system is generally a sediment-carrying zone, in which there is no net deposition or erosion of sediment. However, this masks dynamic movement of sediments and annual erosion may be very high, matched only by rapid redeposition. If the source of fresh sediment is severed, the river may continue to erode its bed without replacing lost sediment. Dam construction on the River Danube in Austria has led to scouring and lowering of the river bed further downstream, threatening wetlands as the water table drops (see Dobson & Frid, 1998). A similar process is occurring in the River Rhine (see **Section 2.4**).

Temperature

Downstream discharge normally derives from deep water close to the foot of the dam. Insulated from the air by surface water layers, this deep water fluctuates in temperature less than surface waters, so is normally slightly warmer than the original river would have been during the winter but cooler during the summer. As river species are generally limited by warm summer temperatures, dam construction in temperate regions allows headwater species to extend their ranges downstream in the cold water plume. In the Gunnison River, a tributary of the Colorado River, a series of dams has allowed the cold water fauna to extend their range 60–70 km downstream, a vertical drop of around 500 m elevation (Ward, 1998).

Breaking river–floodplain interactions

Larger rivers tend to be characterised by seasonal fluctuations in flow, and their inhabitants have life cycles adapted to follow and exploit these. Therefore, breaking the cycle by excessive extraction can threaten these organisms. Over-extraction can reduce the extent to which a river inundates floodplain wetlands, but damming a river to capture and store its water can cause it to stop altogether.

The River Niger in West Africa has been extensively modified by the construction of Lake Kainji, which effectively traps seasonal flood waters. The Niger enters the Atlantic Ocean through a large, complex delta containing many small lakes. These lakes are hydrologically separated from the river for much of the year, but prior to dam closure the monsoon flood restored connections on an annual basis. Among the species that took advantage of this flooding was the sardine *Pellonula leonensis*, which migrated into lakes in the delta to spawn. After spawning, the adults returned to river, while the juveniles fed on zooplankton in the lakes before they, too, migrated into the main river channels. Since closure of the dam the zooplankton of Lake Oguta, a small lake in the Niger Delta, has become dominated by the formerly very rare calanoid copepod *Propodiaptomus lateralis*. This species has benefited from the inability of its main predator, *Pellonula*, to migrate into the lake. The fish, in turn, has become much less abundant since connections to its spawning grounds were severed (Dumont, 1999).

Coastal dams

Several rivers in coastal parts of South Africa have been dammed close to their mouths, with consequent influences on adjacent estuaries. A common occurrence is for reduced freshwater discharge to allow salinity to rise in the estuary, so that it loses its estuarine features. Such conditions can be permanent, as in the case of the Kromme River, dammed several kilometres upstream of its estuary and with an outflow of less than 2% of the mean annual runoff. As a consequence, the estuary has a salinity approaching that of the open sea, and the freshwater inputs serve mainly to stop it from being hypersaline due to evaporation (Bate & Adams, 2000). Elsewhere, extreme drought can cause short-term changes, which may be devastating because communities are unable to adapt to changing conditions. Outflow from Lake Mzingazi ceased in February 1992, allowing saline water to penetrate the lower Mzingazi River from the sea. This water penetrated laterally up to 20 m into groundwater, causing the death of a large number of trees in swamplands (Cyrus *et al.*, 1997).

2.4 Effect of channelisation

Channelisation is the containment of a river within pre-defined limits using artificial structures. As the aim is to improve navigation or to create an effective means of flood water removal, rivers are normally straightened and deepened; as a consequence they are normally reduced considerably in width, and defensive banks destroy riparian habitats. Occasionally, a small urban river may be completely buried, if small enough to fit into a pipe. This is the fate of most of London's rivers, about 30 of which are buried underground. The River Westbourne, for example, surfaces only to fill the Serpentine, an artificial lake in Hyde Park, before continuing, piped, to the Thames.

Channelisation cuts rivers off from their catchments, even where natural vegetation remains. A consequence of this is increased variability in flow, river discharge responding more directly to precipitation events than it would if vegetation were to dampen fluctuations. The Comite River catchment around Baton Rouge, Louisiana, illustrates the effects of land use change. Between the early 1950s and the late 1970s, expansion of agricultural land and rapid urbanisation led to a major increase in the number and size of floods recorded on this river (Table 2.1). The total volume of water passing along the river did not change significantly, but precipitation was no longer able to seep slowly through groundwater and riparian wetlands, so entered the river channel in a series of flushes and floods. Downstream of the city, the turnover rate of water in Lake Maurepas increased by 30% (Stone *et al.*, 1982).

River and coastal dredging is carried out to ensure that shipping channels are deep enough to take ships. Dredging creates instability in the bed of the water body and normally needs to be repeated at frequent intervals, so disturbance is common. Therefore, its effects are generally negative, although in some cases it benefits the local environment by increasing habitat diversity (Attrill *et al.*, 1996).

Inland boat transport requires navigable rivers or artificial canals. Canals may be new cuts, created from scratch and fed by reservoirs, or they may incorporate rivers as a source of water, the canal being so large relative to the river that flow may almost cease. The Manchester Ship Canal in northwest England is such a structure, created

Table 2.1 *Increase in discharge rate and flood frequency of flood waters in the Comite River around Baton Rouge, Louisiana*

	Percentage increase in maximum recorded discharge between the 1950s and 1960s	Number of annual floods		
		1950s	1960s	1970s
Upstream of Baton Rouge	−2	2.4	2.6	2.6
Downstream of Baton Rouge	23	3.8	3.9	5.3

Source: Stone *et al.* (1982).

Figure 2.5 *The Manchester Ship Canal. The entire flow of the torrential River Irwell is diverted into the canal, whose much larger volume causes it to become a very slow flowing system. (Photo by Skyscan/William Cross)*

by trapping the waters of the River Irwell, formerly a tributary of the River Mersey, and extending the torrential river into a large linear lake (Figure 2.5).

Channelisation of the River Rhine

The River Rhine in central Europe has been impacted by a series of structural alterations along its length (Havinga & Smits, 2000). The Upper Rhine formerly consisted of two morphologically distinct stretches (Figure 2.6a). Between Basel (Switzerland) and Rastatt (Germany) it was heavily braided, with numerous constantly shifting channels. The high bed load transport, a consequence of large volumes of sediment from the Alps, combined with the erosive force of the river to ensure constant erosion and infilling of river channels. Ecologically, this was a very rich environment, but human settlements were constantly threatened with flooding or complete destruction, while oxbow lakes and other residual water features meant that waterborne diseases such as malaria constituted a major health problem in the region. To the north of this, as far as Mainz

(Germany), the river was mainly a single channel but with large meanders.

The Upper Rhine was modified by three phases of channelisation within the space of less than a century. The first were initiated by the German hydraulic engineer Tulla; by 1878, they had created a single channel in the braided section and straightened the meanders further downstream (Figure 2.6b). Although not a primary concern, these modifications improved its navigability. Tulla appreciated that, by confining the river to a single channel, its erosive force would be increased, so proposed a floodplain along the channel, into which the river could flood and disperse its energy during high flow events. In the event, however, dykes were built close to the river to increase the area of reclaimed land. As a result, the river began to erode its bed, lowering water tables and, ironically, causing severe water shortages in the surrounding agricultural land.

By the beginning of the 20th century it was appreciated that constant erosion and redeposition of sediment would threaten navigation channels, so the river was narrowed. This ensured adequate depth for large boats but, by confining flow to a narrow channel, further increased bed erosion and lowering of the water table.

In 1928, the final phase of channelisation was initiated, the construction of a lateral canal, the Moderne Oberrheinausbau (Modern Upper Rhine Extension), whose primary aim was to exploit the river for hydroelectric power generation but which also acted as a navigation canal (Figure 2.6c). The river has a mean discharge of around 1100 $m^3 s^{-1}$, of which all but 20–30 $m^3 s^{-1}$ was diverted into the lateral canal; only when the flow exceeded 1400 $m^3 s^{-1}$ did the original river channel receive more than this trickle of water. The original plan was for the lateral channel to extend continuously as far downstream as Strasbourg, but it reduced water tables to such an extent that a series of short canals was constructed instead in the southern section, the water being diverted into the canal, through the hydroelectric dam and then

Figure 2.6 *Changes to the channel structure of the Upper Rhine. (a) The original channel was complex in structure, with braiding in the upper reach and large meanders further downstream. (b) The 19th century Tulla regulations simplified the river into a single straightened channel. (c) During the 20th century, hydroelectric dams and lateral canals were added (from Havinga & Smits, 2000).*

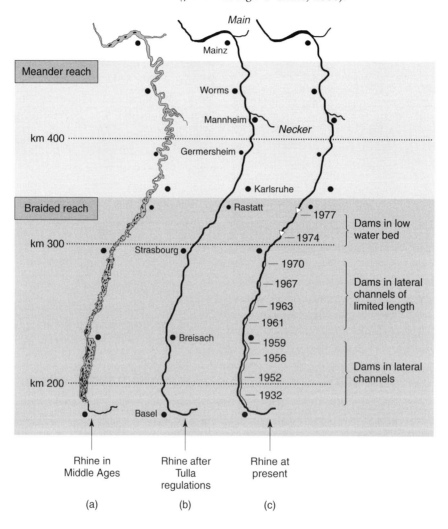

back into the original channel (Figure 2.6c). However, the dams trapped sediment, which was replenished by erosion from the river bed, leading to further severe degradation. The final phase of this development involved constructing dams in the Rhine itself, downstream of the hydroelectricity diversions to stabilise sediment movement. However, downstream erosion is still continuing and is countered by adding gravel into the river.

The channelisation of the Upper Rhine has had enormous consequences for its ecology, although most of its influences are probably not documented. The area flooded by the Upper Rhine has shrunk from more than 1000 km² to 130 km² (Figure 2.7a), the river bed has become severely eroded, and floods have become more severe as the floodable area has been reduced (Figure 2.7b).

Figure 2.7 *Effects of channelisation of the Upper Rhine. (a) Reduction in flooded area since 1800. (b) Increase in intensity of flood discharges. The figure shows the profile of two floods near Worms, Germany, each carrying an identical volume of water. Prior to channelisation, the flood was longer lasting but with a lower peak (from Havinga & Smits, 2000).*

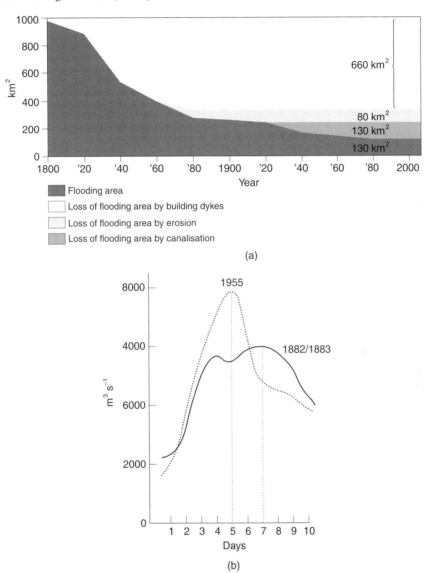

Ecological effects of River Rhine channelisation

Human influence has had some interesting effects upon the flora of the River Rhine. Bryophytes (mosses) are often associated with stable hard substrates in turbulent river channels. Before regulation, the High Rhine, immediately downstream of Lake Constance, was characterised by a succession of waterfalls and rapids, supporting a diverse flora of rheophilous mosses. Below Basel, where the river – now called the Upper Rhine – enters an alluvial floodplain, the substrate was dominated by sand and pebbles, subject to disturbance during high flows and not, therefore, providing suitable substrate for mosses. Straightening the river channel in the Upper Rhine section required creation of solid embankments, and these have since become occupied by bryophyte species typical of the High Rhine, with a consequent increase in moss species richness from 4 to 15. In the High Rhine section, in contrast, the water level has been raised by barrages, to facilitate shipping or generate electricity, inundating the rapids and waterfalls. As a result, bryophytes formerly typical of the area are now very rare. Therefore, this component of the flora has been displaced downstream by human modification of its habitat. The changes have been complicated by increasing eutrophication of the river. Species requiring oligotrophic conditions have been replaced by those preferring eutrophic water. Again, human modifications have had an interesting side effect in that the oligotrophic flora has found refuge in some of the former side channels of the river. Since disconnection from the main river by the canalisation process, these side channels have been fed mainly by groundwater, which is appreciably cleaner than the river itself, as a result of which their floral assemblages show affinities with the High Rhine rather than the adjacent river (Vanderpoorten & Klein, 1999).

Mitigation works are now under way along much of the length of the Upper Rhine. Hydraulic works have been initiated to protect areas downstream of the canalised section from flooding. These involve creation of 'polders', lateral areas into which flood waters can be directed. These systems will also be subject to 'ecological flooding', inputting water to restore and maintain the lateral wetland systems along either side of the main channel. Ironically, this ecological flooding threatens to destroy the remnant floral communities that require clean water, because inputs from the Rhine itself will be of highly eutrophic water.

2.5 Effect of extraction

Extraction removes water and inevitably reduces the volume of water in the source. If water is extracted from a natural lake, then no change in its mean volume may be discernible, but the volume discharging into the outflow river will be reduced.

There are several effects of reduced volume on surface waters.

1 There is less living space for organisms to occupy. Reduced volume of water is an effect in itself, ultimately, in extreme cases, leading to complete drying out and ensuring that pelagic fauna are eliminated. It also results in altered water quality. Temperature fluctuations are no longer dampened and, as greatest water extraction normally occurs during warm, dry periods, temperatures may rise to levels at which they cause physiological stress to organisms. If the water body is receiving any pollutants, their concentration will increase as it is no longer able to dilute them sufficiently.

2 If the water body contracts in surface area, the bed will become exposed towards the edges, reducing benthic area and threatening benthic organisms; only those animals able to migrate into the hyporheic sediments or into the still flowing channel area will have a chance of survival. Peripheral habitats will be damaged; for example, fringe wetlands adjacent to a water body may lose their source of water and dry up.

3 Aquatic organisms that are able to survive exposure to the air will come into competition with terrestrial species. During naturally low seasonal flows, exposed gravel bars in rivers become dominated by terrestrial vegetation, a process exacerbated by exposure of the river channel proper.

In addition to surface waters, groundwater is also exploited as a source of water. In most areas

Figure 2.8 *Annual changes in average precipitation in England and Wales between 1766 and 1985 (from Wigley & Jones, 1987).*

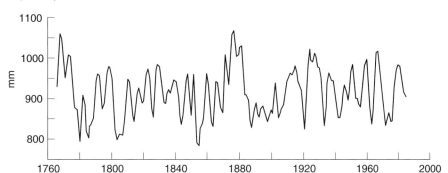

where groundwater is being used for irrigation, it is being extracted more rapidly than it can be replenished. Therefore, water tables are dropping, often by as much as 0.1 m per year in major agricultural areas. As well as increasing the energy required to pump it out, this is clearly unsustainable in the long term, and aquifers will eventually begin to run dry. The main ecological impacts are upon surface water features whose water derives from the groundwater source being extracted, as their water supply is reduced.

2.5a Rivers

River discharge is determined ultimately by rainfall. Rainfall is not constant, but undergoes changes with climatic fluctuations (Bradley *et al.*, 1987). In the USA, for example, rainfall declined between the 1880s and the 1930s; since then, it has been gradually increasing. Elsewhere, rainfall is currently declining; in southeast Asia it rose from the beginning of the 20th century, to reach a high during the 1920s and 1930s, but has gradually declined since then. Overlaid on these long-term changes, measurable in decades, are inter-annual variations (Figure 2.8). Most fluctuations in rainfall identified from past weather records are almost certainly consequences of natural climatic fluctuations, to which rivers have

always been exposed. Nowadays, however, their effects on river discharge are being confounded in several ways by human activity.

1   Extraction for human use is consistently increasing. Therefore, if natural rainfall declines, human extraction takes out a larger proportion of the remaining water. Increasing human use is a consequence of increasing population and, in the developed world, increasing domestic use per capita.

2   Inter-annual fluctuations may be dampened by the storage of excess water in floodplains and other associated wetlands, or through gradual release of excess water by forests in the catchment. Human activity is increasingly destroying these catchment habitats, causing flow to fluctuate more rapidly than under natural conditions.

3   The role of human-induced climate change in altering rainfall patterns is unknown, as long-term changes may turn out to be natural fluctuations. However, there is a strong possibility that anthropogenic activity is contributing to altered weather patterns, and therefore river flow.

The problems of extraction for water levels are compounded by the fact that the greatest demand for water is during the driest periods, when flows will naturally be at their lowest. Some of the largest and seemingly most permanent rivers in the world now run dry along parts of their lower reaches. The Yellow River in China ran dry, for the first time in several thousand years of recorded history, in 1972; now several hundred kilometres are dry for an average of 70 days per year. The Ganges and Indus in south Asia can also run dry for several months per year, while the lower reach of the Colorado River rarely nowadays contains any water.

If fluctuations in flow are dampened, by excessive extraction or by damming, then flood events will no longer occur. Sudden floods are important processes in small rivers, altering the structure of the river channel and moving bed sediments; they ensure that highly competitive species do not completely dominate the community. Without such occasional floods, species diversity will decline as the most competitive species eliminate the others. Very high levels of extraction lead to rivers running at very low levels or even drying out completely, and therefore reduce the availability of water for communities, both human and ecological, downstream.

Streams and small rivers are susceptible to direct influences of local precipitation on discharge, and undergo frequent major fluctuations in flow. Therefore, their biota are resilient to such changes (Morrison, 1990; Ledger & Hildrew, 2001). Indeed, even permanent rivers may occasionally cease to flow under very severe drought conditions, but their fauna and flora are able to persist as egg stages or deep within hyporheic sediments. Such disturbances are exacerbated, however, by extraction, which can make otherwise very rare events commonplace. Recovery is not instantaneous once water levels have returned to normal, and too many low flow episodes will compromise the ability of organisms to recover.

Some species are unable to persist in dry channels because they are obligate aquatic species with no drought refuge. Fish, for example, will be eliminated unless pools remain in the river channel. Once eliminated, recovery of their populations requires the presence of a suitable population to act as a source of colonists and, of course, a direct connection between the source population and the impacted river. Upland stream stretches may be isolated from each other by lowland reaches that provide unsuitable habitats for headwater stream species, slowing down recolonisation rates considerably.

Shrinking lakes

Freshwater lakes are the least adapted to cope with fluctuations in flow. The vast majority of freshwater lakes have a large excess of input over storage capacity, ensuring a permanently flowing outflow river that maintains the water at a relatively constant level. In arid areas, however, extraction of water from the lake or its catchment can lead to a decline in its volume.

2.5b The Aral Sea

No textbook on aquatic ecology would be complete without mention of the Aral Sea disaster (Jones, 1997). Situated in the semi-arid steppes of what is now Kazakhstan, the Aral Sea was, in 1961, the fourth largest lake in the world, covering $66\,000$ km^2 and containing 1064 km^3 of water. It was fed by two rivers, the Amu Darya and the Syr Darya, running from the south and east respectively and bringing in an estimated 56 km^3 of water per year. During the 1960s, a major irrigation scheme was implemented, diverting water away from the rivers and into cotton, as a commercial crop, and rice, as a staple for the population of the region. By 1975, no more than 11 km^3 was reaching the Aral Sea, and both rivers became intermittent, with inputs into the lake being the exception rather than the norm. The Aral Sea has no outflow, losing its water through evaporation. With its sources of water cut off, it began to dry up (Figure 2.9). By 1990 it had been reduced to half of its original area, and it still continues to lose water through evaporation, without replenishment.

Figure 2.9 *Shrinkage of the Aral Sea. As water is diverted from its influent rivers, the volume of the lake continues to decline (from Jones, 1997).*

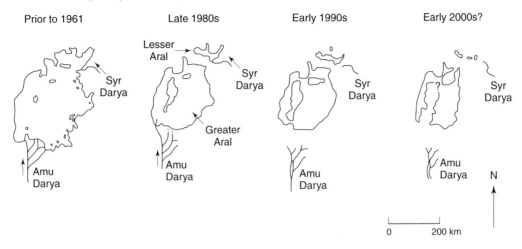

In some ways, the plight of the Aral Sea has analogies with the natural fluctuations in water volume that occur in saline lakes in arid areas. Such lakes lack an outflow, because too little water enters them to fill the depression in which they sit. Therefore, whereas a freshwater lake responds to fluctuations in incoming flow by increasing or decreasing the volume of outflow, and therefore maintains an approximately constant volume, arid zone lakes fluctuate in volume according to the volume of input. This, in turn, is dependent upon precipitation in the lake's catchment. As the volume declines, salinity will rise, because the total quantity of salts remains the same but their concentration increases. Therefore, organisms inhabiting such lakes need to be adapted to high salinity (often several times higher than those encountered in the sea) and, more importantly, to fluctuating salinity. The Aral Sea is an admittedly very extreme example of what is a common natural process and there is, therefore, the possibility that its ecosystems may recover to a certain extent if the process of its destruction is reduced. However, being originally a very large lake, its volume dampened the large fluctuations in salinity experienced by smaller lakes, so its biota may be less well adapted to changes in salin-

ity or to the high salinity to which the remnants of the lake are now subjected. Furthermore, as a large volume of water in an arid zone, it had a major effect on its local climate, evaporation from the lake surface creating cloud cover that created a relatively mild, damp climate. As the lake shrinks, the climate changes, with unknown effects upon its ecology.

2.5c Effects on the irrigated area

Irrigation not only alters the volume of water in natural water supplies, but it also has an influence on water availability and its quality in the area where it is being used.

Water quality

Irrigation is employed to its greatest extent in arid and semi-arid areas, where evaporation rates are high. Therefore, much of the irrigation water simply evaporates, leaving behind its dissolved components. In this way, the soils in irrigated areas become more saline, and any runoff will also carry a high concentration of dissolved components. Groundwater extraction in coastal areas compounds this problem, as fresh water extracted from aquifers below the land can be replaced by saline water derived from marine aquifers. Ironically, salinisation is often associated with excessive waterlogging, caused by leakage from inefficient irrigation channels and year-round application of water. Salinisation and waterlogging have serious detrimental effects on crop production, such that many large irrigation schemes are deemed to be failures within a few years of inception.

Creation of irrigation channels

Irrigation channels, built to transfer water from its source to the irrigated area, provide a novel habitat in arid areas, particularly if they are permanently filled. The water is flowing and therefore, although evaporation may be very high and hence result in highly inefficient transfer of water, the channels themselves remain relatively fresh. The ecological communities that develop within such systems are poorly studied; however, irrigation of new areas in Egypt has been cited as the cause of health problems such as increasing prevalence of waterborne diseases, including bilharzia and malaria. The pathogens causing these illnesses are dependent upon aquatic invertebrates as intermediate hosts; bilharzia, in particular, is associated with freshwater snails that, in turn, require algae upon which to feed. By creating open water, in the form of irrigation channels, storage ponds or even accidental ponds where waste water collects, the spread and establishment of highly dispersive freshwater species, and their human parasites, is encouraged.

2.6 Effects of boat transport

Use of water bodies for transport has a wide range of detrimental impacts upon the environment. The direct impact of the boat in displacing water is a short-term influence, probably with little effect on pelagic organisms, although the wash produced by a moving boat can subject any adjacent shoreline to an erosive force. The propellers used by powered vessels churn up the water and can inflict physical damage on any organisms with which they come into contact. In enclosed water bodies, such as canals and small lakes, propeller damage has been cited as a major factor limiting macrophyte growth: plants growing in the path of vessels are continually subjected to turbulence created by the propellers and even to shredding action if they come into direct contact with a propeller. Propellers presumably also inflict damage to plankton drawn into their path, although this will be highly localised and unlikely to limit populations of organisms unless boat traffic is very heavy. Large, slow moving animals may also be vulnerable to propeller damage. In Florida, many manatees (*Trichechus manatus*) bear the scars of injuries caused by collisions with boats, a problem exacerbated by the fact that the occupants of pleasure boats are often actively seeking manatees (see **Section 8.2b**).

Aquatic animals may also be impacted by noise produced by vessels. This is of particular concern for cetaceans, as most toothed whales use sound for hunting (echolocation) and communication; noise from vessels, oil and gas exploration (see **Section 7.4**) and recreational users might interfere with their normal socialising and feeding behaviour. The most convincing actual evidence comes from the stranding of 12 Cuvier's beaked whales (*Ziphus cavirostris*) in Kypari Siakos Gulf, Greece, on 12–13 May 1996. At this time NATO was conducting trials in the area of a new type of submarine detector known as low frequency active sonar (LFAS). The spatial and temporal correlation of the two events has been used to suggest causality. The probability of the stranding being due to other environmental causes was calculated to be only 0.007. The stranded animals showed no signs of ill health on necropsy; they had been feeding normally until just prior to death (Jones, 1998). This is an example of an extreme type of noise emission not normally associated with vessels, but recent modelling of the effect of icebreakers on beluga whales (*Delphinapterus leucas*) suggests that they may be audible to the whales up to 78 km away, and interfere with communication over most of this range (Erbe & Farmer, 2000). Lesage *et al.* (1999) demonstrated that beluga in the Saint Lawrence Estuary, Canada, responded to the approach of vessels by a progressive reduction in calling rate, an increase in repetition of calls and a shift in calling frequency. The impact of sound pollution upon whales is reviewed in Richardson & Wursig (1997).

More influential than direct physical effects of boat traffic is the input of pollutants into the environment. All types of vessels, from small pleasure boats to ocean-going vessels, are a source of pollution, and differences between different types of boats are generally those of degree. The problems associated with antifouling agents are considered in **Section 6.3c**, while discharges of litter are considered in **Section 6.4**. All vessels which spend more than a few hours away from land act as potential sources of sewage; if this is discharged untreated into the water, problems may result (see **Section 6.2**). Added to these sources of pollution are oil, either as fuel or cargo, and contaminated bilge water, both of which may be spilled either accidentally or deliberately (see **Section 6.3c**).

The damage caused by pollution originating from boats is a function of the size of the input, the length of time over which pollution occurs, and the relative volume of the water body. Small vessels at low density cause only localised damage and indeed can enhance the environment for some organisms. For example, occasional inputs of nutrient-rich sewage discharge, if small in volume, will quickly disperse and enhance phytoplankton productivity, while moderate physical disturbance may keep highly competitive macrophyte species at bay, allowing a diverse flora to persist (Murphy & Eaton, 1983). If boat densities are high, however, disruption becomes too great for organisms with low tolerance to disturbance. Along the River Nile in Upper Egypt, wave action produced by boats is second only to water quality

in determining species assemblages of macrophytes (Ali *et al.*, 1999).

The detrimental influences of shipping considered so far are not unique to vessels. The widespread development of rapid cargo vessels has, however, created an environmental problem peculiar to shipping, the transport of alien species in ballast water. This is considered further in **Section 2.7c**.

2.7 Translocation

Immigration of harmful non-indigenous species from one location to another has been an increased issue in recent years (see Table 2.2 and **Box 2.3**). Some of these introductions have been deliberate, for aquaculture (**Chapter 5**) or sport fishing (**Section 8.3a**), but most are accidental.

2.7a Water transfer schemes

Water may be moved from its source to the area in which it is required using pipes or artificial canals. A less expensive method in non-arid areas is, however, to exploit natural river channels. Water transfer schemes extract water from a river, transfer it through pipes into another river, upstream of the area where it is required, then allow it to flow downstream to the extraction point (**Box 2.1**).

Water transfer schemes nowadays connect formerly separate catchments in many parts of the world. In North America, even the Continental Divide is breached, with 13 schemes in Colorado transferring water from the headwaters of the Colorado River, which flows west, into the Platte and Arkansas Rivers, which flow east into the Mississippi. Most are over distances of only a few kilometres, but several proposals have been made in recent decades for major transfers over long distances, including piping water from the Congo River north to Lake Chad and from rivers in western Canada to southern California. One of the

Table 2.2 *A selection of macrobenthic species which have established populations in the German Wadden Sea and its estuaries in the past 100 years*

Species	Year of first appearance	Possible origin	Likely transport mechanism
Crustacea			
Corophium curvispinum	~1920	Caspian Sea	Migration/drift/encrustations
Elminius modestus	1952	Australia	Ballast water/encrustations
Eriocheir sinensis	~1910	China	Ballast water
Gammarus tigrinus	1965	North America	Released
Gastropoda			
Crepidula fornicata	1935	North America	In aquaculture products
Potamopyrgus antipodarum	~1900	New Zealand	Ballast water/encrustations
Bivalvia			
Corbicula fluminalis	~1920	North America	Ballast water/encrustations
Crassostrea gigas	1983	Japan	Imported as aquaculture product
Ensis americanus	1978	North America	Ballast water/encrustations
Petricola pholadiformis	1904	North America	Ballast water/encrustations
Polychaeta			
Nereis virens	~1920	?	?
Marenzelleria viridis	1983	North America	Ballast water/encrustations
Marenzelleria wiremi	1932	North America	Ballast water/encrustations
Tharyx killariensis	~1970	?	?
Tunicata			
Aplidium nordmanni	1985	Netherlands	In aquaculture products

Source: Nehring (1998).

most controversial such ideas was a plan, now shelved, to divert northward flowing rivers in Siberia south to the Caspian Sea.

Interbasin transfer may have detrimental effects on the receiving river. Water quality may differ between the source and the sink river system, while transferring large volumes of water disrupts natural hydrological patterns (**Box 2.1**). In some situations, inter-basin transfer may be carried out with the precise intention of dampening variation in discharge within the river channel. During drought conditions, for example, water is transferred from the Kielder system in northeast England to the nearby River Wear in order to maintain minimum flows. One justification for this is to maintain populations of salmon parr, whose shallow gravel bed habitat may otherwise decline in area by up to 70% as the river recedes (Gibbins & Heslop, 1998).

Water transfer can alter the chemistry of the recipient river, and facilitates colonisation of new river systems by alien species. It may also allow intermixing of genetically different populations of the same species (**Box 2.1**).

2.7b Movement in canals

Canals provide aquatic environments where previously they may have been absent. However,

Box 2.1 Inter-basin transfer in southern Africa

Southern Africa is a relatively arid area, and most of the rain that does fall evaporates or enters groundwater. In South Africa, for example, only 8.6% of rainfall therefore reaches rivers. Furthermore, one third of the country receives only 1% of the runoff. In view of this low input and poor distribution of water, large numbers of water transfer schemes are either operational, under construction or proposed in the region. Most of these are situated within South Africa, or involve transfer of water to South Africa from adjacent countries. By 2017, it is expected that IBTs in the region will be carrying 4.82×10^9 m^3 y^{-1}, equivalent to nearly 9% of the total annual runoff for South Africa. Construction of the proposed Zambesi aqueduct would almost double this volume of water (Petitjean & Davies, 1988).

Despite the large number of IBTs in the region, and the volume of water moved, very little monitoring of their environmental effects has been carried out. However, important ecological information is available from two of the currently operating IBTs.

The Orange River Project
(Petitjean & Davies, 1988)

The Orange River Project captures water in the Henrik Verwoerd Dam on the Orange River (which flows west into the Atlantic Ocean) and transfers it, via an 83 km tunnel, to the Great Fish River (which flows south into the southern Indian Ocean). It is transported to a holding reservoir, from where it is piped into the Little Fish River. It is then extracted and passes along a canal to the Sundays River system (which also flows into the southern Indian Ocean). The water is used mainly for irrigation in the region around Port Elizabeth, while flood control and hydro-electricity generation are further benefits of the system.

The Orange River Project has had several effects upon the Great Fish River. Its hydrology has been radically altered by the influx of water, its flow above the reservoir being 500–800% higher than previously. Below the dam, the mean flow remains unchanged, but the river is now perennial where it formerly was

Current and proposed inter-basin transfers in southern Africa.

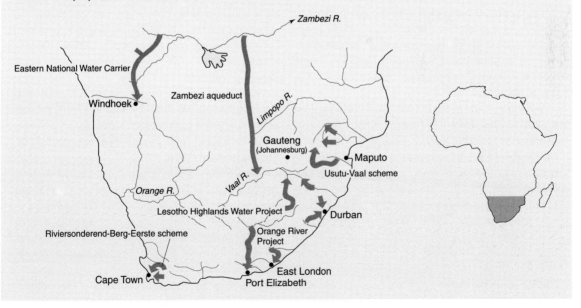

Box 2.1   continued

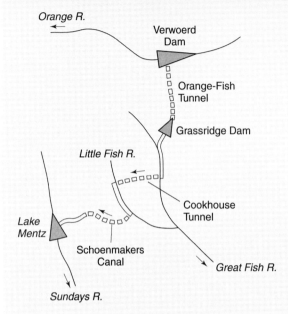

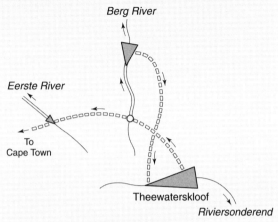

seasonal. Water chemistry has been altered, too, the salinity of the recipient river being lower than before water transfer commenced. These physico-chemical changes have resulted in changes to the benthic invertebrate fauna, the most important being the major increase in abundance of the blackfly *Simulium chutteri*, a blood sucking cattle pest, and the decline of *S. adersi* and *S. nigritarse*.

A further result of the IBT has been the migration of fish species from the Orange River catchment. The transfer tunnel contains a series of baffles, grilles, valves and high pressure points. Despite this, four species have been recorded in the Great Fish River for the first time since the scheme was opened (smallmouth yellowfish *Barbus aeneus*; Orange River mudfish *Labeo capensis*; catfish *Clarias gariepinus*; rock barbel *Gephyroglanis sclateri*), while individuals of three other species already present in the recipient river (mud mullet *Labeo umbratus*; mirror carp *Cyprinus carpio*; chubby-headed minnow *Barbus anoplus*) are believed to have moved between the catchments, resulting in a merger of formerly distinct genetic populations.

The Riviersonderend–Berg–Eerste Scheme
(Snaddon & Davies, 1998)

The Riviersonderend–Berg–Eerste Scheme captures water in the Theewaterskloof Reservoir on the Riviersonderend (which flows east into the southern Indian Ocean) and transfers it, via a 25 km tunnel, to the Eerste River (which flows northwest into the Atlantic Ocean), from which it is piped to Cape Town to provide domestic water. Between the source and recipient rivers the tunnel passes over the Berg River (which also flows northwest into the Atlantic Ocean), into which some of the water is transferred for irrigation and trout farming. Transfer only occurs during summer and early autumn, when the natural flow of the Berg River is very low. A dam further downstream on the Berg River captures water, from which it is regularly abstracted and transferred back into Theewaterskloof.

Transfer of water significantly alters hydrology of the Berg River, with discharges below the outlet in late summer up to 4500% higher than those upstream. As a consequence, invertebrate species diversity is lower downstream of the outlet than upstream. However, some species thrive in the continually high discharge, including hydropsychid caddis larvae, which are probably feeding on the large numbers of zooplankton extracted from the Theewaterskloof and transported to the river; their densities below the outlet in late summer are 20 times higher than upstream.

normally they connect natural water bodies, thus providing a means by which obligate aquatic species can colonise water bodies from which they would otherwise be excluded.

In Great Britain, canals now link the relatively diverse eastern catchments with the more impoverished western and northern ones, and have allowed alien species such as zebra mussel (see **Section 2.7c**) to expand their ranges. Completion of the Rhine–Danube canal in Germany has created a direct freshwater connection between North Sea and eastern Mediterranean catchments. The predatory mysid crustacean *Hemimysis anomala* was possibly one of the first species to take advantage of this link, reaching the Rhine basin from its original home in the Caspian Sea in the late 1990s; it has now spread into the Scheldt and Meuse catchments, where it has decimated populations of Cladocera in some reservoirs (Dumont, 1999).

The most famous canal-mediated migration of all is that associated with the Suez Canal, known as 'Lessepsian migration'. A marine canal, flowing entirely at sea level, its opening in 1869 provided the first warm water connection between Indo-Pacific and Atlantic marine biota since closure of the Isthmus of Panama around 2 million years ago, and therefore a great potential for faunal exchange. To date, however, movement of species has been almost exclusively one way, as the canal flows from east to west (Por, 1978).

2.7c Ballast water

Cargo vessels, when unladen, take on ballast to ensure that they lie deep in the water, thus improving their stability. When they reach their destination, the ballast is jettisoned and the cargo takes its place, providing the same stabilising function. Until the end of the 19th century, ballast was mainly in the form of sand or pebbles, but solid ballast was gradually replaced by water, which is now used by the great majority of ocean-going cargo vessels. Normally a vessel will take on water ballast at or close to the port at which it discharged its cargo, so most ballast derives from shallow marine or estuarine water. The water itself contains pelagic organisms, and it may include a large amount of sediment, containing benthic species.

Almost as soon as use of water as ballast became commonplace, marine organisms began to appear many thousands of kilometres from their natural ranges. In 1903, for example, a dense phytoplankton bloom in the North Sea was blamed on a diatom normally confined to warm waters in the Indian and Pacific Oceans (Anderson, 1992). However, such translocations have become much more common in recent decades, as vessels have become larger and faster. Larger sizes mean that more ballast is needed, so greater volumes of water are carried, while increased speeds ensure that organisms taken on board are confined to the dark and often contaminated conditions within ships' holds for shorter periods before being discharged, hence improving their chance of surviving the voyage. If they are discharged into conditions similar to those from which they were extracted, then they may persist. Eno *et al.* (1997) list 15 species of seaweeds, five species of diatom, one species of flowering plant and 30 invertebrate animals as being non-native marine species which have become established in UK waters (Table 2.3). They attribute 18% of introductions definitely to ballast water and a further 12% to either ballast water or via fouling on hulls. Many other species will have been transported but unable to establish.

Most transplanted species do not survive in the new environment. However, those that do persist often thrive upon being released from their natural controls by competitors and predators. They foul pipes, screens, conduits, boat bottoms, floats, buoys, rocks, submerged objects and predate or compete with native animals and plants. Their presence causes detrimental influences on native communities and, in many cases, can also result in

Table 2.3 *Non-native marine taxa found in British waters*

Division	Class	Species (or subspecies)
Non-native marine flora which have been recorded in British waters		
Bacillariophyta	Coscinodiscophyceae	*Thalassiosira punctigera*
		Thalassiosira tealata
		Coscinodiscus wailesii
		Odontella sinensis
	Bacillariophyceae	*Pleurosigma simonsenii*
Rhodophyta	Rhodophyceae	*Asparagopsis armata*
		Bonnemaisonia hamifera
		Grateloupia doryphora
		Grateloupia filicina var. *luxurians*
		Pikea californica
		Agardhiella subulata
		Solieria chordalis
		Antithamnionella spirographidis
		Antithamnionella ternifolia
		Polysiphonia harveyi
Chromatophyta	Phaeophyceae	*Colpomenia peregrina*
		Undaria pinnatifida
		Sargassum muticum
Chlorophyta	Chlorophyceae	*Codium fragile* subsp. *atlanticum*
		Codium fragile subsp. *tomentosoides*
Plantae	Magnoliopsida	*Spartina anglica*
Non-native marine invertebrates in British waters		
Cnidaria	Hydrozoa	*Gonionemus vertens*
		Clavopsella navis
	Anthozoa	*Haliplanella lineata*
Nematoda	Dracunculoidea	*Anguillicola crassus*
Annelida	Polychaeta	*Goniadella gracilis*
		Marenzelleria viridis
		Clymenella torquata
		Hydroides dianthus
		Hydroides ezoensis
		Ficopomatus enigmaticus
		Janua brasiliensis
		Pileolaria berkeleyana
Chelicerata	Pycnogonida	*Ammothea hilgendorfi*
Crustacea	Maxillopoda	*Elminius modestus*
		Balanus amphitrite
		Acartia tonsa
	Ostracoda	*Eusarsiella zostericola*
	Eumalacostraca	*Corophium sextonae*
		Eriocheir sinensis
		Rhithropanopeus harrissi
Mollusca	Gastropoda	*Crepidula fornicata*
		Urosalpinx cinerea
		Potamopyrgus antipodarum
	Pelecypoda	*Crassostrea gigas*
		Tiostrea lutaria
		Ensis americanus
		Mercenaria mercenaria
		Petricola pholadiformis
		Mya arenaria
Tunicata	Ascidiacea	*Styela clava*

Source: Eno *et al.* (1997).

Box 2.2 The Colorado River

The Colorado River has been heavily impacted by dam construction and water abstraction. One of North America's longest rivers, it flows for 2250 km from the Rocky Mountains to the Sea of Cortez in Mexico, all but the final few kilometres being in the USA. Prior to dam construction, the Colorado River was a warm river with high sediment loads and major seasonal fluctuations in discharge. Since completion of the Hoover Dam in 1935, a further 27 dams have been constructed to provide water for irrigation and domestic use. As a consequence, fluctuations in flow are now severely dampened and sediment is trapped in impoundments,

Location of major dams and reservoir dams on the Colorado River system.

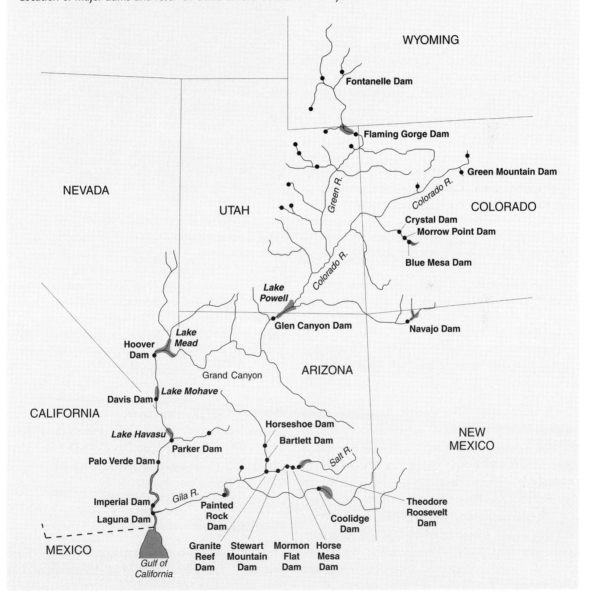

Box 2.2 continued

while multiple releases from deep reservoirs have also lowered the average temperature of the water. The diverse native fish fauna, adapted to warm, muddy waters and tolerant of fluctuating discharge, has been largely replaced by introduced species such as rainbow trout (*Oncorhynchus mykiss*) and largemouth bass (*Micropterus salvoides*), which are better adapted to the cold, clear water (Briggs & Cornelius, 1998).

Flood dampening and sediment capture in the Colorado River has led to changes in river channel structure, often favouring introduced species including invasive riparian plants. In the Grand Canyon, early signs of this included the gradual erosion of sandy beaches along the river and the build up of deltas where tributaries discharged into the main river and deposited their sediment loads. In March 1996, therefore, a controlled flood event occurred by opening sluices on the Glen Canyon Dam, immediately upstream of the Grand Canyon. Although much lower in magnitude than floods that had naturally occurred on the river before impoundment (**Fig. 2.2**), it substantially restructured the river channel, re-creating many of the sand aggregations that had been eroding, and with little apparent effect upon its biota (Collier *et al.*, 1997).

The lower Colorado River delta, which formerly covered over 780 000 ha, is gradually eroding now that the river only flows into it during occasional periods of high discharge. The wetlands that formerly covered this area are very much reduced in extent and suffer from saltwater intrusion from the sea. One of

the main sources of water maintaining the remaining wetland areas is waste water from irrigation projects. Among these is the Ciénega de Santa Clara in Sonora, Mexico, a 20 000 ha wetland maintained by runoff from two irrigation canals. Only 200 ha in extent in the early 1970s, this wetland increased to its current size following completion of the Welland–Mohawk Canal in 1978. Ironically, this wetland is now threatened because the canal may be diverted elsewhere (Briggs & Cornelius, 1998).

economic damage. The North American comb jellyfish *Mnemiopsis leidyi*, accidentally carried to the Black Sea, took less than a decade to replace most of the native fish species as the dominant predator, mainly by effectively predating their eggs and larvae. Probably originating from a single ballast discharge in the early 1980s, by 1990 its biomass in the Black Sea had reached an estimated 900 million tonnes (Pearce, 1995); it has contributed to the collapse of fisheries in the sea (Shiganova & Bulgakova, 2000; Shushkina *et al.*,

2000). In the UK, many oyster fisheries have been adversely affected by competition from the slipper limpet (*Crepidula fornicata*) and predation by the oyster drill (*Urosalpinx cinerea*), both introduced from the USA.

Dinoflagellates

Translocations of large invertebrates such as those described above are often well documented, but little is known about the scale of transport of

Box 2.3 The Great Lakes

The Great Lakes of North America support one of the highest concentrations of introduced species in the world. The first recorded introduction was the sea lamprey (*Petromyzon marinus*), which arrived from the eastern seaboard, probably by the Erie Canal, some time in the 1830s. There are now around 140 established introductions and the number of recorded species continues to rise, although their taxonomic composition changes with time.

Shipping is at least partially responsible for over 50% of these introductions, bringing organisms mainly from Europe and the east coast of the USA. Macrophytes were widely introduced during the 19th century as seeds or fragments in solid ballast. Introductions of planktonic algae became more prevalent during the 20th century after use of water as ballast was widely adopted, and particularly since the opening of the Saint Lawrence Seaway in 1959 allowed larger ocean-going vessels access to the lakes and more rapid transit times. The increase in number of macrophytes during the 20th century is due to the increasing prevalence of escapes from aquaria.

Twenty-four exotic fish species now occur in the lake, their origins matching those of other taxa. Unlike invertebrates, plants and algae, however, a large proportion of fish releases was deliberate. Deliberate release of fish, for sport or commercial harvest, peaked during the late 19th century but is no longer carried out, probably because the dangers of such translocations are now appreciated, but also because all the potentially valuable fish species have now been introduced. The last species to be deliberately released was the kokanee salmon (*Oncorhynchus nerka*) in 1950.

In a few cases, human activities have enabled organisms to expand into new areas, not by creating artificial connections or by direct translocation but by modifying the environment of natural water bodies through their activities. In the Great Lakes region there are now several fish species which originate in the prairies in the west of the Mississippi catchment, and whose arrival in the Great Lakes cannot be related solely to transport routes. These species, including the suckermouth minnow (*Phenacobius mirabilis*) and

Timeline of introductions to the Great Lakes (from Mills et al., 1993).

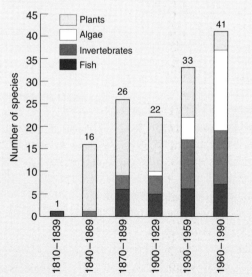

Entry mechanisms of exotic species into the Great Lakes (from Mils et al., 1993).

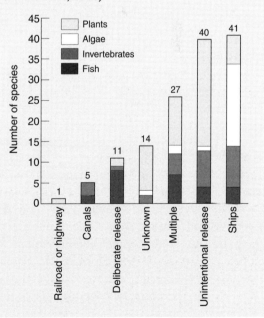

Box 2.3 continued

Origin of exotic fish species in the Great Lakes.

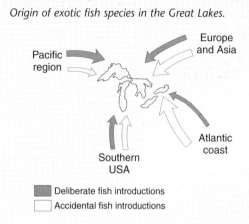

Pacific region

Europe and Asia

Atlantic coast

Southern USA

■ Deliberate fish introductions
□ Accidental fish introductions

orange-spotted sunfish (*Lepomis humilis*) favour turbid, organically rich streams typical of the Mississippi prairie region. The eastward movement of these species occurred as prairie and forest in the Midwest was converted into agricultural land, and runoff from farms transformed streams: where they had formerly been clear with gravely and sandy beds they became turbid and silt-dominated. As their habitat expanded, the suckermouth minnow and orange-spotted sunfish also extended their ranges, until eventually they reached canal systems connecting the Mississippi and Great Lakes catchments.

Source: Mills *et al.* (1993)

bacteria and dinoflagellates. These two groups often cause major nuisance problems – the former by causing diseases in native stocks which have no history of exposure and therefore lack resistance, and the latter by 'blooming', unchecked by natural competitors and grazers. Blooms of dinoflagellates often lead to problems of toxicity in shellfish and aesthetically unpleasant scum on beaches. Dinoflagellates associated with toxic blooms, such as *Gymnodium catenatum* and *Alexandrium tamarense*, have become established in southern Australia, having been introduced in ballast from Japan. One ballast tank from a recently arrived ship was estimated to contain over 300 million viable cysts of *A. tamarense* (Anderson, 1992).

Zebra mussel and quagga mussel

Probably the most infamous translocation in recent decades is that of the zebra mussel (*Dreissena polymorpha*) (Figure 2.10) and the closely related quagga mussel (*D. bugensis*). A native of the Ponto-Caspian fauna, found in southeast Europe and the Caspian Sea region, *Dreissena* is tolerant of both fresh and brackish waters. It spread throughout Europe during the

Figure 2.10 *Zebra mussel. (Photo by Still Pictures/ Ed Reschke)*

19th century and is now widely established in canals and shallow lakes. However, it was not perceived as a major problem until it appeared in the Great Lakes of North America. Like many Ponto-Caspian species now occurring in North America, it was probably not transported directly from its native range, but from an established population in northwest Europe. It was first recorded in North America in 1986 and has since spread rapidly throughout the Great Lakes and into adjacent river systems (Mills *et al.*, 1993). It has even managed to spread to lakes not directly connected to the Great Lakes drainage, through its attachment to small pleasure boats transported by road; under cool, humid conditions, mussels can remain alive for several days out of water, so this gives the potential for wide dispersal by road. Its presence in the Great Lakes is having enormous ecological and economic repercussions. *Dreissena* attaches itself to solid substrates and then filters water, capturing and eating particulate matter including plankton. In shallow lakes it can reach extremely high densities, forming a continuous mat over the substrate. In Lake Ijselmeer, Netherlands, it is so abundant that the entire lake (5 km^3 of water) can be filtered every 12 days during the mussels' most active periods (Reeders *et al.*, 1989). North America possesses a rich native bivalve fauna, none of which have the ability to attach to hard substrates. Therefore, the introduction and rapid spread of *Dreissena* has resulted in major changes to solid substrates. Among the surfaces it prefers are shells of other bivalve species, preferring to attach to live mussels rather than dead shells (Schloesser & Nalepa, 1994), and it is therefore threatening the survival of several species in areas in which it has established. It has caused a major biofouling problem, attaching to structures such as piers and boat hulls and even the inside of water pipes, which it can effectively clog. Its large numbers have led to changes in phytoplankton composition as filtering intensity has increased.

Dreissena has had a positive effect upon some benthic species in the Great Lakes. It is large-bodied and provides physical structure for attached benthos such as hydroids. Tightly packed individuals within a colony provide structural shelter from predation and trap aggregations of detritus, including faeces deposited by the mussels, both of which enhance the environment for amphipod shrimps and gastropod snails. Furthermore, it has increased abundance of bivalve predators, including some leeches and flatworms (Planaria) (Stewart *et al.*, 1999).

Controlling ballast water translocations

The increasing number of introductions in ballast water has led to demands for management to control the immigration of harmful species from one location to another. In pursuit of this, the US Congress passed the Nonindigenous Aquatic Nuisance Prevention and Control Act of 1990. Also in 1990, the Australian Quarantine Inspection Service (AQIS) introduced voluntary ballast water quarantine guidelines as a result of concerns about the introduction of exotic marine organisms, especially the toxic dinoflagellates, into Australian waters. In November 1991, the International Maritime Organisation (IMO) published voluntary guidelines for controlling the discharge of ballast water, while the IMO's Marine Environment Protection Committee (MEPC) has also implemented creation of legally binding provisions on ballast water management.

In practical terms, there are two options (Bolch, 1993; Boylston, 1996). The first is to reduce the chances of carrying potentially harmful organisms in the first place. This can be done most simply by delaying taking on ballast water until the vessel is far out to sea, or by discharging and replacing water at sea. Coastal organisms are unlikely to persist if released into the mid ocean; oceanic water contains very little sediment and lower densities of plankton than that from the coastal zone. There are, however, problems with exchanging water at sea, in that the vessel will lose stability during the exchange process.

The second option is to treat water before it is discharged, either within the vessel itself or after pumping it into an on-shore treatment facility. This can be done by heating or by exposure to ultra-violet radiation, to sterilise the water. Chlorination is effective, but results in contamination of the sea when the water is discharged. Although successful at destroying potential colonists, these processes are expensive.

2.8 Sustainability

Water is a continually renewing resource and the human processes described here have little influence upon the renewal mechanism. Non-consumptive exploitation, particularly for transport, is therefore a sustainable use; the negative influences, including habitat destruction and translocation of alien species, do not influence the ability of water bodies to transport vessels. Extraction, too, although depleting the resource, does not influence the processes by which it is renewed, although large scale extractions such as that around the Aral Sea may eventually influence rainfall in the headwaters of the rivers from which extraction occurs.

The issue of sustainability is, therefore, difficult to address with respect to the uses described here. What they do influence, however, is exploitation of living resources described in subsequent chapters. Furthermore, deterioration of water quality, habitat diversity or water volume may impact the ability of the system to contribute to ecosystem services. Sustainability is therefore an issue which needs to be considered beyond the confines of the direct uses of water.

Some of the uses described here are not sustainable. If water is extracted from the headwaters of a river, this leaves less to extract further downstream, so extraction at one point will inevitably influence the catchment further downstream. Over-use of water is currently a major human and ecological issue; irrigation, the main user, is dominated by inefficient mechanisms in which much of the water is wasted and very often the land is degraded. As human populations increase and the demand for food becomes greater, continually increasing the area under irrigation and extracting more water are not sustainable options. The technology for efficient extraction and application of irrigation water, and to ensure maximum use of rainwater and minimal wastage, currently exists. The challenge is to apply it.

2.9 Summary

- Water is used for domestic and industrial consumption, irrigation, power generation and transport. Each of these uses normally requires infrastructure, leading to major changes in the physical structure of water bodies.

- Alterations to the aquatic environment include construction of dams to impound reservoirs, channelisation to control flooding and improve rivers for navigation, creation of artificial canals, both to transport water and to carry boats, and transfer of water between basins.

- Effects of water extraction and its infrastructure can be divided into two main groups: disruption of linkages through dam construction and loss of water volume through extraction.

- Enhancing transport links may result in major modification of river channels, although new canals may become positive environmental features in due course.

- Damming a river breaks longitudinal links in a river system and causes changes both upstream and downstream.

- Channelisation destroys much of the habitat heterogeneity supported by a natural river channel. It also leads to greater intensity of floods, with consequent erosion of the river bed.

- Extraction of water exacerbates the effects of natural changes in river discharge. Excessive extraction can cause rivers to run dry or be so low that their ecological integrity is compromised. Lakes in arid zones may shrink.

- Boat transport has some direct effects on organisms, including collisions and noise pollution, but its main effects relate to pollution. Canals constructed for passage of boats provide new habitats, often in areas where natural habitats are severely reduced in extent.

- Interbasin transfer and long distance movement of boats both facilitate translocation of exotic species.

Part 2 Aquatic Organisms

Chapter 3 Biological Basis for Harvesting Living Organisms

3.1 Introduction

People have been harvesting the aquatic environment from earliest times (Desse and Desse-Berset, 1993). Even up until the beginning of the 20th century, it was the perceived wisdom that, compared to natural mortality, human activities were small-scale and the aquatic environment would always provide a bountiful harvest. The folly of this for freshwater systems quickly became apparent and by the 1920s there was good evidence of the impact of human activities on marine populations. This gave impetus to attempts to understand the nature of harvesting, in order that appropriate management strategies could be applied to ensure sustainability. As the human population increases, resources derived from aquatic habitats have become ever more important, particularly as a source of protein, and the need to manage stocks more imperative.

The biological basis for harvesting is straightforward: living organisms reproduce and their capacity for reproduction exceeds the ability of the environment to support all of the offspring. This excess can be removed without any overall effect on the size of the prey population. Populations of prey organism can be harvested indefinitely if the rate of removal of individuals is matched or exceeded by the population's ability to reproduce. If this rate is exceeded, however,

the population will begin to decline and the harvesting will become unsustainable. Therefore, effective harvesting requires an understanding of the population dynamics of the exploited species.

In this chapter we look at the biological processes which underpin the harvesting of populations. In doing so we consider the basics of biological production, simple population dynamics, and the effect of harvesting on populations.

3.2 Production

Harvesting exploits the productivity of a system: the rate at which **production** occurs. This determines the energy – and therefore the biomass – available within a system.

3.2a Sources of energy

In any living organism we can distinguish a number of basic functions that must be supported, including the basic metabolic needs – respiration, the growth of new tissue (either to increase the organism's size or to replace expired tissues) and investment in reproduction.

Respiration provides the energy for metabolic activity to take place. Most organisms can only practise **aerobic respiration**, 'burning' organic material in the presence of oxygen to liberate energy. In **anaerobic respiration** the oxygen in the 'burning' may be replaced by methane or hydrogen sulphide.

In order to be available for respiration, the energy required must be in a form metabolically available for living organisms: organic matter. Production is the synthesis of new living organic matter, either from inorganic materials – **primary production**, or from other living or dead organic matter – **secondary production**.

Organisms that are able to synthesise material through primary production are termed **autotrophs**. They are mainly photoautotrophs, which are able to use light energy for photosynthesis; these include green plants, algae and cyanobacteria. Certain bacteria, known as chemoautotrophs, are able to utilise chemical reactions as the driving force for production. They exploit energy liberating (exothermic) chemical reactions to fix carbon dioxide into carbohydrate. For example, hydrogen sulphide (H_2S) is normally rapidly oxidised in the presence of molecular oxygen. However, under certain circumstances they can be found in close proximity, including the warm waters surrounding hydrothermal vents in the deep sea, and at the interface of oxygenated water and anoxic waters in deep lakes. Under these conditions, chemoautrophy can occur.

Autotrophs also have the ability to synthesise amino acids. Amino acids require carbon, oxygen, hydrogen and nitrogen and smaller amounts of sulphur and phosphorus. The carbon, oxygen and hydrogen are usually obtained from carbohydrates synthesised by one of the above reactions, while the nitrogen, phosphorus and sulphur are taken up as inorganic salts.

Among **heterotrophs**, including all animals and decomposers, requirements for both energy and structural materials must be met from ingested organic matter. Therefore, they are dependent, either directly or indirectly, upon the production of autotrophs.

3.2b Determinants of primary production

The major requirements for primary production are light, water and carbon dioxide for photosynthesis and nutrients (nitrogen and phosphorus) for protein synthesis. Obviously in an aquatic system water will not be limiting, while carbon dioxide is also rarely limiting. The determinants of primary production, therefore, are light and nutrient availability. Light in water bodies derives from the sun, and therefore is at its maximum near the water's surface. The depth of water that is illuminated enough for photosynthesis to occur is referred to as the photic zone. It is generally considered to be down to a depth where the light value is about 1% of surface light, a depth referred to as the **compensation depth**, as the rate of primary production compensates for the material used in respiration.

Light intensity declines with depth, in a process called **attenuation**. It has two main causes: **absorption** of light and **scattering**. Light is absorbed by water molecules, dissolved organic molecules, phytoplankton and by suspended particulate matter. Scattering is merely the reflection of light by suspended matter. Obviously the more turbid the water the more light is scattered and absorbed. This attenuation yields a decaying curve (Figure 3.1a):

$$I_d = I_0 \, e^{-kd}$$

where I_d is the light at depth d, I_0 is the light at the surface (depth 0) and k is the extinction coefficient. It varies with the wavelength of the light and depends on the presence of dissolved or suspended matter.

We can measure k directly with a light meter or it can be estimated as:

$$k' = \frac{1.7}{D_s}$$

where D_s is the Secchi depth (i.e. the depth at which a Secchi disc lowered from the surface disappears from view; Figure 3.1b).

Figure 3.1 *(a) Extinction of light curve. (b) Secchi disc being used to estimate the extinction coefficient.*

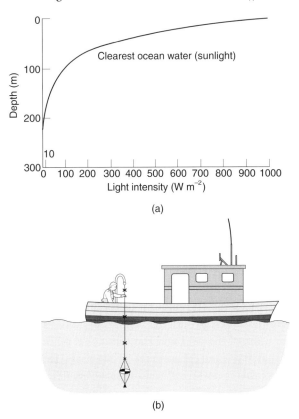

(a)

(b)

Figure 3.2 *Penetration of light into very turbid estuarine waters (a), coastal waters (b) and very clear tropical ocean water (c). Note that not only does the total amount of energy decrease with increasing turbidity but also the wavelength of maximum energy (maximum penetration) shifts from the blue to the green part of the spectrum.*

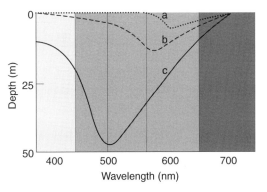

In clear waters blue-green light (wavelength $\lambda \approx 480$ nm) penetrates to the greatest depth; in turbid waters blue light is selectively scattered and the wavelength of maximum penetration moves towards the red, and into the green part of the spectrum ($\lambda \approx 550$ nm) (Figure 3.2).

3.2c The fate of primary production

Fixed carbon is used in the production of more cells, whether of multicellular plants or algae or of new single celled algae. Much of this, in turn, is eventually converted into particles of detritus. These particulate components are together described as particulate organic matter (POM) or, if only the carbon fraction is referred to, particulate organic carbon (POC), and form the basis for food webs in all ecological communities. However, while phytoplankton and zooplankton are living particles of organic matter, the terms POM and POC are generally used to refer to the non-living particulate matter, the detritus. In addition to POC, however, phytoplankton excrete organic materials in a dissolved form, or dissolved organic carbon (DOC). Early estimates were that DOC represented 50–90% of the total carbon fixed by phytoplankton; these are now known to be overestimates caused by problems with the methodology – present estimates are that around 5–20% of the total carbon fixed by phytoplankton is excreted as DOC. Most of the excreted DOC is in the form of glycolate, an early photosynthetic product, and also sugars, sugar-alcohols, and vitamins. DOC production is greater at high and low light levels. In culture, when populations are growing rapidly, very little production is lost as DOC; however, when populations go into the stationary phase, DOC production increases. Very few DOC consumers are able to exploit DOC directly, but it is the basis of the microbial loop (see Dobson & Frid, 1998).

Production of Particulate Organic Matter

Zooplankton feeding leads to fragmentation of phytoplankton cells, releasing DOC and POC. The ingested material is eventually excreted as faecal pellets – a form of POM; zooplankton moults and carcasses also contribute to the total pool of POM.

In the oceans about 75% of the POM produced in the photic zone is re-utilised above the thermocline; on average only 1% reaches the sediments, the rest being consumed by detritivores and filter feeders in the water column. There is a fairly constant ratio of carbon as:

- DOM – 1000;
- POM – 125;
- Phytoplankton – 20;
- Zooplankton – 2;
- Fish – 0.02.

POM and DOM therefore represent a large reservoir of carbon. In the world ocean 90% of POM is of phytoplanktonic origin, although seagrasses, salt marsh plants, mangroves and macroalgae may be locally important in coastal waters. About 70% of detritus is composed of highly refractory material, for example, skeletal remains and diatom frustules.

3.2d Global patterns of productivity

If harvesting is to be optimised, then it needs to take place where productivity, and therefore the potential biomass of the target species, is greatest. Within marine environments, productivity is highly variable, both spatially and temporally, but clear patterns may be discerned. The mechanisms and resulting patterns are explained in detail in Dobson & Frid (1998), but a summary of the major patterns is presented here. In the oceans, productivity is largely determined by the availability of light and nutrients. In equatorial regions productivity is limited by the availability of nutri-

ents and is generally low except at the weak upwelling associated with the equatorial counter current and at the western margins of continents (Figure 3.3). In the former case the shear stress between the currents flowing in opposite directions pulls nutrient-rich deep water to the surface, while at the western margins of the continents the prevailing winds set up a slope in the sea surface which drives a current system flowing westwards away from the land mass. As the water moves away from the coast, nutrient-rich deep water wells up to replace it. In temperate areas, upwellings are also important in controlling productivity. In stable areas the bloom of algae associated with spring increases in light levels exhausts the nutrients trapped above the thermocline; productivity therefore falls until the autumn, when mixing brings nutrients from below the thermocline into surface waters and allows a second bloom before light levels become the limiting factor over winter. In high latitudes, productivity is limited by low light for most of the year and the summer is marked by a massive burst of primary production. In some areas this is sustained until light levels fall, but in other areas it may outstrip the available inorganic nutrients, causing productivity to decline during mid to late summer.

The advent of satellite remote sensing has allowed a massive increase in our ability to gain synoptic data on levels of primary productivity (estimated from a complex calculation involving levels of irradiance at three or more wavelengths, but basically a measure of the chlorophyll content of the waters). Satellites also cover marine, freshwater and terrestrial environments, giving a true global picture (Figure 3.4a). What this reveals is the same basic pattern as shown by the earlier spot measurements (Figure 3.3), but has also highlighted the high degree of spatial heterogeneity, especially in the tropics and in frontal regions where currents meet (for example, off the Grand Banks) (Figure 3.4b). Much of this structure is due to meso-scale features such as eddies and cold-core rings (Richardson, 1983; Mann & Lazier, 1991).

Figure 3.3 *Distribution of primary productivity: (a) in the oceans, showing the high values associated with the western margins of the continents and the equatorial counter currents; (b) in terrestrial and freshwater systems.*

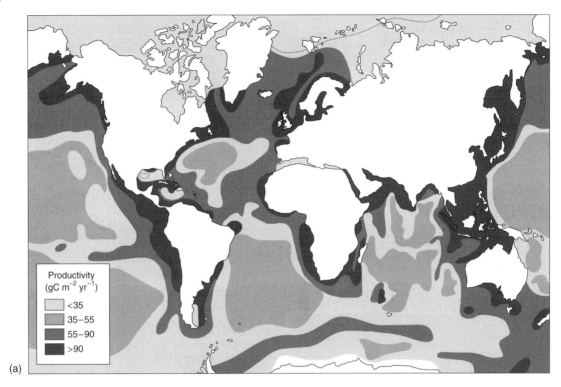

(a)

Productivity
(gC m^{-2} yr^{-1})

<35
35–55
55–90
>90

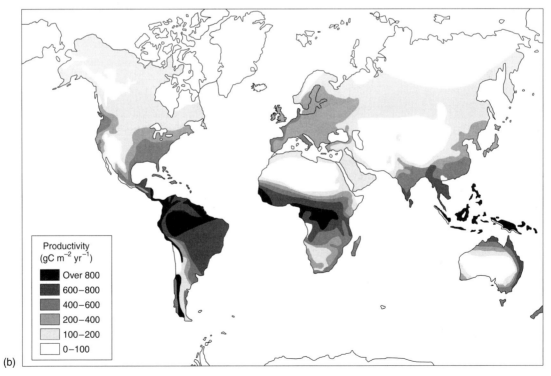

(b)

Productivity
(gC m^{-2} yr^{-1})

Over 800
600–800
400–600
200–400
100–200
0–100

(a)

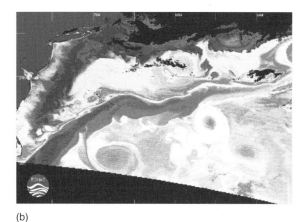

(b)

Figure 3.4 *(a) Global patterns of marine productivity as derived from estimates of chlorophyll abundance as seen in satellite images. (Image NODC/NOAA) (b) Mesoscale features of ocean circulation, such as these eddies off the Gulf Stream, may be very important in contributing to global productivity. This satellite image shows the east coast of the USA in the top left and 'cold core' rings spinning off from the Gulf Stream and into the central Atlantic. It is also possible to see some evidence of 'warm core' rings to the north of the Gulf Stream in the top right of the image. (Image University of Miami/MODIS Terra)*

However, their presence means that global productivity estimates based on spot measurements may be under-estimates (possibly by as much as 20%) of actual global production.

Productivity of freshwater systems is light-limited in temperate and high latitudes in the same way as marine systems, but also reflects climatic conditions and inputs from the terrestrial system. Fresh waters show similar global patterns to terrestrial systems, with latitudinal zones and east–west differences across continental land masses.

They are, however, heavily influenced by more local conditions so that, for example, a single lake district may contain oligotrophic (low productivity) lakes at high altitude and eutrophic (high productivity) lakes at low altitude, production being determined largely by the size of the lake's drainage basin, from which nutrients are derived. Rivers may further complicate general patterns by transporting water and nutrients from high productivity areas into generally lower productivity regions. For example, the semi-arid subtropical regions of Africa, typified by low terrestrial productivity, are the location of some of the continent's largest and most productive wetlands, nourished by major rivers flowing out from the equatorial region.

3.3 Energy transfer

Individuals must make decisions about how they utilise the resources they consume – energy and structural materials. These are not conscious decisions but are made automatically as a result of generations of selection. Once adulthood is reached resources can be channelled into additional somatic growth (which might lead to reduced risks from predators, increased ability to overwinter or to survive periods of harsh environmental conditions, and better access to mates) or into reproduction. This is not necessarily an either/or choice; in most cases some somatic growth continues while resources are allocated to reproduction. Both these processes contribute to growth.

The bulk of the resources obtained are used for maintenance. If resources are low or if conditions are harsh, requiring more energy/resources for basic maintenance, then there is a further decision to balance the needs of the current situation against the future. In some cases resources may be directed into maintenance, i.e. pollutant detoxification, at the expense of growth. The **scope for growth** method of assessing pollution effects (see **Box 9.1**) is based on this. It seeks to measure the reduction in resources available for growth through the stress imposed by the pollutant. In extreme circumstances an organism may pump resources into reproduction, in order that its genes may be perpetuated, even if this means it succumbs.

Production is determined by the amount of resource, such as organic carbon, that an organism assimilates into its body tissues. Among secondary producers, not all matter consumed is assimilated; some will be undigested, either because it is not amenable to digestion (e.g. structural tissues such as diatom frustules) or because the residence time in the gut is too short to allow full digestion. This excess is passed out of the body, a process referred to as **egestion**. Incorporating these processes together allows construction of a simple budget:

$$A = C - E \tag{1}$$

or:

$$A = I - E \tag{2}$$

The amount of material assimilated (A – the ration) equals that consumed (C, which equals ingested, I) minus that egested (E). The assimilated fraction (A) is that available to meet the energetic and material needs of the organism.

The amount of material assimilated (A) is divided between the amount used for production of new tissue (P) and the amount used for growth of reproductive (gonadal) products (G), while some is continually lost through respiration (R) and excretion (U):

$$A = (P + G) - (R + U) \tag{3}$$

The more of this material that is used for respiration to provide energy or which is excreted as waste, the less there is available for production.

Assimilation and transfer efficiency

The energy incorporated into living organisms is potentially available for consumers in the next trophic level. However, only a small proportion of the potential is successfully assimilated by consumers (Figure 3.5). Typically, predator biomass is rarely higher than 10% of that of their prey, while assimilation by herbivores and detrital feeders is even less. The remainder is lost through R and U in equation (3), is lost through E in equations (1) and (2), or is simply not consumed and passes as detritus to detritivores and decomposers upon the death of the organism. Therefore, the further up a food chain an organism is situated, the less common it will be because the energy available

Figure 3.5 *Transfer of energy between trophic levels, demonstrating the rapid decline in available energy with increasing distance from the source (basic food chain, plus assimilation efficiencies between levels). In this example, 10 t of phytoplankton is required to support 6 kg of fish. From a human perspective, the optimum level to harvest would be at level 1 (for energy) or level 2 (for protein), but for logistical and cultural reasons, harvesting is most likely to occur at level 4.*

Level	Food chain	Biomass (kg)	Assimilation efficiency
4	Fish	6	
			12%
3	Predatory zooplankton	50	
			12.5%
2	Grazing zooplankton	400	
			4%
1	Phytoplankton	10 000	

declines. For this reason, the most efficient use of the available energy is to harvest organisms low down in a given food chain.

3.4 Population growth

All populations have the potential to undergo growth as a result of the reproductive success of the individual members. In general a population has the capacity to produce many more individuals than are needed to replace the existing population. These excess individuals suffer mortality through predation, disease or starvation. The size of a population is therefore a dynamic balance between the causes of mortality and the variable success of the individuals in reproducing. We can represent this dynamic process in the form of simple mathematical growth models.

3.4a Exponential growth model

We can visualise the population at some time in the future (N_{t+1}) as the number of organisms alive now (N_t), plus those born (B) minus those that die (D) in the interim:

$$N_{t+1} = N_t + B - D \tag{4}$$

However, B and D are not likely to be constant, but are a function of N_t, because more births and deaths are likely in a large population than in a small population.

If we also express B and D together as the net population growth or reproductive success: $R = B - D$, then we can say that:

$$N_{t+1} = N_t R \tag{5}$$

This means that the population in any given generation is a product of the population in the previous generation and the reproductive success.

The model represented by equation (5) applies to a population that breeds in discrete breeding seasons. To make it generic we can say that:

$$N_t = N_0 R^t \tag{6}$$

The population in generation t is the initial population multiplied by R to the power of t, the elapsed time, in generations. If births exceed deaths, then R will be positive, so the population will increase. If the reproductive output per individual remains constant, the reproductive rate of the population as a whole will accelerate as population size increases. Therefore, plotting N_t against t gives an exponential curve (Figure 3.6a), the actual shape of the curve depending on the value of R. This is obviously unrealistic, as populations cannot continue to grow forever. As numbers increase, a limiting factor will check population growth.

3.4b Sigmoidal growth model

In considering the exponential growth model it was assumed that D, the number dying in a given time interval, was simply a constant fraction of

Figure 3.6 *(a) Growth of a population with discrete generations over time. Exponential growth is shown by the left-hand line, while sigmoidal growth, produced by the constraining of the population to a carrying capacity, is shown by the right-hand line. (b) Density dependent mortality in trout fry keeps numbers approximately constant (data from Le Cren, 1973). (c) Numbers of macro-benthic organisms at a station off the coast of Northumberland. Numbers in March remained approximately constant from 1973 to 1981, irrespective of variations in the abundance the previous September (due to recruitment variation). This implied strong density dependent mortality over winter (data from Frid et al., 1996).*

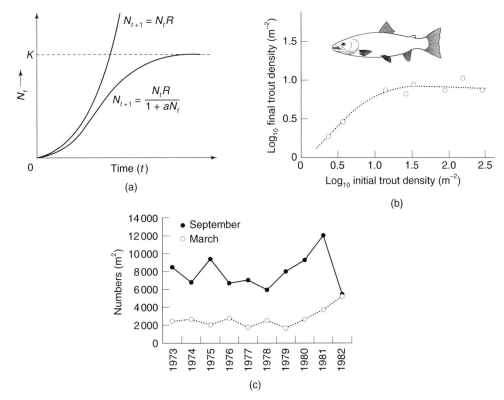

(a)

(b)

(c)

the number alive. In practice, there are two types of mortality, **density independent** and **density dependent**.

A certain proportion of the population will die from natural causes in any given time interval, independent of the total population size. This is referred to as density independent mortality and does not act as a check on total population numbers. Density dependent mortality occurs when the proportion dying varies, depending on the density of the organisms. Therefore, as the population gets larger, so a larger **proportion** suffers mortality. Initially, death rates are low

because resources are plentiful. As the population increases, however, the proportion of essential resources, such as food, per individual declines. Therefore, starvation is unlikely to be a mortality factor for a low population, but becomes important when the population rises. This provides a mechanism to limit the population size.

If we begin with the simple exponential growth model for a population breeding continuously:

$$N_{t+1} = N_t e^{rt} \qquad (7)$$

Note that the net reproductive rate R has been replaced by a term e^r, where e is the mathematical constant and r is the intrinsic population growth rate. This is simply the result of integrating the discrete breeding model into a continuous breeding population.

If we wish to produce a check in this model, the simplest way is to restrict the population to some maximum size, called K, the **carrying capacity**. The carrying capacity is the maximum population that can be supported by the available resources. Inserting K into equation (7) gives:

$$N_{t+1} = N_t \, e^{r(1-N_t/K)} \qquad (8)$$

This is the Logistic Equation, which produces a sigmoidal growth curve (Figure 3.6).

The Logistic in its differential form becomes:

$$\frac{dN}{dt} = rN\frac{(K - N)}{K} \qquad (9)$$

The term in parentheses regulates population growth rate by intraspecific competition, the mechanism by which populations are held at their carrying capacities. As with all models it is a simplification of the real world situation; in reality competition is probably only one of a range of factors influencing population densities. There are, however, many studies which have demonstrated that, while the size of population varies, often fluctuating considerably from year to year, there is a tendency for densities to vary only within certain limits (Figure 3.6c).

3.4c The recruitment curve

The population growth curve has three components (Figure 3.7). Initially, population density changes slowly. As the population increases, however, its rate of increase will also rise, so the middle part of the curve represents rapid growth rate, because the population is well below its carrying capacity and density dependent mortality is absent. As the population approaches carrying capacity, density dependent mortality will increase and the rate of

Figure 3.7 *A sigmoidal growth curve, demonstrating different levels of recruitment at different population densities. Maximum recruitment occurs in the middle section, when the population is high enough to produce many offspring but recruitment is not yet influenced by the limiting factor to population growth. Keeping the population in this region maximises recruitment.*

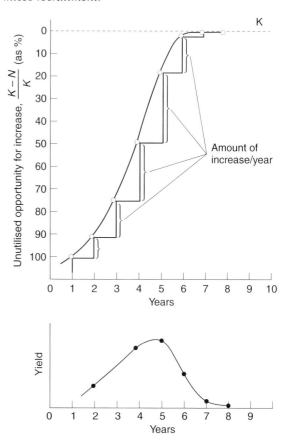

population growth will decline, causing the curve to level out.

Effective harvesting can exploit this pattern by reducing the population well below its carrying capacity, inducing a high reproductive rate. By reducing numbers below the carrying capacity, harvesting can actually stimulate extra production by increasing fecundity (and therefore moving the population into the steep section of the curve in Figure 3.7).

3.5 Maximum sustainable yield

The ultimate question for fisheries managers is how much can be taken without the stock collapsing? The optimum harvest is the maximum number that can be removed without reducing the population in subsequent generations, and is known as the **maximum sustainable yield** (MSY).

There are two broad approaches to determining MSY. The simpler approach is the Surplus Yield group of models, which uses a 'biomass production' model as its basis and so subsumes a whole range of biological and ecological dynamics into the single model. In the Dynamic Pool models at least some of these underpinning processes are modelled explicitly, giving greater realism but much higher levels of complexity. Both approaches have spawned a variety of models and a detailed consideration of this diversity or the mathematics of the models is beyond the scope of this work. We illustrate the two approaches by a consideration of one of the classic models in each class and discuss the general lessons which can be gleaned from each group of models before considering virtual population analysis, the approach used in many contemporary fisheries management schemes to set quotas.

3.5a Types of MSY model

Surplus Yield models (logistic-type models)

The starting point for this group of models is the idea that a population, when below its carrying capacity, will be continually increasing. Fishing can be regarded as continual removal of this production. So the model seeks to use an appropriate model of population growth towards the carrying capacity and applies fishing mortality to it in order to keep the population away from the carrying capacity.

The most widely applied model of this type is the Schaefer model (Figure 3.8) (Schaefer, 1954).

Figure 3.8 *The Schaefer Surplus Yield Model. The equilibrium line represents, for a given stock density, the recruitment rate (i.e. the reproduction rate) required to replace losses and keep the population constant. Therefore, to the left of point K, more individuals are produced than are needed to maintain the population. At point F, 'a' is the number required to maintain the population but, being well below K, the recruitment curve is much higher; 'b', therefore, is the excess production. At point G, the recruitment required to maintain the population is very high ('c'), but the actual recruitment is appreciably lower because K has been exceeded and the population can only fall. See text for further details.*

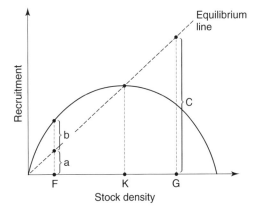

Schaefer's model uses the simple logistic equation to provide the biomass growth function. MSY can be determined by considering the recruitment curve of a population.

Where the recruitment curve in Figure 3.8 is above the equilibrium line (to the left of K), the difference between the lines represents the excess that can be harvested. This difference is at its maximum at $K/2$; therefore, in theory, if the population is kept at half of its carrying capacity ($K/2$) or, in fisheries terms, half the unexploited biomass ($B_\infty/2$), this will produce the MSY.

Applying the surplus yield approach is attractive as it is very simple. In reality there are a number of problems, particularly those that occur when detailed background information is lacking, a problem further explored in **Section 3.7a**. Outputs are very sensitive to the quality of the inputs, while in many ways it lacks biological realism. The evolution of fishing gears and the general increasing trend in power and efficiency of vessels and gears always confound estimates of fishing effort.

Dynamic Pool models

Dynamic Pool models, such as the General Dynamic Pool Model of Beverton and Holt (1957), provide much greater realism than Surplus Yield models but require much more in the way of input data. Dynamic pool models explicitly include the processes which alter fish stock biomass – recruitment, growth of biomass, natural mortality and fishing mortality. They also take account of the age structure of the population. The number of recruits per fish may vary with the age of the fish, as will its growth rate, its susceptibility to natural mortality events (e.g. predation) and vulnerability to fishing.

The basis of the Dynamic Pool model is five linked equations which generate a function describing yield from the fishery.

- The first three equations merely integrate across the cohorts (age classes) so that total numbers, total numbers caught and total biomass are the integrals over all cohorts.

- The fourth estimates the number of fish of a particular age in the stock by deriving it from the rates of natural and fishing mortality and the stock size at some later age.

- The final equation is an integral equation to express the total yield from the fishery. This equation is difficult to solve mathematically without making assumptions which may compromise its reality, but can readily be solved iteratively by computer.

Modern developments of the Dynamic Pool approach have replaced the integration across the whole population with separate consideration of each cohort and subsequent summation. This is more realistic, easier to understand, but more time consuming to calculate, although the use of computers negates this disadvantage.

The general Dynamic Pool model allows incorporation of any pattern of growth, natural mortality and fishing mortality. The patterns can be derived from fisheries observations or empirical models. The results can be expressed as yield per recruit relationships, or if recruitment is included directly, the results can be predictions for specific years. The gaining of the necessary recruit data can be from plankton surveys of fish larvae and young fish surveys in nursery areas, by back calculation using VPA (see below) or by using stock–recruit relationships such as the Ricker curves. However, the model is restricted by the availability and quality of the input data/model functions.

Virtual population analysis (cohort analysis)

Virtual population analysis (VPA) or cohort analysis seeks to estimate fishing mortality (F) and the numbers at age in a stock from catch data only; in doing so it must assume that natural mortality (M) is constant. For example, consider the number of five year old fish, which is the number of fish reaching the age of four the previous year minus the number of four year olds dying (total mortality Z_4). The total mortality is composed of fishing mortality (F_4) and natural mortality (M_4), but we assume M_i is the same for all age groups and is known. The number of fish of a particular age class caught is a proportion of the total number dying which are killed by fishing. These two relationships can be combined into a single equation containing M, numbers at age plus 1 and catch and fishing mortality at age. If M, numbers at age plus 1 and catch at age are known, then F can be calculated. The actual equation is complex and so solving for F is done by computer iteration, beginning with the oldest cohort (because number at age plus 1 is then zero), giving a value for F for the oldest cohort. This can then be used to calculate the numbers in the cohort one year younger. The procedure can then be repeated, using the appropriate catch data, to get a value of F for the next cohort and so on until all the cohorts have been modelled. Alternatively an initial 'guess' of F for the oldest cohort is seeded into the iteration and analysis proceeds from there.

The biggest assumption of VPA is that M is constant and known. M can be calculated if total mortality and fishing mortality are known, but if F is known, then VPA is unnecessary! It is relatively easy to include age specific rather than constant natural mortalities into the model, but it is not easy to incorporate interannual variability in M or density dependence.

Unfortunately, in using VPA to set quotas (see **Section 4.7**) the least reliable estimates of F are those for the later years. Considerable effort has gone into developing more accurate VPA, including multi-species VPA, which contains inter-specific interactions such as large cod predation on small whiting, for example. However, it remains unlikely, given the multiplicity and complexity of the factors controlling fish stock sizes, that we will be able to develop models which are effective in driving 'real-time' fisheries management.

Figure 3.9 *The optimum yield model. Increasing yield requires an increase in harvesting effort. Optimising yield per unit effort means taking a slightly lower harvest than the maximum sustainable yield.*

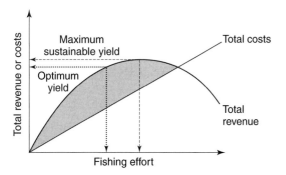

3.5b Optimum yield

MSY gives the maximum biomass that can be taken per generation on a sustainable basis, but fishery management tends to operate towards a slightly different target – optimum yield. This is a balance between trying to harvest the maximum possible and the increasing effort that is required as the catch increases (Figure 3.9). Optimum yield is slightly lower than the MSY, because of the disproportionate increase in effort required for even small increases in yield as the revenue curve levels off.

3.6 Fisheries models

Many types of animals are harvested from marine and fresh waters (see **Chapters 4, 5**). However, by far the most important are fish, whose life cycles necessitate further modelling in order to optimise harvesting. While logistic and similar models may predict the size of a population, fisheries rarely deal with the whole population. In many species the exploited population comprises a series of sub-units which are effectively isolated from each other, either by physical barriers or behavioural mechanisms, such as the different herring stocks in the North Sea (Figure 3.10). These sub-populations are referred to as **stocks**. Different stocks experience different environmental conditions and so vary in parameters such as growth rate, fecundity and carrying capacity. It is therefore inappropriate to try to use a single model for all stocks.

Stocks may migrate around their range seasonally or may move following prey aggregations or particular hydrographic features, such as fronts or temporary upwelling systems. Many species migrate to breeding grounds so the larvae are released in a suitable habitat for the fry or in a location that allows planktonic larvae to drift with the currents to their nursery grounds (Figure 3.11).

Most exploited fish species breed once a year and the populations comprise individuals of a

Figure 3.10 *Spawning grounds for the four North Sea herring stocks. Each stock is semi-isolated and differs genetically, morphologically and in parameters such as fecundity and growth rate.*

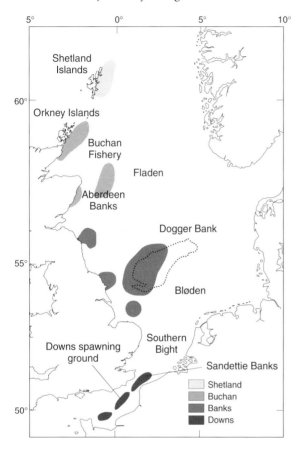

Figure 3.11 *Spawning grounds and larval drift in North Sea plaice.*

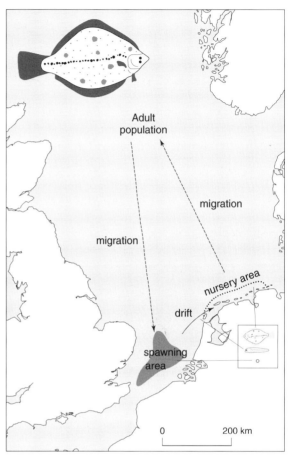

number of age classes, usually referred to as cohorts. Individuals in a particular cohort vary in their fecundity and their likelihood of suffering mortality (natural and harvesting). To reflect these differences, models need therefore to include information on the age/size structure of the population.

3.6a Fish growth – the von Bertalanffy model

The different stages in a fish life cycle tend to grow in different ways, so it is inappropriate to seek one model to describe the entire life cycle. Fish growth is expressed in changes in both length and weight with time. Weight is useful as it can be directly converted to yield from a fishery, but as marine fish are often gutted and de-headed at sea, length is a more easily measured parameter in a fishery. To allow conversion between the two, standard length–weight relationships are produced. These generally take the form:

$$w = a \, l^b \qquad (10)$$

Where w is the weight, l the length, b is ≈ 3 and a is a constant. This relationship can be logged to yield a straight line relationship:

$$\ln (w) = \ln (a) + b \ln (l) \qquad (11)$$

which allows simple regression techniques to be used.

The exact value of b varies between species and also seasonally (for example, as the gonads develop). It has been used as an ecological measure of the 'health' of fish as the 'condition factor' k, where:

$$k = w/l^3 \qquad (12)$$

This assumes that $b = 3$, so that $k > 1$ indicates good condition (weight is greater than predicted for length) and $k < 1$ indicates poor condition.

The most commonly used model of fish growth is the von Bertalanffy equation (von Bertalanffy, 1957). This starts at the time the fish recruits to the stock (see below) and assumes that a fish grows towards some maximum size (l_∞, w_∞), with the growth rate slowing as the maximum size is approached. The von Bertalanffy equation is:

$$l_t = l_\infty (1 - e^{-k(t-t_0)}) \qquad (13)$$

Or, expressed in terms of weight:

$$w_t = w_\infty (1 - e^{-k(t-t_0)})^3 \qquad (14)$$

with l_t and w_t being the length and weight at time t, while k, the growth parameter, describes the rate at which the length or weight approaches the maximum. The parameter t_0 represents the age at which length is zero; as such it has no biological reality and can be regarded as a mathematical consequence of the structure of the equations.

Nowadays, the von Bertalanffy equation is only one of many models available to describe growth applicable to fish, and fisheries scientists will fit a number of models to the data and then use the best fit as a basis of further population modelling.

3.6b Recruitment

Models such as the von Bertalanffy apply only to fish that have recruited, meaning that they have grown large enough to leave the nursery area(s) and join the main adult part of the stock. This is the portion of the stock that is available for exploitation, so recruitment marks the time in a fish's life when it becomes vulnerable to fishing. It is the continual recruitment of new individuals that allows a fishery to be sustained. The recruitment process is complex as it is the combined effect of spawning, egg development, hatching, larval growth and survival, metamorphosis and juvenile growth, and survival and migration to the stock. While an understanding of recruitment is crucial to our ability to predict adult stock sizes, and hence manage a fishery, these early life stages are the most difficult to study. Many attempts have been made to develop simple models of recruitment, but it is probably the case that for each species, recruitment has its own very specific characteristics. The following relationships must therefore be regarded as generalities which are useful abstractions for understanding the processes but do little to allow us to develop species specific models. Modern fisheries management uses very complex models, but their development can be traced back to the simple ones described here.

Most fish hatch into a planktonic larval stage, which feeds on the available plankton, often initially on phytoplankton before switching to zooplankton as it grows. During this phase the larvae are drifting with the water currents and mortality rates can be very high. The planktonic larvae metamorphose into juveniles and at this stage their increasing swimming ability allows them to select an environment; for example, juvenile flatfish adopt the flattened body form and migrate to the seafloor from the water column. These juvenile fish seek out nursery habitats which are usually distinct from that of the adults. Plaice, for example, use shallow sandy bays and estuaries in contrast to the open sea used by adults, and many pelagic freshwater fish spend their nursery phase in reed beds at the margins of lakes. As the fish get older their preferred habitat changes and they may begin to occupy the same environment as the adults but without undertaking any spawning migrations.

In many species there is evidence of strong density dependent mortality in the juvenile stages such that this is the crucial life history stage regulating population numbers (Southwood, 1978). Egg mortality is generally low, of the order of 2–5%, but mortality in the planktonic and juvenile phases can be high – only 1 in 10 000 haddock survive their first year of life. While much of this mortality may be density dependent, a component is density independent and derives from variations in the physical environment, including weather conditions and the degree of synchrony between the timing of fish breeding and the annual cycle of plankton production.

3.6c Stock–recruitment relationships

Of great interest to fisheries managers is the relationship between the number of recruits and the size of the stock from which they derive, the so-called stock–recruit relationship. This takes two basic forms: a curve to an asymptote (Figure 3.12a) or a hump-shaped curve skewed to the left (Figure 3.12b,c). In all cases however there is considerable variation around the smoothed curves. In the case of plaice, the relationship is one of increasing recruitment with increasing stock size – more spawners produce more recruits up to a certain level, followed by an asymptote – further increases in the number of spawners does not lead to any additional recruits. This can be explained by strong density dependent mortality amongst the juveniles – food limitation in the nursery areas. In contrast, in the gadoids (cod and haddock) and pelagics (sardines and herrings) the relationship is hump-shaped. Recruitment increases with increasing stock size, but above a critical level declines with increasing stock size. This is usually the result of cannibalism (cod) with the adult stock predating juveniles, or competition for planktonic food (juvenile haddock, pelagics).

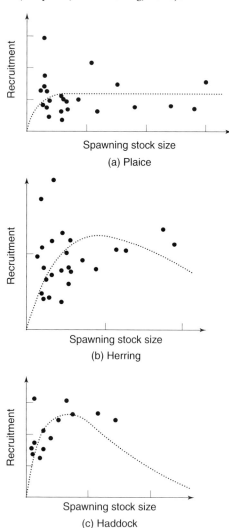

Figure 3.12 *Stock–recruit relationships for (a) plaice, (b) herring and (c) haddock. Note the variability in the data around the fitted relationships: (a) fitted Beverton–Holt curve, (b) and (c) fitted Ricker curves (adapted from Cushing, 1977).*

Two families of model describe the two fundamental types of stock–recruit relationship. The Ricker curve is a model producing a hump-shaped stock–recruit curve, while the Beverton–Holt equation yields an asymptotic relationship. Both models use only two parameters; while more complex models are available they require more parameters to drive them.

The Ricker family of curves derive from the model:

$$R = \alpha S \, e^{-\beta S} \qquad (15)$$

where R is recruits, S the spawning stock size and α and β are the parameters of the curve. α is density independent, while β is the density dependent parameter; increasing α makes the curve higher and more humped, while increasing β decreases the height but makes the curve more peaked.

From the Ricker equation we can see that maximum recruitment (R_{max}) occurs when the stock is at $S_{max} = 1/\beta$ and the size of $R_{max} = 0.3679\alpha/\beta$. The stock will exactly replace itself when the stock size (S_r) = $\ln{(\alpha)}/\beta$.

The Beverton and Holt family of asymptotic curves is derived from the model:

$$R = 1/(\alpha' + (\beta'/S)) \qquad (16)$$

α' and β' are different parameters to the α and β in the Ricker model. Decreasing α' increases the asymptote and reduces curvature, while decreasing β' causes the asymptote to be reached more quickly. The stock size for exact replacement, $S_r = (1 - \beta')/\alpha'$ and $R_{max} = 1/\alpha'$ which occurs when the stock size is infinity.

The Beverton–Holt model applies when there is a ceiling on abundance of recruits, such as juvenile food supplies or habitat space. Ricker curves apply when there is strong density-dependent mortality such as cannibalism on fry, when increased larval densities result in increased time spent in a vulnerable stage, or when a high density of fry causes a time-lagged increase in predator numbers (by attraction or reproduction). Highly humped curves are produced when there is scramble competition for food, habitat space or oxygen. Low humped curves will occur with interference competition, e.g. predator refuges on coral reefs, trout territories in riffles.

3.7 Problems with applying theory

3.7a Lack of information

In order to harvest a population sustainably, we ideally need the following information:

- the rate of reproduction, or recruitment of individuals;

- growth rate;

- natural mortality;

- harvesting mortality;

- the effect of harvesting on natural mortality.

Agriculture and forestry operations have little difficulty in obtaining this information, and so can work at a high level of efficiency and confidence in the prediction of crop size. Unfortunately, such data are very hard to determine for organisms harvested from the wild, and particularly so when they are under water, making observations difficult. For a fishery, the only details known with any accuracy (assuming the people carrying out the harvesting tell the truth!) are harvesting mortality (the catch) and fishing effort. There is a straightforward relationship between these two parameters (Figure 3.13): increasing fishing

Figure 3.13 *The effort–yield relationship. E_{MSY} is the fishing effort required to give the maximum sustainable yield (MSY).*

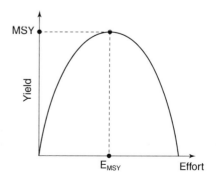

effort increases the yield, but if effort is too great, then the population declines because it is being over-harvested and yield drops. Therefore, harvesting aims for effort that maximises yield. Unfortunately, many fishery yield predictions are made from empirical information about previous years, whereas natural populations are influenced by many parameters that may reduce stocks considerably, so a sustainable effort one year may be disastrous the next if stock recruitment is lower than normal. Therefore, among real populations, E cannot simply be a fixed value, but must be continually adjusted, depending upon current conditions.

The problem of predicting MSY and other parameters for fisheries is normally compounded by the fact that they have been harvested for many years before such management is considered necessary. Therefore, the population is no longer operating under natural conditions, but there is no information about how that harvesting has influenced it. When people's livelihoods are at stake, it is almost impossible to introduce a moratorium to allow the population to return to a level at which natural processes can be studied, although the International Whaling Commission almost succeeded with its worldwide ban on whaling since 1977.

3.7b Unpredictability

MSY estimates inevitably make many assumptions about the exploited stock. One common assumption is that, once the carrying capacity (K) has been reached, the population will remain stable, so long as K does not change (Figure 3.14a). However, the population may oscillate regularly (Figure 3.14b) or even chaotically (Figure 3.14c), quite apart from extrinsic factors that may alter K.

If a population of a harvestable species remains constant prior to harvesting, then it is possible to predict the level to which it will fall once constant harvesting effort commences. However, processes dampening fluctuations may only operate at the

Figure 3.14 *Oscillations in populations around K. On reaching K, a population may (a) stabilise, (b) fluctuate around K in a predictable manner, or (c) fluctuate around K in an unpredictable, chaotic manner.*

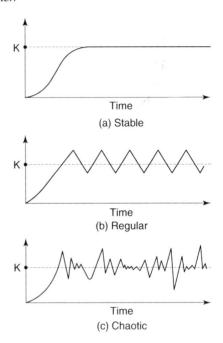

higher level, and once harvesting has commenced it may not equilibrate at the lower level (Figure 3.15).

3.7c Multiple harvesting

If more than one species is harvested from within a single system, their altered population densities may interact with other species, particularly if they are connected by a direct trophic relationship or are competitors. MSY can be calculated for each species individually, but exploitation of the other species will require their recalculation.

Exploitation of krill (*Euphausia superba*) in the Antarctic Ocean has been proposed, but krill is the major food resource of baleen whales. Therefore, it is possible to calculate an MSY for krill and another for whales, but the MSY for

Figure 3.15 *Possible effect of harvesting on population stability. (a) Before harvesting, the population is predicted to retain its stability and equilibrate at a low level. (b) In practice, lowering the population density may remove the process that dampens fluctuations, and population density may become chaotic.*

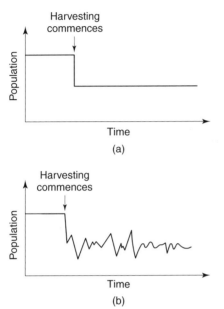

whales assumes a certain density of krill for them to eat. Therefore, if krill is directly harvested, this will reduce the amount available for whales, requiring MSY to be recalculated. In theory, a situation such as this requires calculation of a constantly changing MSY for each species, based upon harvesting levels of the other. The difficulties of calculating an MSY for a single species, based on data that are difficult to obtain, or on guesswork when data are simply unavailable, are complicated many times, even by adding a single extra species into the equation.

The North Sea is a classic example of a harvested multi-species pool. There is direct exploitation of a large number of the ecosystem components (Figure 3.16). As each component is also ecologically bound to several others, the ecological ramifications of changes in harvesting practice on one component are therefore very difficult to predict. This is further complicated by the fact that within a component, a number of taxa may be exploited which are also tightly linked ecologically.

3.8 Unsustainable harvesting

Ideally, harvesting should be sustainable, but very often a stock is harvested beyond its ability to replenish itself, a process that is inevitably unsustainable. This may occur by accident, but under some circumstances it makes economic sense, particularly if sustainable harvesting results in economic loss in the present for the sake of preserving the stock for the future.

3.8a When the MSY is not economic

Manageable harvesting to exploit the MSY usually requires sole ownership or tight control over harvesting rates. If a stock is exploited by many parties and one party takes too much, it will increase its profit but only at the expense of the others, who will have less to harvest between them. Of course, taking too many individuals from a population is unsustainable in the long term, but short-term economic gain may have more benefits to the individual concerned than long-term sustainability.

If the MSY does not provide an economic return (for example, the stock has suffered over-exploitation in the past or its value has declined), then the best option, from an economic perspective, is to over-exploit, to gain some benefit in the short term. If a party has invested heavily in equipment, for example, a boat which requires a catch of 2000 t per year to pay for itself and a useful life of 20 years, then it makes economic sense to catch 2000 t per year, even if the MSY is less. However, after exceeding the MSY for several years, it may become difficult to find 2000 t per year in later years; so it makes even more sense to harvest as much as possible, as quickly as possible.

Figure 3.16 *A simplified (highly aggregated) North Sea food web. Half of the food web's components are subject to direct human exploitation, including macro benthos (crabs), benthic predators (whelks and lobsters), demersal fish, pelagic fish, small fish and juvenile pelagics (industrial fisheries) and predatory fish.*

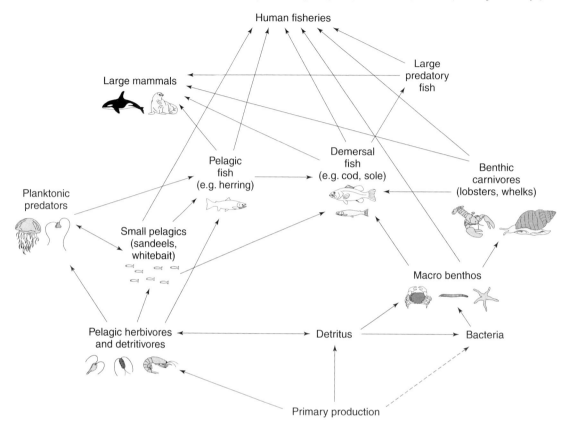

Whale harvesting displays many of the features described above. Over-hunting whales is a serious problem from a conservation perspective, but is probably an example of good economic sense!

3.8b Switching target species

If a stock is over-exploited, its increasing rarity means that increasing effort is required for the same return. In some situations (for example, rhinoceros horn), the product is very specific and increasing rarity raises its value, so that harvesting continues to be economical. Most aquatic resources are, however, neither specific in the same way

as rhinoceros horn nor particularly valuable, so eventually the cost of harvesting exceeds the return and the fishery will cease to operate or switch to another prey.

One of the most celebrated examples of prey switching is that of the great whales. Whaling is a costly exercise, requiring the maximum return. As the effort of catching an individual whale is relatively constant, whatever its size, it makes economic sense to catch the biggest individuals, in order to maximise the return per effort. However, if large whales become rare, it becomes more economical to hunt slightly smaller ones, which give less return per unit effort of catching, but which are commoner and therefore easier to find.

Figure 3.17 *Catches of baleen whales in the Southern Hemisphere between 1910 and the introduction of a ban in 1977. Apart from the interruption caused by the Second World War in the early 1940s, hunting effort remained high, but the dominant species caught changed (from Allen, 1980).*

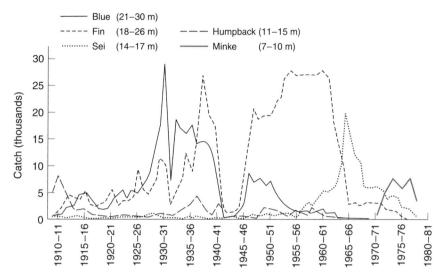

Exploitation of whales around Antarctica commenced around 1905, but only took off with the development of factory ships, introduced in 1928. During the late 1930s, 46 000 whales were taken per annum, following which there was a gradual decline until by the late 1960s fewer than 10 000 per annum were caught. In terms of the product sought, the species of whale was unimportant, because all contain the blubber and meat that was sought, and ten different species were hunted. Most are now very rare but, significantly, not a single one was hunted to extinction. Figure 3.17 illustrates the whale catch during the 20th century, and illustrates the reason for their survival, albeit in low numbers. At first, hunting was selective, targeting the blue whale because it is the largest species (21–30 m in length). As blue whale numbers declined, harvesting switched to the fin whale, the second largest species (18–26 m). When this declined in numbers, effort switched to the sei whale (14–17 m) and then, in turn, to the minke whale (7–10 m).

3.9 Summary

- Harvesting is the removal of organisms for human consumption. The term normally refers to sustainable harvesting, in which the harvest potential is maintained from year to year.

- Harvesting is possible because reproduction normally exceeds the carrying capacity of the environment, resulting in an exploitable surplus of biomass. Harvesting capacity is determined by the productivity of the system, which is in turn normally a function of primary production and the efficiency of energy transfer between trophic levels.

- Effective harvesting requires an understanding of population dynamics of the species to be exploited. When reproduction exceeds mortality, a population will grow exponentially until a limiting factor to its further growth comes into play. The population will stabilise at its carrying capacity, with further reproduction matched by mortality.

- Harvesting attempts to achieve a maximum sustainable yield (MSY): the greatest biomass that can be removed without reducing potential for harvesting the succeeding generation. In practice, increasing harvest effort incurs extra cost, so there is an optimum economic yield below the MSY.

- Calculating the MSY, or optimum economic yield, requires an accurate model of population dynamics, including the relationship between current stock and future recruitment. Various fisheries models are available, based upon recruitment dynamics and length–weight relationships of the fish being caught.

- Predicting effective harvesting strategies suffers from lack of detailed information, particularly for marine fisheries in which the only information available is often the fishing effort and the biomass caught during the previous season. Furthermore, populations may behave unpredictably in response to harvesting. If two or more species are harvested from the same system, this may induce complex interacting effects.

- In certain circumstances, sustainable harvesting may be less economic than unsustainable harvesting, leading to over-exploitation.

Chapter 4 Harvesting for Economic Gain

4.1 Introduction

Living organisms have the inherent ability to increase in number, allowing scope for harvesting the excess (see **Chapter 3**). In this chapter we consider the harvesting of living aquatic organisms for economic gain. This does not simply mean for sale, but includes subsistence harvesting, and we only exclude harvesting when the primary aim is recreation – a process considered in detail in **Chapter 8**.

The range of aquatic organisms harvested is considerable, ranging from reeds and seaweeds, through shellfish and fish to beavers, otters, seals and whales. Economically fish and shellfish harvesting far exceeds the utilisation of other taxa. In 1994 the world fish catch was estimated to be 121 million tonnes (FAO (the Food and Agriculture Organisation of the United Nations), 1998a).

In this chapter we consider not only the range of aquatic organisms harvested, but also how they are harvested and the effects on the aquatic ecosystem of the harvesting process. We include both the effects on the populations harvested (a direct effect), the impacts of the harvesting process and the altered ecological balance caused by man becoming an additional predator in the system (i.e. direct and indirect effects). We also consider some of the approaches used to manage these impacts.

Harvesting methods and definitions

A wide range of techniques is employed for the harvesting of living aquatic resources. Below we provide an outline of some of the most common (for a more comprehensive account see Sainsbury, 1986). Some, such as fish trawls, are used widely, while others are localised to certain geographic regions, for example, the fishing cormorants of China, or restricted to certain taxa, such as explosive harpoons for whales. Harvesting techniques also vary in their selectivity; hand gathering of shellfish and direct hunting of whales and seals are highly selective and take only the targeted individuals, which will be both the correct species and the correct size or age. Other methods, including most nets, are much less selective and take the intended species, but also capture other taxa; they may also take a proportion of individuals of a size, or other characteristic, which makes them unusable. In these situations we can recognise the following:

- **The target(s)**: those individuals of appropriate size/age/sex/condition and species which are sought by the harvesters.

- **The by-catch**: other species captured incidentally.

- **The discard fraction**: that part of the catch not wanted due to its size (under legal landing size, under marketable size) or species composition. This fraction of the catch is often 'discarded', i.e. returned to the environment.

The by-catch therefore comprises the discard fraction plus a 'marketable fraction', comprising individuals of species which are not the target of the fishery but do have an economic use. For example, in the North Sea demersal fishery, cod and haddock are the principal targets but turbot, dogfish, plaice, sole and monkfish are also captured by the trawl and brought to market.

The term **artisanal fisher** is often encountered; we use it here to indicate a small-scale fisher who uses traditional methods. This definition does not preclude the commercial sale of the resulting catch.

4.2 Fish and shellfish fisheries

4.2a Introduction

When we consider the harvesting of aquatic living resources, it is fish and shellfish harvesting which generally comes to mind. Man has harvested fish and shellfish from rivers, lakes and the coastal seas since prehistoric times. There is, therefore, a wide range of local and cultural influences on fishing activities, some of which have a biological basis, such as closed seasons to protect breeding grounds.

Harvesting can be motivated by a need to gain the material for the fishers and their dependants – subsistence fishing – or for monetary gain by the sale of the catch. There is no real difference in the ecological processes associated with these two motivations, but the technologies used, the geographical locations/ecological environments exploited and the scale/intensity of the activities do often vary between the two categories.

Marine ecosystems fishing, which represents the biggest anthropogenic impact (Dayton *et al.*, 1995), is both widespread and has been ongoing since pre-history (Desse & Desse-Berset, 1993).

Fishing impacts the populations of target species, other species captured incidentally, and potentially all the other species in the community which interact with them (Dayton *et al.*, 1995). In addition bottom fishing gears impact on the seafloor, causing mortality and injury to surface living and shallowly buried fauna (Tuck *et al.*, 1998), and altering physical habitat features (Auster *et al.*, 1995), sedimentation (Churchill, 1989) and nutrient cycling (Mayer *et al.*, 1991). In the short term such disturbed areas may attract mobile scavengers and predators (Kaiser and Spencer, 1994). It is difficult to assess the impact of these changes at the scale of the ecosystem (Thrush *et al.*, 1998) and the lengthy time scales over which exploitation has occurred (Frid *et al.*, 2000).

Fisheries are most commonly divided into demersal and pelagic fisheries on the basis of fish habits. Pelagic species are those which live in the water column; they tend to be active, fast swimming species such as tuna, mackerel, herring and sardines. Pelagic fisheries employ long lines, seine and gill nets and mid-water trawls. Demersal fisheries target fish that live on or adjacent to the seafloor, such as cod, haddock and the flatfish such as plaice. They are captured by bottom trawls, traps and bottom-set fixed nets, seine nets and long lines.

4.2b Global fisheries

In 1995, 1.26 million (decked) fishing vessels were recorded as participating in global fisheries (FAO, 1998b). The gross registered tonnage (GRT) of these vessels was recorded as 24 million tonnes. Both the number of vessels and the GRT were approximately double those recorded in the 1970s. An estimated 11% of the world's fishing fleet are trawlers, the remainder using fishing techniques such as purse seines, gill nets, long lines and traps. The majority of fishing effort is targeted on marine systems, but in some areas freshwater fisheries are highly important (see **Box 4.1**).

Box 4.1   Freshwater fisheries in Africa

Commercial harvesting of fish from wild stocks is dominated by marine fisheries. In many parts of the world, however, freshwater fisheries are significant sources of food for subsistence. Africa, in particular, has a large freshwater fishing industry, with approximately 1.8 million tonnes landed per year, compared with only 70 000 tonnes derived from aquaculture (see **Box 5.2**). Even amongst many countries with coastlines, including Ethiopia, Kenya, Tanzania and Madagascar, inland fisheries dominate wild catches. Normally inland fisheries are based upon multiple species of fish, with over 50 species exploited in the River Niger and 120 species in Lake Malawi.

The demand for wild caught fish continues to increase in Africa as the population rises. Therefore, attempts are continually being made to increase yields. The first major innovations were in the types of gear used. Until the middle of the 20th century, most fishing was practised using traditional techniques and locally fabricated gear, including baskets, seine nets and spears. These techniques are often highly selective, reducing the problem of unwanted by-catch, either of inappropriate species or undersized individuals. Following introduction of the nylon gill net into the continent, however, it has become the dominant gear in most African fisheries, and has allowed many, particularly those in the Great Lakes of East Africa, to increase in size into major commercial ventures. The increased efficiency and less discriminate catch introduced by gill nets have contributed to the overexploitation of many fishery resources. Further declines in abundance of commercial species followed the introduction of bottom trawling. In Lake Victoria, bottom trawling began commercially in 1965, precipitating a decline in numbers of many haplochromine cichlid fish species. In Lake Malawi, introduction of a commercial bottom trawl fishery in 1968 led to a rapid decline in harvestable species, with 20% of species disappearing by 1974.

Declines in catches of native species can be offset by introducing new species, or by exploiting artificial lakes. Introduction of alien species was a common management strategy during the mid-20th century. In Lake Victoria, for example, a severe decline in fish landings occurred following introduction of gill nets and outboard motors, compounded by eutrophication caused by agricultural expansion in the catchment. Therefore, during the 1950s and 1960s, several large-bodied species of fish were introduced in an attempt to improve yields. These included tilapiine species (*Oreochromis niloticus*, *O. leucostictus* and *Tilapia zilli*) and Nile perch (*Lates niloticus*). *Lates* and *O. niloticus* now dominate commercial fisheries in most of the lake, and their combined catches have increased the total harvest of the lake several-fold, albeit based upon a top predator (*Lates*) whose current populations are unlikely to be stable.

From an ecological perspective, the introduction of more efficient harvesting methods has had more profound effects than simply altering the abundance of fish species. The African Great Lakes each support several hundred endemic species of haplochromine fish, many of which are particularly vulnerable to unselective fishing, such as trawling. In Lake Victoria, the effects on endemic species have been compounded by introductions, and particularly that of the predatory *Lates*, with the result that many of the fish species endemic to the lake are probably now extinct and all but a handful of the other native fish species have declined in numbers. Fisheries management in lakes which have so far escaped addition of introduced species, such as Lakes Malawi and Tanganyika, faces a major dilemma – whether to maintain the high diversity of endemic fish species or add new species, which may increase yields high enough to satisfy demand.

A possible option for increasing protein yields without precipitating extinctions is effective exploitation of fish stocks in the many artificial lakes that now cover much of tropical Africa. Creation of a lake by impounding a river has a negative effect upon river fish (see **Section 2.3**), but provides an opportunity to create new assemblages in the lake. Furthermore, as a new landscape feature, an impounded lake has no native fauna to be disrupted and so introductions can be less controversial. Natural colonisation from the river system can be augmented by introducing key species of potential economic importance, as has occurred with the sardine *Limnothrissa miodon*, introduced to

Box 4.1 continued

Lake Kariba (Zambia/Zimbabwe), and the tilapiines *Oreochromis macrochir* and *Tilapia rendalli*, introduced to Lake McIlwaine (Zimbabwe). *Limnothrissa* was introduced to Lake Kariba in 1967–68, following closure of the dam in 1958. The introduction was a success, probably because the native fish species in the Zambesi River were confined to parts of the lake adjacent to the shore and not exploiting the pelagic zone with its high abundance of zooplankton. *Limnothrissa* is a specialist pelagic zooplanktivore in its native Lake Tanganyika, and adapted well to the vacant niche in its new home. By 1985, up to 1000 tonnes of this species were being harvested from Lake Kariba per month, comprising 85% of the total catch. In Lake Tanganyika itself, this species is economically unimportant.

Sources: Lévêque (1997); Talling & Lemoalle (1998)

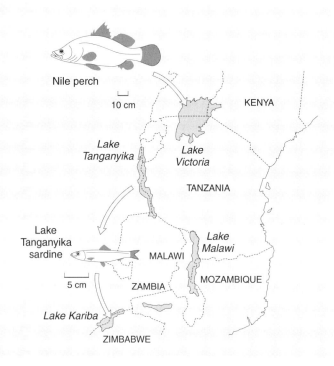

In the capture fisheries, global landings of pelagic species form the largest species grouping (21% of global tonnage landed), demersal gadoids (cod, haddock, whiting, etc.) form 10%, whilst shrimp and prawns represent around 3% of global fish production (FAO, 1995).

Some 35% of the world's total annual fish catch is derived from non-tropical continental shelves (for example the North Sea, see **Box 4.2**), even though these only occupy approximately 5% of the total sea area. Pauly & Christensen (1995) estimated that 36% of non-tropical continental shelf primary productivity was required to sustain these levels of exploitation, an amount that is likely to result in substantial changes in these ecosystems (Beddington, 1995).

A wide range of harvesting techniques is used to capture fish and shellfish. An appreciation of

Box 4.2   The North Sea fisheries

There are significant offshore fisheries in the North Sea which lie outside coastal waters, beyond the six-mile limit. Vessels operating in these offshore waters tend to be larger than those in the inshore fleets, and typically may have lengths up to 80 metres and engine powers up to and in excess of 3000 HP. There are four main offshore fishery sectors: the demersal round fish fisheries, the flatfish fisheries, the pelagic fisheries and the shellfish fisheries. Within the North Sea, a variety of fishing gears are used: beam trawls, pair trawls, seine nets, gill nets, otter trawls and traps (Anon, 1998). The major commercial demersal fish species caught in the North Sea are cod, plaice, sole, *Nephrops*, haddock, whiting and anglerfish (MAFF, 1997; Anon, 1998; European Commission (DG XIV), 1998).

The northern area (ICES area IVa) is regarded as the most productive part of the North Sea, with UK landings from this area valued at £192 million. Reported UK landings from the central and southern areas (ICES areas IVb and IVc) were £84 million and £17 million respectively (MAFF, 1997). The estimated total landings and values from the North Sea are detailed in Table 1.

An estimated 21 300 fishermen and 9800 vessels (Table 2) are registered in ports bordering the North Sea (MAFF, 1997; EU (DG XIV), 1998). The UK, Denmark, Netherlands, as well as Belgium, France and Germany have significant stakes in the North Sea fisheries (Table 3).

In general terms, the demersal fishing effort patterns in the North Sea show signs of change, with more effort gradually shifting towards beam trawling, gill netting and the use of light trawls, whilst seine netting, long lining and otter trawling are all in decline (Anon, 1998). The landings of demersal fish for human consumption from the North Sea have shown a steady

The North Sea (ICES areas IV a, b, c) showing the ICES divisions used as the basis of stock management.

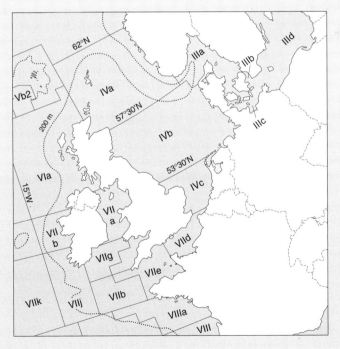

Box 4.2 continued

Table 1 *Estimated value of North Sea landings in 1998*

Species	1998 TAC/landings (tonnes)	Gross value (£)
Cod	140 000	88 million
Plaice	87 000	66 million
Sole	19 100	63 million
Crangon	28 550	50 million
Haddock	88 500	42 million
Nephrops	15 200	42 million
Angler fish	22 070	38 million
Whiting	60 000	25 million
Pelagics	177 500	24 million
Miscellaneous other demersal	100 000	68 million
Miscellaneous other shellfish	58 000	55 million
Total estimated value (rounded)	800 000 tonnes	£550 million

Source: MAFF (1997), EC (DG XIV) (1998), Anon (1998).

Table 2 *Details of the fleets from the North Sea nations*

Country	GRT (tonnes)	No. of vessels	No. of fishermen
UK (Registered in North Sea ports)	110 504	1086	8090
Belgium	23 262	156	720
Denmark	97 455	5165	5491
Germany	78 799	2406	4142
Netherlands	180 607	1009	2834
Totals	491 000	9800	21 300

Source: MAFF (1997), EC (DG XIV) (1998), Van Marlen *et al*. (1998).

Table 3 *The North Sea total allowable catch (TAC) national allocations of selected fish species (1998)*

Country	% of total North Sea TAC allocated to country			
	Cod	Whiting	Plaice	Sole
Denmark	19	11	19	4
Germany	12	3	6	7
Netherlands	10	6	37	75
Belgium	3	3	6	8
UK	43	44	27	4
France	4	16	1	2
Non-EU	9	17	4	0

Source: EC (DG XIV) (1998).

decline over the past 25 years. For example, the 1996 North Sea landings as a percentage of 1970 landings were: cod 47%, haddock 14%, plaice 64%, whiting 43% and saithe 67%. Overall the decline is described as severe, with aggregated North Sea demersal landings at 38% of the 1970 level (Anon, 1998). Only the landings of sole have shown no decline over this period (Anon, 1998).

their ecological impacts requires an understanding of their design and mode of operation, which is described below.

4.2c Pelagic fishing processes

Long lines

Long lines are the most basic design of fishing gear, consisting of a hook on a fishing line. The hook may be baited or may be attached to a lure designed to resemble some prey of the target species. Pelagic long lines are used for tuna, swordfish, sharks and other oceanic predatory fish. Long lines may be suspended from buoys and left to drift in the water for a period, or may be towed, often in an undulating way ('jigged') to resemble the movements of the prey species (Figure 4.1a). Long lines are generally regarded as quite selective, taking the target species and those which feed in the same area and on the same prey. Unfortunately in some regions this can include significant numbers of seabirds (particularly albatross and petrels), sharks, turtles and juveniles of the target species. The by-catch of birds is dependent on the time the hooks spend in near surface waters (many birds congregate to feed as the baited hooks are deployed). Conservation measures have therefore been aimed at reducing this time. This can be achieved by the more rapid sinking of the lines or the use of the so-called 'smart hooks' in which the point is only opened once a certain depth is attained. Alternatively, or in addition, bird deterrents such as noise, towed buoys and streamers of floats and lines suspended from long poles, may be deployed from the sides of each vessel.

Gill nets

Gill nets are suspended in the water column in such a way as to produce a vertical curtain of netting. The top of the net is supported by floats, while the bottom is weighted (Figure 4.1b). These nets are deployed into areas where fish swim. The fish swim into the nets and become trapped by their projecting gill covers. The nets are made as inconspicuous as possible, and are now commonly made of mono-filament nylon, which is extremely strong and almost invisible underwater. Gill nets are highly size selective but are very efficient at catching all organisms larger than the mesh size, including marine mammals and birds which then drown.

Seine nets

The basic concept of a seine net is to enclose a group of fish. In pelagic seine netting a shoal of fish is first located (by sonar, spotter plane or an observer at the masthead of the vessel). The vessel approaches the shoal and then remains stationary, while the end of the net is towed in a circle around the shoal by a small ancillary vessel – the seine skiff (Figure 4.1c). Once the net is deployed in a circle, a line is pulled which closes the bottom; the net is then winched back on board, effectively decreasing the circumference of the enclosed area. As the fish become concentrated in this area they can be captured using simple scoop nets to transfer them to the vessel or they may be drawn up in the seine net as it is hauled on board. Seine nets can have diameters of over 600 m.

Seine nets are used to capture pelagic predatory fish, such as tuna, mackerel and salmon, as well as herring and Antarctic krill. They are very effective at capturing all the organisms enclosed by the net; therefore, in addition to the targeted shoal they capture other species, either shoaling with the targets or attracted to feed on the shoal. There has been much concern about the capture of dolphins, which shoal with tuna, by the tuna seine fishery. This has led to use of cetacean scaring devices (acoustic and visual) and the lowering of part of the net during the hauling-in period to allow dolphins to escape.

Figure 4.1 *Pelagic fishing techniques: (a) long lines; (b) gill nets; (c) seine net; (d) mid-water trawl.*

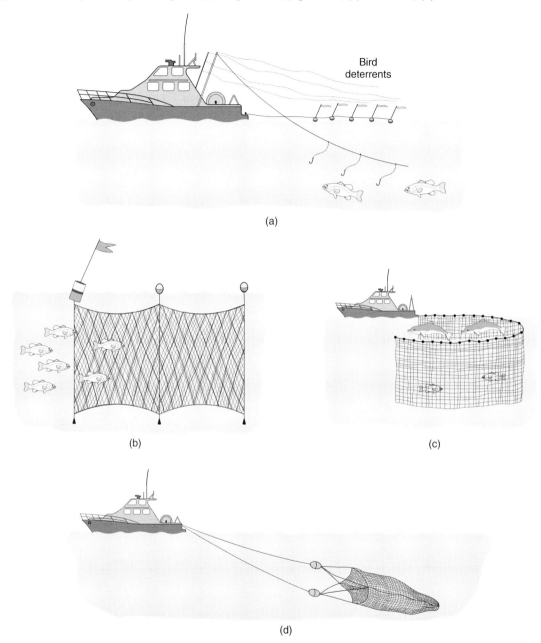

Mid-water (pelagic) trawls

Trawls are basically bags of netting which taper from a wide mouth (the larger the mouth, the greater volume fished) to the 'cod end'. The bottom of the mouth of the net is weighted, while the head rope has floats attached; this holds the net open vertically. As it is towed through the water, the hydrodynamic effect of the water passing over the 'otter boards' opens the net in the horizontal plane (Figure 4.1d). Floats may be attached to the cod-end to support the weight of the catch and prevent it from dragging the net down. Towing a mid-water net requires more power from the vessel (and so fuel and hence cost) than a net of the same size fished on the bottom.

4.2d Demersal fishing

Long lines

Bottom set long lines are deployed on the bottom in such a way that the baited hooks rise a short distance clear of the seafloor (Figure 4.2a). On the east coast of England, long lining was a traditional means of capturing cod and haddock inshore during the winter. The hooks were baited with mussels and limpets collected on the shore or with fish offal or unmarketable material from an earlier catch. They were then deployed for a few hours or overnight before being hauled. The availability of fish from the trawl fishery and the labour intensive nature of collecting and cutting bait and then baiting the lines meant that this fishery became uneconomic. In recent years the availability of mechanical baiting machines and the ability of long lines to exploit areas closed to towed gears – either by regulation or because of the nature of the seafloor (wrecks and rocky reefs) – has seen an increase in this sector.

Beam trawl

The first records of beam trawls can be traced back to the 12th century. In 1376 the abbot of a monastery in southern England petitioned the king about this new fishing device. He was concerned about it destroying the local beds of oysters harvested by the monks – this is the earliest record of concern about the environmental impact of a fishery being expressed (de Groot, 1984). A beam trawl consists of a bag of netting, tapering from the mouth to a cod-end (Figure 4.2b). The top of the mouth of the net is attached to a rigid beam (originally of wood, but now often of steel) which in turn is held above the seafloor by runners mounted at each end. The bottom of the net is weighted to give a good contact with the seafloor. There may also be one or more 'tickler chains' to flush fish or shellfish out of the sediment. In some areas mats of chain – a chain matrix – are used both to drive the targets out of the sediment and to protect the bottom of the net from chaffing. Beam trawls are used for the capture of flatfish and shrimps/prawns in estuaries, enclosed areas such as bays and in inshore waters around sandbanks, due to their greater manoeuvrability compared to otter trawls.

Otter trawl

Probably the most commonly used fishing gear in the world is the otter trawl (Figure 4.2c). Developed from the beam trawl, the otter trawl uses the motion of the vessel through the water to keep the net open in the horizontal plane. The foot-rope is weighted and so sits on the bottom, while the headline rises up supported by floats. The two otter boards (or 'doors') are dragged over the seafloor, and as they are pulled forward their shape causes them to attempt to shear away laterally (in the way that moving air causes a kite to rise) and so open the net. The passage of the otter boards over the seafloor creates a cloud of sediment that helps herd the fish so they remain in front of the net. One or more chains, known as 'tickler chains', may also be placed in advance of the foot-rope to flush fish or shellfish out of the sediment. Video recordings from cameras mounted on the fishing gear show that fish tend to be herded by the otter boards and ticklers but remain swimming in the path of the net until they become fatigued; only then are they carried back into the net and captured. Shellfish are captured immediately as they cannot sustain a fast swimming speed.

Figure 4.2 *Mobile demersal fishing techniques: (a) long lines; (b) beam trawl; (c) otter trawl; (d) scallop dredge.*

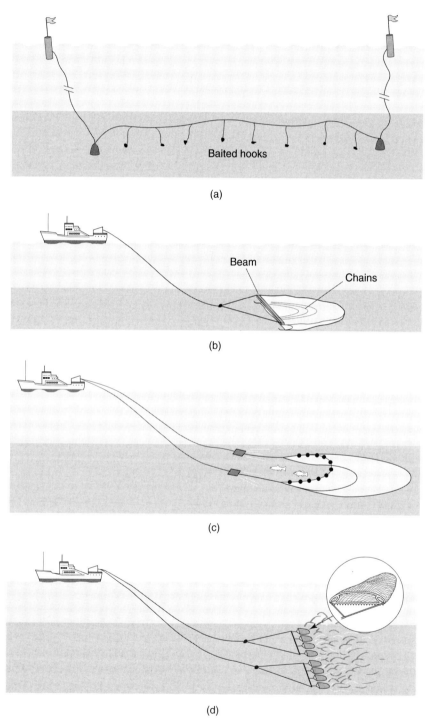

Scallop dredges

Scallops are large bivalves that burrow shallowly or sit on the surface of the seafloor. They occur in sediments ranging from mud to mixed mud and cobbles. They are fished by dredges which consist of a bag of netting, often enclosed in a bag of chain mail for protection, mounted on a frame whose lower edge is 'toothed' to dig into the sediment and extract the scallops (Figure 4.2d). Scallop dredges are usually fished in groups, attached to a beam which is towed from the vessel. In order to penetrate the sediment and to be robust to withstand the impacts with cobbles, scallop dredges are heavy. This means that the vessels need to be powerful to tow them and the physical impact of the gear on the seafloor is large.

4.2e Set/fixed gears and traps

Bottom set nets

Gill nets and trammel nets can be set close to the seafloor to target demersal species. These are often set near wrecks and rocky reefs where towed gears cannot operate. The gill nets work in the same way as pelagic gill nets, but trammel nets consist of three layers of net, a fine mesh sandwiched between curtains of coarser mesh. Fish pass in through the outer coarse mesh and then become entangled between it and the finer mesh net (Figure 4.3a). Trammel nets are less selective than gill nets, and bottom set nets often take a by-catch of invertebrates, which are attracted to the catch.

Fish weirs and traps

Fish weirs and traps seek to use the normal behaviour of the target species to entrap it (Figure 4.3b). In some cases bait is used, but in many cases it is curiosity or a shelter-seeking behaviour that attracts the target into the trap. There is a massive range of designs used worldwide, with many regional differences in details. Some designs are highly selective, including the urns used to capture octopus in the Mediterranean.

Creels and pots

Many types of invertebrate scavenger are taken in creels and pots, including lobster, crawfish, crabs, whelks (a scavenging gastropod mollusc) and *Nephrops* ('scampi' or Norwegian lobster). The design of pot varies depending on the target species and local tradition (Figure 4.3c). In all cases the principle is the same; the pot is baited with some decomposing material, the targets follow the chemical/olfactory gradient to the prey and enter the pot through the narrow opening. Once inside, the geometry of the pot prevents, or at least restricts, their ability to escape. Pots are moderately good at restricting the entry of non-target species; as the material is not damaged by the capture process when the pots are hauled, there is a very high survival of discards.

4.2f Hand gathering

Hand gathering covers a massive range of operations. They are however all highly selective. At its simplest, the only equipment is a basket to contain the catch – gathering gastropods on the shore, for example. Hand raking the sediment to extract bivalve clams is similarly simple in the equipment required. The hunting of individuals of the target species, whether by spear fishing, explosive harpoon (whaling) or by rifle (seals, walrus, beaver, etc.) is highly selective, with minimal direct impact on any but the targeted individual. This of course does not preclude population and ecosystem changes arising through altered abundance/size range/sex ratio, etc. of the targeted species.

Figure 4.3 *Fixed, or set, bottom fishing gears: (a) a bottom set 'trammel' net; (b) fish traps from various parts of the world: (i) eel trap, Germany, (ii) Madeira, (iii) Thailand, (iv) Gilbert Islands, (v) Lake Chad; (c) pots and creels for (i) crab, (ii) lobster, (iii) whelk.*

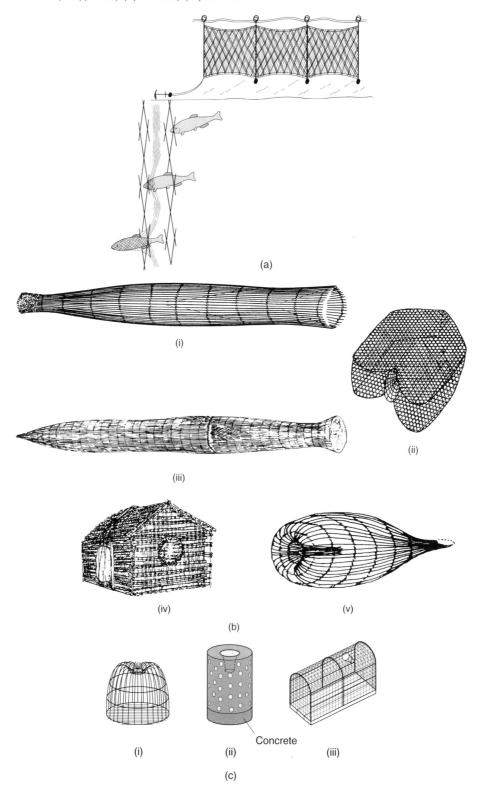

In some areas scuba divers are used to collect echinoderms and molluscs to supply the curio trade or for consumption, or corals for jewellery and ornaments. This can have low impact or be highly destructive, depending on the techniques used. For example, in the Mediterranean, divers using compressed air powered tools to extract molluscs from the reef matrix have destroyed large areas of limestone reef.

4.2g Dynamite/poisons

In habitats where fish are difficult to net, other techniques have been employed to give a large catch for little effort. In some parts of the world, fish are captured in a highly destructive way by poisoning or by explosions. An explosion in the water column will cause a pressure wave to travel outwards, killing or shocking fish nearby. A proportion of these will float to the surface to be collected by the fishers or swimmers, using masks, to get a greater proportion of the fish. However, large proportions of the killed fish are never collected and the explosion also causes mortality to invertebrates and damage to the physical structure of the habitat – especially if used in coral reefs. Coral reefs which have been repeatedly dynamite fished become reduced to rubble, which is too unstable to support new coral recruits and becomes dominated by low grade, opportunistic algae. This in turn supports a totally different fish community to that which inhabited the reef prior to destruction.

Poisons, including bleach, DDT and cyanide have also been used to kill or stun fish to allow them to be collected. As many fish recover from cyanide poisoning (at least in the short term), this technique has been widely used in tropical marine and fresh waters to collect fish for supply to the aquarium trade. These poisons, while not physically degrading the habitat, cause massive and indiscriminate mortality to a wide range of organisms, and so have very large ecological effects.

4.3 Impacts of fisheries on ecosystems

4.3a Impact on target species

The most direct effect of fishing is the changes in abundance of target species and the size structure of their populations. Most fisheries are size selective. Initially large individuals are targeted, but as these are exploited their abundance decreases, so progressively smaller individuals become included in the catch. Up to a point, the replacement of old, large, slow growing individuals with younger, faster growing ones is efficient in terms of conversion of the available production into catch (human food). Old fish allocate a larger proportion of their food intake to maintenance rather than growth and so do not make it available for harvest (see **Chapter 3**). This shift in the size of fish exploited also tends to occur alongside shifts in the species exploited. This can lead to the progressive exploitation of lower levels of the food chain (Pauly *et al.*, 1998) as well as a shift in the size frequency distribution of the fish assemblage (Gislason & Rice, 1998).

It is widely accepted that fishing mortality of most, if not all, commercial species is too high, being above the rate that is sustainable. In 1998, the FAO reported that over 66% of the world's major wild capture fisheries were either fully exploited, over-fished or depleted (FAO, 1998b). The North Sea spawning stocks of cod, haddock, whiting and plaice are all currently reported to be at low, or record low, levels (see **Box 4.1**).

4.3b Impact on non-target species

Fisheries kill organisms other than the target species, the by-catch. Some by-catch may be marketable and be landed, but the remainder is discarded at sea. By-catch mortality can be unsustainable for a non-target species for two different reasons. Firstly, direct exploitation may be too high. Species such as elasmobranchs (sharks and rays) and cetaceans (whales and dolphins) may only be able to withstand much lower mortality rates than the fishing mortality for directed fisheries. Secondly, because the management regime typically sets single species TACs (total allowable catch – the 'quota'), a fishery targeting a mix of commercial species may therefore continue fishing, thus generating additional mortality on commercial species as long as not all TACs are taken (see **Section 4.7a**).

In addition to the 121 million tonnes of fish harvested in 1994, estimates of the global incidental discarded by-catch were around 27 million tonnes (Alverson *et al.*, 1995), although this figure has since been revised down to 20 million tonnes per year (FAO, 1998a). Shrimp/prawn fisheries tend to have the highest discard rates, with an average of 84% of the catch (by weight) being discarded, compared with an average of 26% for fisheries in general (Pascoe, 1997).

The total amount of fishery discards in the North Sea alone is around 262 000 tonnes of roundfish, 299 300 tonnes of flatfish, 15 000 tonnes of elasmobranchs and 149 700 tonnes of benthic invertebrates per year (Garthe *et al.*, 1996). These discards equate to about 22% of total North Sea landings.

Beam trawl fleets were identified as discarding at the highest rate in the North Sea. Among flatfish beam trawl fisheries, Garthe and Damm (1997) estimated that 6.6 kg of fish were discarded for each kg of sole landed. These estimates suggest that the white fish beam trawling fleet in the North Sea discarded 18 000 tonnes of round fish and 182 000 tonnes of flatfish each year. Van Beek *et al.* (1990) showed that, for Dutch beam trawlers in the North Sea, from 1976 to 1990, 49% of the plaice caught were discarded. In another example of discarding, a total of 69 000 tonnes of North Sea whiting were caught in 1996, of which an estimated 28 000 tonnes (41%) were discarded. This was by no means an unusual occurrence, as records show that whiting discards may often

be greater (in tonnes) than the quantities landed (Anon, 1998).

There is therefore a clear case for the better targeting of fishing activities on target stocks, with a need for greater selectivity to prevent the incidental capture of other species and the capture of unmarketable individuals of the target species.

4.3c Indirect effects

Effects on food webs

Exploitation of fish stocks has altered the abundance of fish in the seas and, frequently, the size composition of the fish populations (Pope *et al.*, 1988; Pope and Macer, 1996). Aquatic communities frequently exhibit size-structured food webs; excessive removal of one size class of organisms is likely to have knock-on effects, upon both its prey species, in the size class below, and natural predators, in the size class above. Removal of a predator releases organisms in the trophic level below from a limiting factor and allows them to increase in abundance, in turn reducing biomass of their prey species in the next trophic level. This phenomenon is often referred to as the 'trophic cascade' (Carpenter *et al.*, 1985), because an influence at one level cascades through indirect interactions throughout the community. The existence of trophic cascades is well established for limnetic systems (Carpenter and Kitchell, 1985; Carpenter, 1988). In the marine environment the greater difficulty of carrying out the necessary experimental studies means that the existence of trophic cascades has not yet been established, although a number of properties of marine communities suggest that they are also likely to suffer trophic cascades (ICES, 1998).

The indirect effects of fishing can feed back to the fished populations. For example, Greenstreet *et al.* (1997) estimate that benthos supply 67% of the annual food requirements of groundfish (those species living near the seafloor and targeted by trawls and other bottom gears), so any increased mortality of benthos could reduce the food available to support populations of target fish species. The selective pressure that direct fishing mortality imposes on the benthos appears to induce a shift in communities towards one dominated by fast growing, small bodied opportunistic species, increasing overall benthic productivity. This change may, in turn, be responsible for the increased growth rates seen in flatfish in recent times (de Veen, 1976; Rijnsdorp & van Leeuwen, 1996). These effects are further compounded by the provision of damaged and dying benthic animals left in the tracks of fishing gears, which may encourage growth in the populations of species with scavenging tendencies (Kaiser & Spencer, 1994, 1996). Extra provisioning resulting from discarding practices may have similar effects (Ramsay *et al.*, 1996).

By combining data sets on the abundance, size frequency and size specific diet of North Sea fish, it has been possible to evaluate predation pressure over the period 1970–93 for the eight most abundant demersal species (Frid *et al.*, 1999). In spite of the declines in target fish populations (gadoids and plaice), the overall level of predation on the benthos increased from around 23 million tonnes per year in 1970 to 29 million tonnes per year in 1993, an increase of over 26%. In addition, there are indications of a decrease in the proportion of crustaceans and molluscs in the diet and an increase in the importance of echinoderms (primarily brittlestars). The principal factor influencing fish stock size of exploited species is fishing, and the expansion in the non-target dab population in the North Sea may be due to competitive or predatory release. There is therefore a case for believing that the observed changes in benthos consumption have resulted from the increase in fishing mortality on the target species. Given that demersal fish biomass has decreased, the increase in predation on the benthos may seem surprising. However, fishing has removed the larger gadoids, whose diet was principally piscivorous, and

allowed expansion of flatfish and young gadoids which prey upon benthos to a greater extent. Differences in diet of the various species would also appear to have influenced the composition of the benthos consumed.

Alteration of physical habitat

The changes in the physical habitats brought about by dynamite fishing are the clearest example of fishing-induced habitat modification. However, all mobile bottom gears impose some degree of alteration on the habitat. This might range from minor short-term changes due to the resuspension of sediments, through alterations in particle size (by winnowing the finer sediments away, or removal of boulders) to loss of biogenic structures such as tube worm or sea-fan beds.

Nutrient cycling

Fishing may also bring about changes in nutrient cycling as a result of physical disturbance of the sediment–water interface. It also adds readily consumed organic matter (discards and offal) to the system, transfers carbon and nutrients from the marine to the terrestrial system, and brings about changes in the food chain arising from manipulation of the density and size structure of the target populations (ICES, 1998). The benthos play an important role in mineralisation of organic matter and release of nutrients to the water column (Rowe *et al.*, 1975), the rate of which is critically dependent on the oxidation state of the sediment (Sørensen, 1978; Prins & Smaal, 1990). Physical disruption by towed bottom gears redistributes sediments and temporarily alters the redox state of the system. Removal of fish biomass is effectively removing production from the marine environment and transferring it to the terrestrial system (for the consumed portion) and into the scavengers (birds and aquatic scavengers) for the discards and offal (Camphuysen *et al.*, 1995).

Litter from fishing operations

A large proportion of the 6.5 million tonnes of litter entering the seas can be traced back to fishing boats, including ropes, floats/buoys, netting, and general refuse including food packaging materials. Much of this behaves as other litter, contributing to the loss of aesthetic value on the shoreline when it is cast up, but has only limited ecological effects, the principal ones being following ingestion by turtles and entanglement of birds and marine mammals (see **Section 6.4**). A special category of effect arising from lost or discarded fishing gear is, however, 'ghost' fishing.

Ghost fishing is the continued operation of fishing gear lost or discarded by fishers. The catch made by these gears imposes mortality on the impacted populations but with no benefit to the fishers. The period of ghost fishing is thought to increase if the net is lost in deeper water (Furevik & Fosseidengen, 2000). In many cases, ghost fishing exerts a higher mortality on non-target species that when the gears are fished commercially. For example, in 1981, 350 dead seabirds were found in a 15 km drift net recovered floating in the North Pacific. This figure is likely to underestimate the actual mortality as some of the victims may well have fallen off, decayed away or been taken by scavengers.

Ghost fishing has become an issue in recent years with the development of increasingly long-lived materials for the manufacture of fishing equipment. Previously, the natural materials used in the ropes and nets would biodegrade, reducing and eventually eliminating their ability to fish. Lost bottom nets tend to get 'balled up' (rolled into a ball) as currents move them over the seafloor, and while the ball of netting may snag or crush epibenthos it is no longer ghost fishing. In contrast, lost set nets around reefs and wrecks may continue to fish for several months. Midwater set nets also continue to drift and fish for long periods and the decaying remains of earlier

victims may attract fish or birds which subsequently get trapped. Lost lobster and crab pots tend to go through a cycle of continual re-baiting: the entrapped organisms die and begin to decay, acting as bait and attracting further individuals, who initially feed on the earlier victims, but then starve to death and in turn become decomposing bait for the next set of victims. In some areas, it is now a requirement that lobster pots are fitted with an 'escape panel' that is fixed closed with a degradable fastener. This means that should the pot be lost, the catch will be able to escape and no further organisms will be trapped.

4.4 Collecting intertidal shellfish

4.4a Recreational and commercial harvesting

In the developed world, intertidal organisms are collected in large numbers as a recreational harvest, either for direct consumption or for bait (see **Section 8.3**). However, the effects of this harvesting are the same as those of economic harvesting and in some situations, it is difficult to distinguish between the two. Large scale harvesting of shellfish from shores in South Africa and Chile is clearly providing a major contribution to the diet of the people involved, whereas similar exploitation in Western Europe is normally assumed to be mainly recreational. For those on low incomes, however, it may be an important source of protein or of money if the harvest is then sold. Little is known of the scale of such activities. In Britain, McKay & Fowler (1997) report upon the commercial collection of the common periwinkle *Littorina littorea* (L.) in Scotland, and indicate a peak in this activity in the late 1970s with a harvest of around 3000 tonnes per year, which has declined to nearer 2000 tonnes currently. The authors, however, advise caution with regard to some of their data, as reports of the intensity of collection were to some degree anecdotal. There appears to be little or no research into collection of other intertidal organisms for human consumption such as mussels, or into the levels of collection for personal consumption, on European rocky shores.

The majority of work regarding the collection of animals from rocky shores has been carried out in South America, South Africa, and Australia. In some regions animals are collected on a subsistence basis and as such may represent an important source of dietary protein which may become increasingly pressurised as coastal populations increase (Hockey *et al.*, 1988). In other regions collection may occur for sale, or for use as bait items in local fisheries, in which case demand may be influenced by the market price, the availability of alternative sources of income, and the demands of local fisheries. The removal of animals from intertidal areas may have far reaching consequences at both population and community levels. Despite the relatively recent research into human collection from rocky shores, it should be noted that such harvesting activities have occurred in some regions for hundreds if not thousands of years (Hockey *et al.*, 1988; Keough *et al.*, 1993; Underwood, 1993a). Therefore, the present community structure may still bear the hallmarks of past exploitation (Moreno *et al.*, 1986).

Despite differences in the motivation for collection, there are several types of response which commonly occur in communities subject to harvesting activities. It is clear that by the removal of animals from a given population, harvesting has the potential to affect the density of target species. Furthermore, if harvesting is carried out in a size selective manner, the likely response will be a shift in the modal size of the target population (Hockey & Bosman, 1986; Lasiak, 1991; Underwood, 1993). Other effects are often dependent upon the nature and intensity of the interactions between the members of a community and may occur either between or within trophic levels.

4.4b Abundance/density effects

One of the most commonly reported effects of exploitation upon rocky shore communities is the reduction in abundance or density of target populations. In a series of human exclusion studies carried out on the Chilean coast, various community parameters were examined at collection sites and compared with non-harvested reserve areas (Godoy & Moreno, 1989). Population densities of target species in each of these studies were significantly higher in reserve areas than in collected areas, as a result of either density increases within reserve areas, or declining densities in harvested areas. In their study in central Chile, Castilla & Duran (1985) recorded a dramatic increase in the density of the exploited gastropod *Concholepas concholepas* within two years of a reserve area being established (Figure 4.4a). The abundance of *C. concholepas* showed considerable temporal variation, although some of this variation was due to annual vertical migration up the shore from the sub-tidal region. Despite this variation a clear upwards shift in abundance of the protected population was seen after the setting up of the reserve area (Figure 4.4b). Moreno *et al.* (1984) noted similar effects upon the population density of keyhole limpets (*Fissurella* spp.) during a four year exclusion experiment in southern Chile. This indicates that some target species may be able to recover quite rapidly, even after being subject to high levels of exploitation.

Several studies carried out in South Africa comparing harvested and non-harvested areas of the rocky shore, report differing effects of human predation upon the densities of target populations. Dye (1992) showed that density of the brown mussel *Perna perna* in exploited areas in Transkei reached only 50–60% of the cover found in protected areas. A similar effect on *P. perna* populations was also reported by Lasiak (1991). The results of both these studies from Transkei concur with those of the Chilean studies discussed above.

Figure 4.4 *(a) Abundance and size distribution of* Concholepas concholepas *in harvested (H) and non-harvested (NH) areas of Chile. (b) Effect of the designation of a marine reserve on* Concholepas concholepas *abundance (from Castilla & Duran, 1985).*

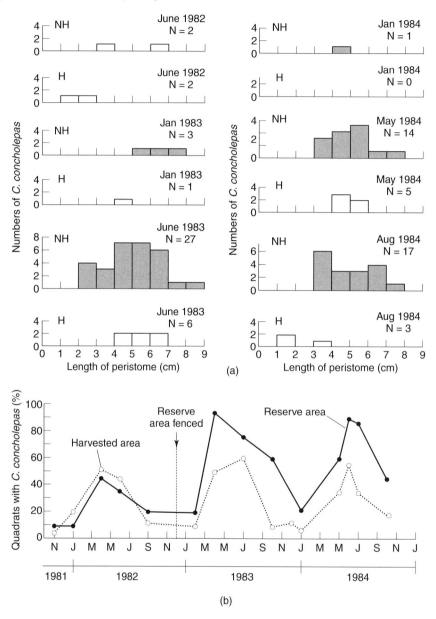

However, a study to the north of Transkei in Kwazulu-Natal revealed populations of *P. perna* which showed no decline in abundance, despite supporting higher exploitation rates than those in Transkei. It was suggested that populations of *P. perna* in an adjacent nature reserve assisted in maintaining recruitment levels, and also that *P. perna* grew faster and reached sexual maturity

98

at a smaller size than in Transkei, thus allowing some reproductive effort to occur before the mussel reached a harvestable size. Transkei, by comparison, had limited inaccessible refuge habitat (Kyle *et al.*, 1997). Keough *et al.* (1993) found that three of four species of harvested molluscs showed no differences in abundance between harvested and non-harvested treatments.

4.4c Size–frequency effects

Several rocky shore studies have shown that human collection may result in changes to the size–frequency distribution of target populations. Collection by humans is frequently size-selective towards larger individuals, resulting in a reduction in modal size within a population, and also decreased reproductive potential in those species where larger individuals are proportionately more fecund (Underwood, 1993a). In some species, sexual maturity does not occur until a certain age or size is reached. If size at sexual maturation is less than the minimum collection size, some reproductive activity can take place before reaching a harvestable size. If the opposite is true, a size-selective harvesting regime may result in the population consisting largely of non-breeding juveniles as an increasing proportion of the more fecund adults are removed (Keough *et al.*, 1993; Underwood, 1993).

Moreno *et al.* (1986) demonstrated the potential effects of size selective harvesting upon a population. Two populations of the muricid gastropod *C. concholepas* were examined, one of which was protected and the other harvested. There was a marked increase in the modal size of animals within the reserve area, while in the harvested area the size distribution of the animals remained essentially unchanged during the same period. The analysis of the size of shells discarded in shell middens by collectors clearly showed that animals as small as 40 mm were being collected, despite the legal minimum size for collection being 100 mm.

An investigation of the effects of harvesting upon two species of limpet in Transkei recorded a shift towards a smaller modal size in both populations, but without any discernible differences in abundance between harvested and non-harvested areas (Hockey & Bosman, 1986). The authors also proposed that the long-distance dispersal of larvae and juveniles employed by the limpets was able to mask declining reproductive fitness within a given population if assessment was based upon abundance alone. However, it should be noted that whilst local recruitment may be maintained by larval supplies external to the population, reduced larval production in the harvested population may result in undersaturation of settling larvae in another area, thus affecting community structure in that locality (Gaines *et al.*, 1985; Lewin, 1986).

4.4d Within trophic level effects

The removal of predation effects from a given community may not only affect interactions between trophic levels, but also result in changes in competitive interactions within a trophic level due to 'cross-linkage' effects (Paine, 1980).

Godoy & Moreno (1989) examined the relationship between two limpets, the exploited keyhole limpet (*Fissurella picta*) and the non-harvested *Siphonaria lessoni*, in exploited and non-exploited areas in southern Chile. In a previous study it had been observed that individuals of *S. lessoni* were larger in areas where *F. picta* was harvested and that both these limpets foraged on the same area of the shore (Moreno *et al.*, 1984). They found that considerable dietary overlap occurred in these two species, and suggested that the reduced size of *S. lessoni* in non-harvested areas was linked to the presence of higher densities of *F. picta*. A size difference of 4 mm between treatments in individuals of *S. lessoni* over a period of eight months was recorded. This, using a size–fecundity relationship, implied a 38.3% reduction in egg production. There was no difference in density of *S. lessoni* between treatments, and the effects of the presence of *F. picta* were manifested in an overall decline in production of *S. lessoni*. Within the reserve, *S. lessoni* may achieve some refuge by virtue of small size, which reduces the effects of predation and desiccation.

4.4e Community level effects

Hockey & Bosman (1986) suggested that at the community level human collecting constituted a form of disturbance, and that the observed comparative increase in species diversity in collected areas could be explained by Connell's 'intermediate disturbance hypothesis' (Connell, 1978). However, Castilla & Duran (1985) reported a greater level of species diversity at the non-harvested site. This apparently contradictory finding is due to the relative roles within the communities of the collected species. In the study of Castilla & Duran (1985), exclusion of human collection effects permitted the increase in numbers of a predatory gastropod, which acted as a keystone predator in the system. The subsequent increase in predator numbers resulted in the decline of the competitively dominant space-occupying mussel, allowing the increased settlement of sessile species, thus increasing diversity. In the case of Hockey & Bosman (1986), human collection targeted sessile filter feeders, and both grazing and predatory gastropods. Collection therefore operated more in the manner of a non-selective disturbance to the community, acting simultaneously on several trophic levels, rather than the addition of another top predator.

A problem with adopting species diversity or richness as a measure of community structure is that neither index conserves species identity. Lasiak & Field (1995) demonstrated no significant effect of collection upon either species richness or diversity on rocky shores in South Africa. However, when the same data were analysed using hierarchical cluster analysis and multi-dimensional scaling ordinations (see **Section 9.2c**), both of which incorporate species identity, there were clear differences between exploited and non-exploited sites.

Disturbance

Whilst it may be disputed as to whether the direct impacts resulting from harvesting may be described as a form of disturbance, indirect effects such as trampling and boulder turning almost certainly can. Considerable disruption to communities may occur from the actions of collectors due to trampling effects (see **Section 8.3**), and the effects of boulder turning whilst searching for individuals of target species (Addessi, 1994).

The communities on the top and underside of boulders in the intertidal zone are often distinct from one another, each being comprised of organisms adapted to life in these different microhabitats; the inverting of boulders may thus subject these animals to high mortality (Lewis, 1964). In examining the human impact upon the rocky intertidal in San Diego, California, Addessi (1994) found a significant positive correlation between the number of overturned rocks and levels of human activity on the shore. The density of all species present was reduced in the most heavily visited areas; a combination of collection, crushing, and exposure to desiccation was suggested as the most probable cause.

4.5 Other living organisms

4.5a Seaweed

In some areas seaweeds have been harvested for human consumption since prehistoric times. These include laver (*Porphyra*), dulse (*Dilsea*) and carragheen (*Chondrus crispus*) amongst the Celtic peoples on the western coast of the the UK and Ireland, and a variety of species including nori

(*Porphyra*) in Japan. Seaweeds have also been exploited for fodder for domesticated livestock and as a source of fertiliser for use in agriculture. In more recent times, seaweeds have been used to extract alginates, including the agar used in microbiology, and gelling agents in a variety of food uses, for example, cakes, confectionery and ice-creams.

The traditional and food uses tend to employ hand gathering, but industrial processing for alginates uses specialist harvesting vessels, which tow rakes and/or cutters through the seaweed beds, with the weed being deposited into a hopper on deck.

The impacts of hand gathering are linked to the scale of collection. In most cases this is small scale, and as the species targeted for human consumption tend to be short-lived ephemeral species, colonisation following harvesting is rapid and no long-term effects are apparent.

The kelp forests which are the target of most of the industrial scale harvesting are subject to natural denuding events, such as storms. The clearance of patches in the kelp forest is therefore a natural event and is important for maintaining their diversity (Dayton & Tegner, 1984; Dayton, 1985; Dayton *et al.*, 1992). However, the total area cleared and the spatial pattern of the harvested patches differs from the naturally generated patches. Concern has been expressed at the consequences of large scale harvesting on the integrity of the kelp forests, and the impact on other species, in particular the fish which use the forests as nursery habitats and species of high conservation status such as sea otters.

4.5b Wetland plants

Wetland plants such as reeds have been harvested for a variety of uses for thousands of years. The range of plant species in wetlands with potential economic use is enormous. Table 4.1 gives an example of uses to which wetland plants are put. As wetlands are normally highly productive, but dominated by annual plants, there is scope for

a continuously renewed harvest. Furthermore, in contrast to most aquatic organisms, the crop is visible to the harvester prior to exploitation, allowing more precise management of the resource than is possible with, for example, fish or shellfish. In Europe, one of the most commonly used types of roofing materials for buildings was, until relatively recently, thatch made from common reed (*Phragmites australis*). Even today, reed harvesting employs around 1000 people, either permanently or seasonally, in the UK (Hawke & José, 1996). Most of the harvest comes from sites maintained for conservation purposes and harvesting is an integral component of the management of the sites, demonstrating that it need not have detrimental effects on the exploited habitat. Elsewhere in the world, wetland plants are still very important sources of raw materials. Inhaca Island in Mozambique contains several small wetlands dominated by common reed, and one with a small amount of papyrus (*Cyperus papyrus*); reeds are widely used for thatching, while papyrus stems are used for walls of huts or for binding thatch (Kalk, 1995).

Wetland plants can be harvested indefinitely at a low level, without damaging the ecosystem. Large scale removal can, however, have detrimental effects upon both the plants themselves and on other organisms dependent upon their presence. Large reedbeds are important habitats for many organisms, and a well managed wetland may provide harvestable resources, such as waterfowl, shrimps and fish, in addition to the plants themselves.

4.5c Sea urchins, sea cucumbers, palolo worms, jellyfish

Although fish and shellfish dominate the world marine harvest, a number of other groups are also harvested in certain locations. While the scale of these harvests is small globally, they can have local impact.

Table 4.1 *Uses of wetland plants in Kashmir*

Wetland Plant	Use
Acorus calamus (sweet flag)	Medicine
Carex spp. (sedge)	Cattle fodder
Ceratophyllum demersum (rigid hornwort)	Compost, ornamental
Echinochloa crusgalli	Cattle fodder
Euryale ferox	Medicine
Hydrilla verticillata (waterweed)	Compost, ornamental
Lemna spp. (duckweed)	Compost, medicine, biogas
Myriophyllum spicatum (spiked water milfoil)	Compost, ornamental
Najas graminea (naiad)	Compost, ornamental
Nelumbium nucifera (lotus)	Human food, honey, cattle fodder, medicine, ornamental, religious ceremony
Nymphaea spp. (water lily)	Human food, honey, cattle fodder, medicine, ornamental, religious ceremony
Phragmites australis (common reed)	Thatching, furniture, cattle fodder, paper pulp
Populus spp. (poplar)	Furniture, timber, firewood
Potamogeton spp. (pondweed)	Compost, ornamental
Sagittaria sagittifolia (arrowhead)	Cattle fodder
Salix spp. (willow)	Furniture, cricket bats, firewood
Salvinia natans (floating fern)	Compost, biogas, paper pulp
Scirpus lacustris (club rush)	Cattle fodder, medicine
Sparganium erectum (bur-reed)	Thatching
Spirodela polyrhiza (great duckweed)	Medicine
Trapa natans (water chestnut)	Human food
Typha angustata (bulrush)	Thatching, stuffing, cementing material, paper pulp, medicine, ornamental
Utricularia flexuosa (bladderwort)	Medicine, ornamental

Source: Pandit (1999).

Sea urchins are exploited for human consumption, their gonads being considered a delicacy in many parts of the world, while sea cucumbers and jellyfish are consumed in parts of the western Pacific and southeast Asia, generally in a dried form. Urchins and sea cucumbers are usually collected by hand gathering, but increasingly destructive techniques may be used to extract urchins and cucumbers from crevices. While traditional levels of exploitation were probably sustainable, the massive growth in the luxury market, particularly in Japan and Thailand, has put increasing pressure on stocks. Many reef areas in Indonesia have been stripped of their sea cucumbers by migrant fishers seeking them in order to supply the Japanese market; these reefs are often physically degraded by the breaking open of crevices to extract the target organisms.

The main jellyfish species harvested are the medusoid phase of Order Rhizostomeae, as they are relatively solid (having a dense mesoglea) and so suitable for drying. In general jellyfish are a high status food, and so command a high price. The world harvest of jellyfish in the late 1990s exceeded 500 000 tonnes per annum (Kingsford *et al.*, 2000), making it greater in tonnage than the world catch of scallops and lobsters. The principal fisheries are Indonesia, China and Thailand, supplying markets in Japan, Taiwan, Korea and the USA, as well as the considerable local demand in China and Indonesia. Jellyfish are relatively delicate and fishing methods most commonly employ dip nets and seines targeted at visible swarms of medusae.

The complex life cycles of the Cnidarians, with an alternation between a sessile polyp phase and a motile medusae generation, provide special challenges to the management of these fisheries. However, the biggest challenge is simply the very high price the product commands, which will encourage fishers to circumvent any management measures if they can. To date work has been advanced to look at the possible management options, but the lack of biological data and a general perception of the unimportance of jellyfish fisheries has not seen any management schemes advanced for the principle fisheries (Kingsford *et al.*, 2000).

Palolo worms (*Eunice viridis*) are benthic polychaete worms which occur in the shallow waters around the South Pacific islands of the Samoa archipelago. The worm's spawning is synchronised to the third quarter of the moon in October and November (Caspers, 1984; Bentley & Pacey, 1992). As the moon rises, the worms emerge from their burrows and the epitoke, the portion of the body containing the gonads, breaks free and swims up to the surface, where they form a swarm. The epitokes break open, the sperm and eggs mix, and fertilisation occurs. The islanders have known about this phenomenon since prehistoric times and a traditional feast has developed, based on the harvesting of epitokes. Islanders in canoes use scoop nets to collect the epitokes from the surface and they are then fried as the basis of the festival meal. The inefficiency of harvesting with scoop

nets from the surface of a moonlit sea ensures that sufficient escape to liberate their gametes and provide for the next generation.

4.5d Whales and dolphins

Many coastal communities traditionally exploited whales for meat and as a source of oil. In many cases, exploitation was limited by the available technology to stranded individuals and smaller species. In some areas the small toothed whales were protected by cultural prohibitions – in ancient Greece and Indonesia, for example, dolphins were regarded as saviours of drowning mariners and therefore not to be harvested.

With increasing sophistication of vessels, more species could be exploited and over larger areas. Some species of whale disappeared from European waters as early as the Middle Ages due to harvesting pressure. In 1660, Dutch whalers were firing their harpoons from blunderbusses (von Brandt, 1984). The explosive headed harpoon was invented in the mid 19th century and quickly came to dominate commercial whaling fleets. This, and advances in ship design and endurance, opened up the Southern Ocean to whaling, where exploitation was dramatic. The primary targets were the large whales, as the whale oil and whale bone were of greater value than the meat. By the time of the first scientific studies in the Antarctic, some whale populations were already in dramatic decline (**Section 3.8b**).

Whales are extremely vulnerable to overfishing, as their inherent rate of reproduction is so low. While the effects on the target populations are dramatic and well documented, there are also concerns about the possible wider impacts on the ecosystem. In the Antarctic it appears that, following the reduction in populations of krill-eating whales, the populations of crab eater seals and several species of penguins increased, in response to increased availability of krill. The timing of these changes provided circumstantial evidence that led some authors to suggest that there had been a competitive release of the smaller krill feeders. However, the effects of climate change in the 20th century were also most dramatic at high latitudes, with altered seasonal cycles and major changes in ice cover. It is possible that the population changes in penguins and seals were driven by these effects (Fraser *et al.*, 1992). Of course it could well be that the observed changes resulted from a combination of effects. The slow recovery of whale populations since the introduction of restrictions has also been attributed to a phase shift in the ecosystem (whatever the cause), with krill production which had been utilised by the whales now going to other consumers (Laws, 1985).

The International Whaling Commission (IWC) was established in 1948 to set quotas and regulate Antarctic whaling. It has since come under increasing pressure to end commercial (i.e. non-indigenous) harvesting of whales, a move driven by the ethical objections to whaling which have developed in western societies. The harvesting, following a fairly dramatic killing, of an advanced, social mammal has raised moral objections in many western societies. These feelings run so strong that in the USA, for example, legislation requires the government to introduce trade sanctions against nations that fail to comply with IWC regulations.

4.5e Fur-bearing mammals

Fur-bearing mammals have been heavily hunted for centuries, to the extent that many are now endangered or vulnerable. Amongst aquatic species, important targets are otters, beaver and seals, particularly unweaned seal pups which have a thick, soft fur.

Beaver have been hunted for meat, for the medicinal properties of their castoreum (scent gland), but most importantly for their fur. Intensive exploitation eliminated the European beaver

(*Castor fiber*) from most of its native range by the early 19th century, and from more settled areas, such as Great Britain, early in the Middle Ages (Kitchener & Conroy, 1997). As the beaver became rare in Europe, hunting and trapping for the lucrative European market began in North America. The value of the pelts of the American beaver (*C. canadensis*) was such that it was the mainstay of the Hudson's Bay Company, and it has been claimed that their demand opened up the interior of North America for European explorers and settlers more than any other single process (Newman, 1987). Beavers were eliminated from large areas of North America over the space of a few decades. Very often the European traders did not capture them directly, but traded for pelts with native American trappers. Therefore, by the time the scientific exploration of the interior of North America got under way, beavers were already much reduced in numbers. Beavers modify their environment greatly by constructing dams across rivers, and the changes that were wrought to upland rivers in North America during the removal of this major structural force, with its indirect effects on other inhabitants of rivers, may never be known.

4.6 Biodiversity

4.6a Genetic diversity

Fishing is selective – not all individuals are equally likely to be caught or suffer mortality or injury. Fishing has therefore the potential to be a powerful evolutionary force. This can occur through selection for traits which alter the vulnerability to fishing and through the creation of evolutionary bottlenecks.

Severe overfishing of a stock reduces the abundance of individuals and even if mortality is non-selective, the frequency of alleles in the reduced stock is expected to differ from that before exploitation and from the species average. Certain alleles may be eliminated (extirpated) from the stock. If the stock then increases following controls on exploitation, it does so with the allele frequencies inherited from the reduced stock; the genetic diversity of the stock has thus been altered. This arises by chance from the random selection of individual genotypes in the 'bottleneck' stock. However, fishing normally exerts a selection pressure, as a tendency to harvest the largest individuals ensures selection for small size at maturity and early reproduction. In the North Sea there is some evidence that cod now reproduce on average one year younger than they did 100 years ago (Law & Grey, 1989), although the issue of growth rates is more problematic to resolve due to the influence of environmental conditions, particularly temperature, which have also changed.

While the extent to which fish have undergone genetic changes in response to fishing is not yet clear, it is certain that the species size spectra of the fish assemblages have changed as a result of fishing (Rice and Gislasson, 1996). This certainly indicates the presence of a potentially strong selective pressure.

4.6b Species diversity

There are a number of measures widely used to measure species diversity. The simplest is species richness (S) and the next most widely used is the Shannon–Weiner index (H′). As a community is fished, one might expect diversity, as measured by Shannon–Weiner, initially to increase as the abundance of the target species, which usually was dominant prior to fishing, is reduced; the 'evenness' of the community therefore increases. As fishing increases, so the abundance of other species starts to decline as they in turn are targeted, at which point diversity starts to decline. This hump-shaped response has been observed in both the Grand Banks and Gulf of Thailand

fish communities (Figure 4.5a and b). However, no such pattern has been observed in the North Sea fish assemblage (Figure 4.5c). A similar lack of change has been reported for comparisons between fish communities within marine protected areas and those outside. In this case, continual migration across the reserve boundaries is the most likely explanation for the lack of difference.

The most dramatic changes may occur when bottom trawl fisheries impact areas of high epibenthic biomass (Collie et al., 1997). Many epifauna are both very vulnerable to the effects of fishing and are important habitat elements for other species. Their loss is both a direct effect on species diversity and causes the loss of the associated fauna (an indirect effect on species diversity). In addition, their loss represents a change in the ecosystem and so also constitutes an impact on ecosystem diversity (Jennings & Reynolds, 2000).

Fishing can clearly impact the abundance of the target species and others taken in the catch. It is therefore responsible for altering the balance of species in the system, even if none are extirpated. For example, recent estimates for the biomass of the entire fish component of the North Sea ecosystem vary, without trend, between 8 and 12 million tonnes (Daan et al., 1990; Sparholt, 1990). Even if the total biomass of fish in the North Sea was somewhat higher in 1900, the fish landed at that time represents removal of around 10% of the total standing crop (Daan et al., 1990). This must be seen as a large alteration of the biodiversity of the system.

Fisheries impose size selective predation on the stocks, causing size spectra in exploited populations to be altered in comparison to unfished populations (e.g. Pope et al., 1987). There is also empirical evidence that the heavier the exploitation of a stock, the steeper the slope of the number per size class versus size relationship (Pope et al., 1987; Rice and Gislason, 1996). ICES (1998) considers this to be a useful indicator of changes in fishing effort, but notes that this relationship

Figure 4.5 *Changes in the diversity (exp H′) of the fish community on (a) the Georges Banks, Canada, (b) the Gulf of Thailand, and (c) the North Sea (redrawn from Hall, 1999; original data from (a) Pauly, 1987, (b) Solow, 1994, (c) Hall & Greenstreet, 1998).*

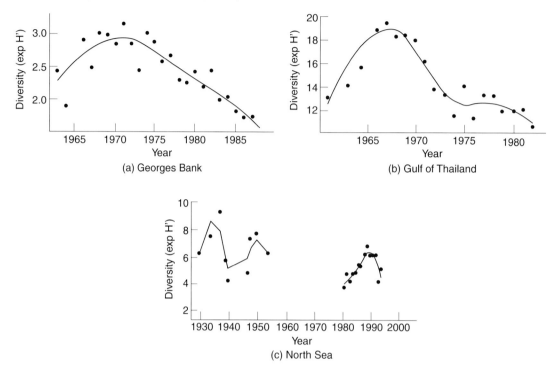

is sensitive to environmental changes which alter the growth rates or the possibility of species replacements. The ecological consequences of these changes are difficult to predict, but include altered predation rates as small and large fish rarely feed on the same prey. Such changes in fish diversity are likely therefore to cascade through other elements of the ecosystem.

It has been suggested that there have been long-term 'changes in benthic communities of the North Sea (e.g. Reise, 1982). Several causes have been proposed, such as the impacts of towed fishing gears, especially on the heavily fished areas of the eastern and southern North Sea, along with the effects of climatic and salinity fluctuations, eutrophication, and changes in zooplankton abundances. However, further evidence is required to determine whether anthropogenic

influences, as opposed to natural or cyclical events affecting the benthos, are responsible for any changes, if, indeed any real changes have occurred.

Overall, it would seem that the ecological effects of fishing extend over multi-decadal time scales and operate at spatial scales, ranging from processes within the trawl tracks (Kaiser and Spencer, 1994) to changes at the scale of the coastal sea (Frid and Clark, 2000; Frid *et al.*, 2000). We can also infer changes across the full spectrum of trophic levels. Phytoplankton will be affected through altered nutrient fluxes, while benthos are influenced by both direct mortality and indirectly through altered predation pressure and competition. The target fish themselves are influenced by direct fishing mortality and indirectly by provisioning from discards and altered competitive regimes. The top predators may also be affected by provisioning of birds (Camphuysen *et al.*, 1995) and altered food resources for marine mammals (Kaiser and de Groot, 2000). The need to conserve biodiversity, and to manage in a precautionary manner, are requirements of international conventions. The extensive changes wrought by fishing on the aquatic ecosystems would therefore suggest an urgent need to explicitly incorporate ecosystem considerations into fisheries management (see **Section 4.7**).

4.6c Ecosystem diversity

Harvesting of a population does not necessarily lead to any changes in the ecosystem and hence ecosystem diversity. However, depending on the type of harvesting gear used, there may be severe modifications to the physical nature of the habitat and to populations of non-target species. If populations of the target species are harvested so intensely that the population size is reduced, this may have implications for the integrity and functionality of the ecosystem (Hall, 1999).

Towed bottom fishing gears have the ability to significantly alter the benthic environment and the diversity of habitats. This can occur through changes in sediment particle size distribution, removal of biogenic structures and altered rates of ecological processes such as nutrient regeneration, bioturbation, etc. The scale of these impacts varies with fishing gear type (scallop dredges have more impact than beam trawls, beam trawls with heavy chains will have more impact than otter trawls). Habitats also vary in their vulnerability. A sandy sediment with no structural fauna may show no effects of the trawl within days (possibly hours) of being impacted. In contrast, a single pass of a fishing gear through a bed of tube worms or erect sponges may change the habitat to one of unconsolidated sediment, which may then prevent colonisation by the original mix of taxa and so persist as an alternative stable state. These impacts

are relatively easy to mitigate against: sensitive environments can only be effectively conserved in areas closed to fishing (see below), while in other areas impacts can be reduced by alterations to the configuration of the fishing gear.

4.7 Sustainability

4.7a Ecological sustainability

With the adoption of the Convention on Biological Diversity, managing the environment in an ecologically sustainable manner has shifted from being an option to a legal necessity – sustainability is now the goal of management policy. Given that reproduction and adaptability are key biological attributes, the real challenge for managing the system is determining the key limits – i.e. what are the ways and rates which can be sustained – and setting in place policies to obtain those goals. The latter is a socio-political issue, while the former is very much a scientific issue and may be the greatest challenge facing ecologists in the third millennium.

There are a number of approaches to limiting the ecological impacts of fishing to sustainable levels. Traditionally the approach has focused on conservation of the target stock with measures such as catch quotas and controls on size/age at capture, i.e. mesh size controls. In recent years the effectiveness of much of the management effort has been questioned – after almost 100 years of fishery management in the North Sea, more stocks were listed as endangered (outside 'safe biological limits') than ever before. This, along with growing recognition of the need to manage fisheries in the context of the wider ecosystem, has led to developments of other management approaches, including effort controls, technological changes to provide better selection for the target and less damage to the ecosystem and closed seasons/areas.

Catch quotas

The most simple and intuitive means of ensuring sustainability of fisheries is to limit the catch to a level which removes only the surplus production (see **Section 3.5**). This is the basis of most fisheries regulations worldwide and also the source of many of the problems of over exploitation. There are two major problems in applying this approach: firstly, the difficulties of calculating, in real time, the levels of this surplus production, and secondly how to match the effort of the fleet to this level of production in space and time.

Estimating the size of stock a year in advance is achieved through modelling the population (see **Section 3.5**) with input parameters based on catch data and fisheries surveys (including young fish and fish larvae surveys). As stock sizes become small, the accuracy of this approach decreases rapidly, both due to the difficulty of accurately measuring the input values and because variations in breeding success and recruitment strength cause greater fluctuations in the stock size. Individual good (or bad) breeding events have much more impact on the total stock size when it is dominated by one or two year classes. As the stock gets smaller, so quotas also get smaller and there is more incentive for fishers to land fish illegally (see below), meaning that the landing data officially recorded are erroneous and their use in the models causes further inaccuracies.

Having predicted the surplus yield available, the **total allowable catch** (TAC), the next step is to allocate this to fishers in the form of an individual TAC or **quota**. This may be done on a vessel by vessel basis, or in international areas by allocation to countries who then divide the national quota between fishers. In doing this, the algorithm used often involves a measure of 'track record', i.e. only fishers who have fished the stock in the past are allocated a quota and the more a fisher has exploited a stock the larger their share of the quota. In the run up to the introduction of quotas, this provides an incentive to individual fishers to catch as much as possible in order to gain a bigger slice of the total quota later.

Problems with quotas

Quotas work well in single species fisheries, especially where individuals are targeted for capture, such as whaling, salmon and sturgeon fisheries. They are difficult in mixed fisheries where the quota of one species often gets filled while other species are still available for capture. For example, in the North Sea demersal fishery, whiting, cod and haddock are often captured together in the trawl. In most years the quota for one species may be filled several months before the end of the season but the other species can still be fished. This results in the fishers continuing to fish but discarding at sea the species whose quota have already been caught (or landing them illegally – so called 'black fish'). However, as most discards are dead the effect on the stock is the same as if the fish were landed, so the quota has failed to control fishing mortality on the stock. In order to address this issue the European Union has in some years introduced 'days at sea' restrictions, which are essentially an effort control measure (see below).

Another aspect of fishers' behaviour that is altered by quotas is their selection of fish to land. As a quota becomes filled, a fisher may go in for 'high grading'. Normally certain sizes of species will command a premium price, so if 2 tonnes of quota are left with time to spare, then on a particular fishing trip the fisher may elect only to land the 1 tonne of the premium fish captured, discarding at sea the lower grade individuals that make up the remainder of the catch of the quota species. This means that it is still possible to land 1 tonne of the high value fish on the following trip.

Quotas require considerable enforcement effort which makes them expensive for the administering authority. Authorities attempt to keep costs down; this often results in under policing and so ineffective management. TACs encourage over capacity and over capitalisation in the fleet – fishers compete for the quota, so will tend to favour larger, more competitive vessels. The rest of the supply chain has to gear up to deal with the quantities of fish arriving when all the vessels are fishing, but for much of the year this capacity is not needed. Thus overheads are higher than they need to be and this is reflected in the price to the consumer.

Fisheries scientists give advice, on the likely size of the available yield, to the managers who set the quotas. The managers have to balance the biological and social aspects of the fishery in setting the quota. Fishers' livelihoods, their families' welfare, and the investment in infrastructure, both in fish capture and also post-capture processing, have to be considered. This results in reductions in quotas always being less than that recommended by scientists. Similarly, if the quota set is so small that the quota for each vessel fails to provide enough legitimate catch for the fisher, then either the fisher will behave illegally or the economics of the whole fleet will collapse. Individual transferable quotas (ITQ) are an attempt to overcome this. Here each fisher is initially allocated quota and can elect to fish up the quota or sell the quota to another fisher. Fishers who buy quota can catch up to the total that they own. ITQs also provide an economic mechanism for managers to reduce the total catch, in that the authorities can buy quota off fishers, effectively buying them out of the industry. If stocks improve, quota can be resold back to fishers.

Size/age at capture restrictions

The simplest technical measure to apply is one based on the size (age) at which individuals are caught – a **minimum landing size** (MLS). This can be set to ensure that all individuals achieve one or more breedings before becoming available for capture. Selection may occur post-capture with small individuals being returned to the wild or, in net fisheries, through alterations in the gear, i.e. the mesh size of the net. Again the biggest

difficulty with this approach is when a mixed fisheries is involved.

In fisheries targeting individuals, it is easy to enforce minimum landing sizes. In net capture fisheries, it is usual to apply a mesh size to the net; fishers still have to comply with an MLS, but the number of small fish caught which have to be sorted and then discarded is reduced. The mesh size used should ensure that most of the time individuals below the desired size escape. However, selection will never be 100%: as a net fills with fish, the fish in the net 'blind' the mesh openings, preventing the escape of smaller individuals, while a heavy catch being towed through the water will stretch the net, causing the mesh to deform.

When fishing for a variety of species which differ in morphology (round fish versus flatfish) and life history (early breeders versus late developers), the selection of an appropriate mesh size is a compromise. The fact that individuals below the species MLS are discarded is of little use if by that time they are dead. There are now efforts directed at developing gears which sort the species in the water by criteria other than size (see below).

Fishing effort reduction

Fishing vessel and fleet capacity is widely considered to be currently excessive and not in balance with the resources. It is generally accepted that many of the issues relating to fisheries effects on ecosystems would be resolved, or at least significantly reduced, if fishing effort occurred at levels which maintained the target stocks within sustainable limits. It is also accepted that this would not provide a solution to all the issues, e.g. loss of vulnerable species such as cetaceans.

The European Commission has in place an existing rolling programme (MAGP), whereby fishing vessel de-commissioning targets are set for each member state in an attempt to reduce the overall fishing effort in EU waters, thus reducing the pressure on stocks. Vessel de-commissioning is likely to be more attractive to those skippers with the least economically successful vessels. For instance, Revill (1996) identified that in a UK fishery consisting of 98 vessels, seven of these vessels were responsible for 49% of the total fleet effort; the combined effort of a further 56 vessels was estimated to be only 6% of the effort.

Such programmes also fail to deal with the issue of 'technological creep'. It is estimated that the efficiency of the fleet increases by 7–8% every year as a result of new vessels replacing old ones and changes in technology on the existing vessels. This implies that even if the fleet size were reduced immediately, there would still be the need to remove a further one vessel in each hundred out of the fleet every 12.5 years.

In the EU an additional restriction has been applied which limits effort during periods of low stock size and hence quota. Known as the 'Days at Sea Regulations', it restricts the number of days per month an individual vessel can fish without removing the vessel permanently from the fleet. It is unpopular with fishers as their costs remain the same and their opportunity to earn is restricted. It also forces vessels to sea in poor weather if quota is not used and there are days remaining in that month. It encourages the race to fish and leads to further high grading and black fish landings from those days when the vessel is at sea.

Greater selection in the water

Fishing gears can be made more 'selective' towards the target species by altering the geometry of the gear. This would reduce post-capture discarding. The mesh sizes of the netting in fishing gear, and in particular the cod-end mesh size, are influential in determining the selective properties of towed demersal fishing gear (Anon, 1996).

The minimum cod-end mesh size is widely legislated for in EU waters and is specific to each target fishery. In many fisheries, however, the demersal fishing grounds are multi-species in nature, and by-catch and discarding are common

resultant features due to poor cod-end selection (Andrew and Pepperell, 1994; Evans et al., 1994). The twine diameter (thickness) of the meshes in the cod-end is also known to affect the selection process in the cod end (Lowry, 1995), as are seasonal processes such as spawning status which may affect the overall shape of the fish (Lowry et al., 1995).

Research into the incorporation of selectivity devices such as square mesh panels, funnels (sieve nets) and grids (Figure 4.6) into towed fishing gears to enhance their overall selectivity is becoming more commonplace (Isaksen et al., 1992; Thorsteinsson, 1992). As a result, the use of selectivity devices within fishing gears is gaining acceptance as a management tool and has resulted in specific legislation being implemented in many instances.

Less destructive gears

Lindeboom and de Groot (1998) reviewed the evidence that benthic disturbance and damage might arise from the passage of towed demersal fishing gears. Subsequent research has focused on minimising the ground contact of the fishing gears (He and Foster, 2000) and the use of less impacting methods to stimulate the target species into the fishing gear, including water jets, electrical pulses, etc.

The incidental impact that 'lost and abandoned' static fishing gears (i.e. gillnets, trammel nets, pots, etc.) are having upon marine life and the benthic communities is also an area of concern. Future developments in this field may lead to formalised 'lost net' retrieval programmes, echo locators, or the development of biodegradable critical net/pot components.

Closed areas

Fishing activity is totally prohibited or restricted in a wide range of areas. In some cases these restrictions may be related to the need to protect

Figure 4.6 *In-water selection devices: (a) sieve net ('veil'); (b) sorting grids.*

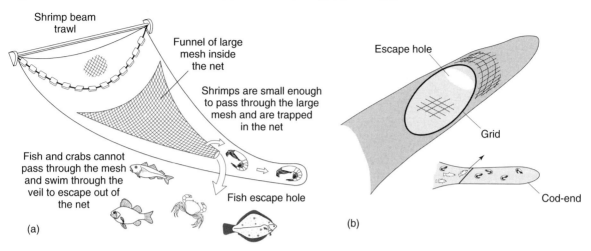

military, oil and gas installations or sites of special scientific or historical interest (Rogers, 1997). In other cases fishing is restricted or prohibited in order to protect the fish stocks (usually juveniles) themselves from over-exploitation. In others the marine protected areas may be established to segregate recreational activities, including tourism, from fisheries. For example, in the North Sea, much of the North Yorkshire inshore coastal waters (inside 3 miles) is closed to all towed forms of fishing in order to protect the juvenile codling and other gadoid species that aggregate in these waters (Rogers, 1997).

The single largest closed fishing area in the North Sea is the 'plaice box' (Figure 4.7). This area acts as a nursery ground for large numbers of juvenile commercially important flatfish species, such as plaice and sole (Anon, 1994). The plaice box is closed to fishing vessels with engine powers greater than 300 HP and therefore precludes the large whitefish beam trawling fleets from access to these grounds. The plaice box has been in existence for over ten years, in spite of which no beneficial effect on the flatfish stocks has been identified. It is postulated that environmental factors may be affecting the structure of the fish stocks and overshadowing the effect (if any)

Figure 4.7 *The North Sea plaice box, where large beam trawlers are prohibited from fishing in order to conserve juvenile plaice and other flatfish.*

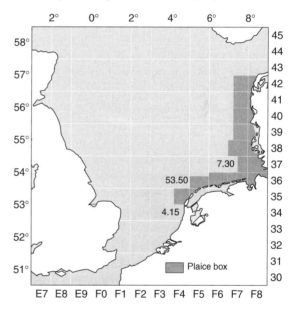

resulting from the existence of the 'plaice box'. In addition, the efficacy of the closed area may be compromised by the continued use of the region by small (<300 HP) vessels.

4.7b Ecosystem sustainability

Ecosystem management schemes are in their infancy, with considerable effort being directed at developing appropriate measures for ecosystem status (health) and function.

One example of an ecosystem level management scheme that has been implemented is the sandeel fishery off the east coast of Scotland and northeast England. A number of internationally important seabird colonies occur in this area, which includes the Isle of May and the Farne Islands. The Isle of May alone hosts around 70 000 pairs of breeding seabirds per year. While these birds range far and wide and take a variety of prey outside the breeding season, sandeels are a very important component of the diet of adults and young during breeding. At this time the birds' foraging is also restricted to sites relatively close to the breeding grounds. In the 1980s a number of inshore areas were exploited for the first time by industrial fisheries targeting the sandeels. There were a number of spectacular breeding failures by the seabirds. For example, 4300 pairs of kittiwakes in the Isle of May raised less than 200 young in 1998 (a pair normally raises one or two chicks from a clutch of three eggs). While the evidence of a fisheries–seabird interaction is only circumstantial, it was sufficient to prompt a precautionary response. Industrial fishing in the 'sandeel box' (which covers the inshore area from eastern Scotland down to northeast England) is closed if the breeding success of kittiwakes in the nearby colonies falls below 0.5 chicks per pair for three successive years. The fishery does not reopen until breeding success has been above 0.7 for three consecutive years. Thus management of this fishery is based on an ecosystem objective (seabird population health), is precautionary (the link is not yet proven), and uses the kittiwake breeding success as a biological indicator of the ecosystem effects of the fishery.

4.7c Economic sustainability

A simple economic analysis of a fishery (Figure 4.8) would suggest that the maximum profit from a fishery is obtained at a fishing effort below that required for MSY. However, this analysis is flawed for three reasons. Firstly, as something becomes less common, the price goes up – caviar being a good example of this phenomenon. This means the price per kg of a fish will be more when the stock is below the level that yields the MSY. Secondly, once a fisher has purchased the capital items required to fish, the cost per unit of extra effort is small (i.e. costs do not follow the dotted line in Figure 4.8). Thirdly, fishers do not operate to maximise profit from the fishery, but seek to maximise their individual profit; in a free access fishery, fish are not owned until after capture, so the first fisherman to catch the fish has an advantage. This leads to a 'race to fish', which drives fishermen continually to seek to improve their catching ability and leads to over capitalisation. This in turn means fishers have to earn more to sustain the loan payments or to get a return on the

Figure 4.8 *A simple steady state fishery model (Schaeffer curve) showing the relationship between maximum sustainable yield (MSY) and maximum economic rent (MER). Total costs (dotted line) are assumed to rise in direct proportion to fishing effort, while the unit value of fish (i.e. price per kg) is constant, so the total value of the catch follows the yield curve. The shaded area is the profit available (economic rent); the maximum profit occurs at a fishing effort below that needed for MSY.*

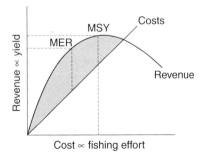

investment. Under these circumstances fishing effort will increase as the individual maximum economic yield will occur at fishing efforts above those required for MSY (Figure 4.8). Therefore, in a free market system a fishery will always tend to over exploitation and over capitalisation.

4.7d The way ahead

Technological advances are in part responsible for the perilous state of our fisheries. Improvements in vessel design, gear efficiency, gear handling, catch processing and navigation have all helped

---

Box 4.3 Lobster fisheries – V-notching and restocking

In England and Wales the Sea Fishery Committees manage the inshore fisheries on a regional basis. These bodies are able to legislate on fisheries matters within their geographical districts and are required to do so to promote the fishery and protect the environment.

Lobsters are a high value catch and are targeted primarily by the use of traps called pots or creels. Lobster pots are recovered by the fisher and the catch sorted with undersize individuals being returned to the sea. These have been shown to have a high survival rate. In many areas fishers are allowed to land 'berried' females, that is females carrying eggs under their abdomen. In other areas these too must be returned to the sea. One of the primary reasons why this rule is not universal is the difficulty of proving that a lobster from which the eggs have been removed was in fact carrying them. There is no point in having a law which cannot be enforced.

In the mid 1990s there was considerable concern about possible over exploitation of lobsters off the northeast coast of England. The local Sea Fishery Committee considered a number of options to try to reverse this trend. The basic challenge was to ensure more lobsters were able to breed and so provide recruits for future years. This could be done in a number of ways. The introduction of a ban on landing berried females was rejected on the grounds of enforcement difficulty, as was raising the minimum landing size. The sea fisheries committees can only make laws covering their district, so a minimum size would apply to lobsters caught in their district, but a fisherman could legally land lobster caught in neighbouring sea fisheries committees districts or further offshore where different sizes would have applied. For this reason the decision was taken to maintain a single

landing size in line with that in the international waters offshore. The third option was to re-seed the area with young hatchery reared lobsters. At the time a lobster hatchery had just started, supplying lobsters to the UK for re-stocking purposes. This was seen as expensive and largely untried. The approach adopted was the introduction of a V-notching scheme. This was the first in Europe but built on techniques developed in Canada.

A new regulation was introduced prohibiting the taking of lobsters with a V-notch cut into the shell of one of the tail fans. Fishermen fished in the normal way and landed their catch, returning under sized and V-notched lobsters. Ashore the sea fisheries committee staff visited fish merchants and purchased female lobsters carrying eggs. These had a notch cut into their shell and were then returned to the sea. The notch persisted for up to two years, slowly growing out as the lobster moulted. The females returned to the sea would release their young before they were able to moult and would probably breed again before the V-notch disappeared. Potentially, therefore, every V-notched lobster could boost the stock through two reproductions.

The scheme was launched in 1998 and in each season since there have been over 2500 lobsters V-notched and released back to the sea. As it takes around seven years for a lobster to recruit into the fishery, it will be some time before the success of the scheme can be assessed in terms of landings. However, the scheme has resulted in around 5000 extra hatchings being available for the larval pool each year. In addition, the programme has gained strong industry support and has been effective in involving the local fishers in a positive way in the management process.

us to impose a greater mortality on fish stocks than ever before, while using fewer vessels and fishers. Technology does not provide a solution to this, but the priority action must be to adjust national fleets so that with the available, and still developing, fish catching technology the level of exploitation reflects the biological reality of fish stock production. This is the challenge for politicians and policy makers.

In realising the potential sustainable yield from fisheries we must also have regard for the sustainability of the ecosystem, both because it ultimately supports the fish stocks and also because of the desire to maintain healthy and natural ecosystems. Closed areas are a very efficient means of protecting key habitats or vulnerable species, but at the scales applied to date they are unable to provide an effective mitigation against the direct mortality of fishing. Given that closed areas often, in reality, merely redirect effort into open areas, which then suffer higher levels of fishing, their role in ecosystem management is restricted to protection of key species or features. Rather than closing areas to fishing, we should seek ways of catching fish that do not do collateral damage to non-target species or ecosystem/habitat features. This will involve a move to more selective and lighter gears and possibly a return to static traps in place of towed gears. This may lead to financial hardships for the fishers in the short term. Such changes will need to be introduced following comprehensive ecological and economic assessments (see also **Box 4.3**). Development of such 'ecologically friendly' fishing gears is the challenge for technologists.

We know much about fish biology, but predicting the size of a stock even a couple of years in advance remains difficult. A reduction in effort will make year to year fluctuations in stock size (and catch and market price) less marked, but it is still important that we develop a better understanding of the relationship between the environment and fish stocks and between fish stocks and the rest of the ecosystem. This is needed to underpin any attempt to provide a holistic ecosystem approach to coastal management. This is the challenge for marine scientists.

4.8 Summary

- Harvesting aquatic organisms has an extremely long history. Fish and shellfish provide an important source of dietary protein in many parts of the world.

- Fishing methods vary with the type of species being pursued; the scale also varies, from hand gathering/hunting, which targets individuals, to large ocean-going vessels using nets many kilometres long.

- Harvesting impacts on the populations of the target organisms and can alter population size structure, sex ratios, gene frequencies and geographic ranges.

- Harvesting frequently affects other components of the ecosystem through changes caused by the alterations in the target population.

- Fisheries provide one of the most obvious examples of a renewable resource and so would seem to be ideal systems for sustainable exploitation. However, most of the world's fisheries are over fished, a consequence of a combination of economics, poor science and weak regulation.

Chapter 5 Aquaculture and Ranching

5.1 Introduction

The term **aquaculture** is used here in a broad way to denote all forms of culture of aquatic animals and plants, in fresh, brackish and marine waters. It is however commonly used in more restrictive ways – to some it refers to culture of all organisms other than fish, to others it means all culture of aquatic organisms except mariculture (culture in sea water). Here we use the word in its widest sense and add additional, qualifying terms where necessary.

5.1a Scale

Aquaculture has an exceedingly long history. Fishing was part of the variety of harvesting methods used by our hunter-gather ancestors. When humans

Figure 5.1 *Egyptian tomb relief, possibly showing Tilapia being harvested from ponds (drawn from a photograph of reliefs in the tomb of Merekuka at Saqquara).*

shifted to a more agricultural system, however, fishing generally remained as a hunting activity. However there is documentary evidence of extensive carp farming in China – Fan hei, a politician turned fish culturist, wrote *Classic of Fish Culture* in 500 BC (Ling, 1977), while Egyptian temple reliefs from 2500 BC are believed to show *Tilapia* being raised in ponds (Figure 5.1). The earliest records of the culture of marine organisms relate to oysters – Aristotle mentions Greek oyster culture, Pliny gives details of Roman oyster farming in 100 BC, while Japanese records indicate oyster farming occurring around 2000 years ago. Indonesian brackish water fish-culture ponds date back to at least the 15th century, by which time monastic pond culture of fish was widespread in Europe (**Section 8.3a**).

Despite its antiquity, aquaculture has mainly remained a small scale enterprise (China and southeast Asia being notable exceptions to this generalisation). Recently, however, advances in biological and related science and technology have overcome some of the barriers to its further development, resulting in massive investment in aquaculture worldwide and consequently a dramatic growth in production since the middle of the 20th century (Figure 5.2). Aquacultural food production now accounts for about 11% of the entire world fish (and shellfish) production of 98 million tonnes, and it is predicted to increase

Figure 5.2 *The increase in global aquaculture production in the late 20th century.*

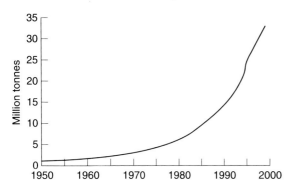

needs of the expanding population have lead to initiatives to seek new sources of food – particularly protein – of which cultured fish is a very efficient supplier.

In contrast, despite recent advances, aquaculture as an economic activity is still risky. The ease of harvesting and the potential constancy of supply appear attractive, but when set against the technical difficulties of providing a suitable environment for optimum growth, most aquaculture operations are economically relatively high risk. For the aquaculturist to make money, the growth rate of the stock must be as efficient and as rapid as possible, while the mortality rate must be low, requiring exclusion of disease and predators. This may therefore involve manipulation of food supply, control of environmental conditions to promote growth and disease/predator controls. How these are achieved varies – both between species cultured but also depending on culture method used, which may depend on the sociocultural setting of the farm.

further over the coming decades. Aquaculture is often relatively straightforward and inexpensive as a subsistence activity, hence its increasing prevalence in the developing world. The recognition by agencies such as the UN Food and Agriculture Organisation of the continuing growth in the divide between world food production and the

---

BOX 5.1   Salmon farming in sea cages

Salmon life cycle

Atlantic salmon (*Salmo solar*) are anadromous fish. In northern Europe the eggs are laid in early summer through to the autumn in gravelly areas of upland streams known as reds. These hatch about six months later and the larvae drift downstream and begin to feed on plankton and insect larvae. Fingerling salmon, known as parr, feed on invertebrates and fish fry in freshwater streams and rivers. After about one year the fish undergo physiological changes and become smoults, at which stage they migrate to sea. The principal feeding grounds of European salmon are in the sub-Arctic waters around Iceland and Greenland. After several years feeding at sea, the fish migrate back to their native river. Navigation is partially by means of an inbuilt magnetic compass, but as they approach their river of origin the final cues are olfactory. Following spawning, the majority of fish drift and swim back to

sea; they return in subsequent years to breed again. This contrasts with Pacific species of salmon, which spawn once and then die.

History of rearing

Sir Isaac Walton, writing in the *Compleat Angler* in 1653, recorded that parr were only found in rivers in which adult salmon ran. By 1830 it had been demonstrated that parr grow to smoults and that ova stripped from adults and then fertilised would hatch and grow into parr. By the end of the 19th century, a number of hatcheries had been built and were raising parr to supplement wild stocks.

In 1968 Unilever Ltd established a hatchery in Inverness-shire (Scotland) and transfered smoults to sea cages in Loch Ailort. This was the beginning of full life cycle farming of salmon.

Box 5.1   continued

Typical arrangement of infrastructure supporting a sea cage as used in salmon farming.

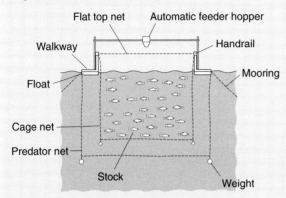

Fish farms and their infrastructure

Fish farms in coastal waters use a variety of designs. However, they have a number of features in common: a usually rectangular flotation collar, a suspended net bag and a mooring system. The flotation collar often supports a walkway.

Individual cages vary in their dimensions, but in the Scottish industry 7 m square has become a common size, although prefabricated cage systems 12 m or 15 m in length are now available. Cages are generally connected to each other to form rafts with a central walkway. The raft is then moored to the seabed with anchors.

Regular feeding, routine observation of the stock and the condition of the nets, and the periodic changing of the nets requires direct access by farm workers. There is therefore a certain amount of small boat activity associated with the operation of the farm. Food supplies, spare nets, ropes, etc. need to be housed in lockable units on the walkways of the farm or in-shore facilities.

Fish farm site requirements

Many factors interact to affect the suitability of a location for an aquaculture site. These include:

■ **Exposure**. Waves and high winds will place a strain on mooring and the joins between cages and will cause structural damage to the farm.

■ **Water exchange**. There needs to be a good exchange of water to remove wastes and keep the region around the fish well oxygenated.

■ **Depth**. Increased depth allows a higher stocking density per unit surface area, and improves the removal of waste due to dispersion in the water column. In the UK, the Crown Estates Commissioners (who lease the seafloor to fish farmers) recommended a minimum depth of 7 m (below low water spring tides).

■ **Salinity and temperature**. These should both be close to the optima of the fish being reared, and not subject to large fluctuations with tide or season.

■ **Pollution**. The water should be free from organic and chemical pollution.

■ **Access**. The site needs to be relatively accessible for stocking, supply of food, etc., and eventual transportation of the harvest.

■ **Security**. The site needs to be secure from the attentions of poachers and vandals.

■ **Shore facilities**. Most farms require areas ashore for the storage of nets, foodstuffs, etc., repair and washing facilities for the nets and eventually processing of the harvest. Movement between shore site and farm is via a boat, so a jetty and mooring facilities are also required.

Fish farm operation

Hatcheries are located in upland areas and either spring fed or supplied with clean stream water. Adult brood fish are gradually acclimatised to fresh water and are then stripped of their eggs and milt, which are combined, usually in the presence of a fungicide. The fertilised eggs are then transferred to hatchery tanks, normally run on a recirculating system with filtration and ultra-violet sterilisation of the water. The larvae may be fed on cultures of plankton or artificial feeds and grown on in indoor tanks, before being transferred to outdoor runs or large tanks at about six months of age.

After a further 12 months some of the larger fish reach the smoulting stage and are ready for transfer to sea cages. There is no need to gradually acclimatise

Box 5.1   continued

A representation of the stages in the production of Atlantic salmon in aquaculture (from Pitcher & Hart, 1983).

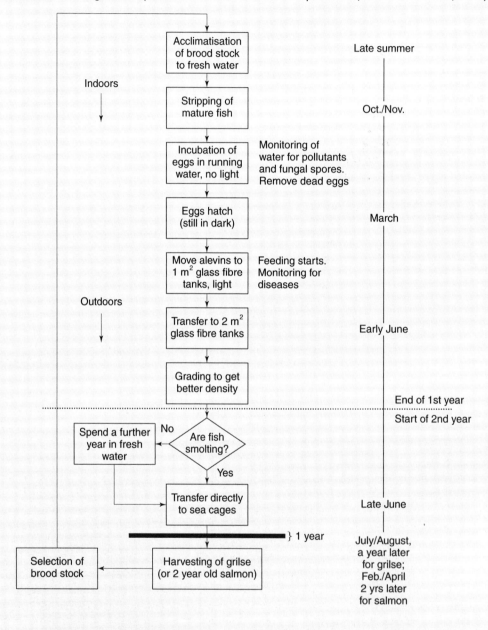

them to more saline conditions. Once at sea the stock are fed on a dry pellet diet. Best quality fish are produced from cages with some water movement (the swimming of the fish improves flesh quality), but excessive swimming wastes energy and slows growth. Usually densities are reduced after the first year, either by harvesting part of the cage as grilse or by transferring them to another cage for further growth.

5.1b General types of aquaculture

We can distinguish a number of categories of aquaculture operation. **Intensive systems** involve raising larvae in hatcheries and then rearing to marketable size in tanks or pens whilst being fed on artificial diets (**Box 5.1**). **Extensive systems** are more 'subsistence' based and grow small fish captured from the wild, often using ponds which may have their fertility boosted by the addition of manure (**Box 5.2**).

Box 5.2   Aquaculture in Egypt

Despite being the origin of the tilapia species that are now successfully reared worldwide, aquaculture is relatively unimportant in Africa. In part, this is due to the scale of harvesting of wild fish, both marine and inland, that is undertaken throughout the continent, making aquaculture uneconomic, but it is also social, in that African cultures have little indigenous tradition of aquaculture (Lévêque, 1997).

The exception to this generalisation is Egypt, a country heavily dependent upon fish as a source of protein for its rapidly expanding population. In 1987, around 250 000 t of fish were produced, of which 141 000 t were harvested from inland fisheries and a further 60 000 t were cultured. As the population increases, more protein will be needed and, with wild stocks being harvested to or even beyond their maximum potential, the only realistic solution to Egypt's protein shortage is aquaculture.

Pond culture is common in Egypt, incorporating native species along with high yield introductions in polyculture systems. Careful combination of species can enhance yields considerably. For example, pond culture of mirror carp (*Cyprinus carpio*) alone yields approximately 380 kg ha y^{-1}, but by combining this species in polyculture with tilapia (*Oreochromis aureus*) and mullet (*Mugil cephalus*), total production can reach 710–800 kg ha y^{-1}, and polyculture with tilapia and other native species can raise total production to 1000 kg ha y^{-1} (Moreau & Costa-Pierce, 1997).

In the Nile Delta region of Egypt, many irrigated areas are now impacted by intrusion of saline water from the coast, making conventional crop growth impractical. However, there is potential to use this water for culture of carp, which are tolerant of brackish conditions. Polyculture in rice fields in the Delta is also a potential major source of protein; rice fields produce around 200 kg carp ha y^{-1}, and their presence enhances rice yields by up to 7% (Moreau & Costa-Pierce, 1997). Currently, total carp production in Egypt is 7300 t y^{-1} (1992 figure), but successful polyculture with rice in 50% of the area covered by rice could increase this production to 30 000 t y^{-1}!

Ever increasing pressure on fresh water for irrigation and domestic use mean that it is not available for aquaculture on a commercial scale. For this reason, Egypt has been at the forefront of research into the use of waste water for aquaculture. Many of the fish species cultured, including carp, are unable to live in water contaminated with untreated or poorly treated sewage, because the high concentration of un-ionised ammonia induces a normally fatal disease known as infectious dropsy. However, tilapia is tolerant of high ammonia levels, while partial treatment of sewage reduces its toxic effects and enables other fish species to thrive, the productive environment allowing them to grow more rapidly than in clean water.

In Egypt, as in much of Africa, the commonest form of sewage treatment is to discharge it into stabilisation ponds (see **Section 6.7a**). By allowing water to remain in these ponds for extended periods, particulates and toxins settle out, while bacterial and algal growth removes excess organic matter and nutrients. Many fish species can thrive in these ponds, exploiting the algae and associated invertebrates. Therefore, a potential solution to Egypt's shortage of both protein and fresh water is to grow fish within stabilisation ponds.

Unfortunately, concerns have been raised about public consumption of fish that may be contaminated with pathogens and toxins (Easa *et al.*, 1995; Khalil & Hussein, 1997). Therefore, a major research

Box 5.2 continued

Survival and growth of cultured fish in water of different quality

	Primary treated sewage	Secondary treated sewage	Clean water
BOD level	178	23	19
Carp survival	0	92	??
Tilapia survival	61	81	83
Tilapia growth (g day^{-1})	1.07	1.11	0.92

Survival is expressed as the percentage alive after 22 days in each water type.
Source: Khalil & Hussein (1997).

programme is under way in Egypt to determine the level of treatment required before fish can be safely reared. The results look promising: residence in stabilisation ponds for only 26 days produces water clean enough to grow tilapia and gray mullet whose quality falls within WHO guidelines and is, in any case, better than that of fish harvested from the wild in Egypt (Easa *et al.*, 1995).

We can further distinguish **open** versus **closed** intensive systems. In the former, stock are raised in water drawn from and returned to the environment, for example, stream fed tanks or cages set within a lake or estuary. In the latter the system is closed off from the environment; stock is reared in tanks fed by a recirculating, filtered and temperature controlled water supply. Hatching and rearing larvae is usually carried out in closed systems, but on growth of stock may use either approach.

In this chapter we also consider **ranching**. This uses the same technology as aquaculture to raise animals in confinement during the naturally high mortality early life stages. They are then released to enhance populations in the wild. The advantage of this system is that savings are made by not having to provide facilities or feed for a prolonged growing on phase. The principal disadvantage, especially in the marine environment, is ensuring the return of the investment accrues to the individual or organisation that made the initial investment in the young. Therefore, it is most common in situations where the harvested individuals are released into enclosed lakes.

5.2 The biological basis for aquaculture

In addition to the economic attractions of aquaculture, fish and shellfish farming compares very favourably with poultry and livestock farming as an efficient way of producing animal protein. The species cultured are poikilothermic (cold blooded), so a relatively small proportion of the food consumed is used for metabolic maintenance (as there is no need to expend energy in maintaining a constant body temperature), leaving a high proportion available for growth. This high food to productive growth ratio, combined with the fact that stock are grown in a three-dimensional water body, means that production per unit area can be much higher than for farms of terrestrial warm-blooded species.

Aquaculture proceeds on the basis of simultaneously increasing somatic growth of the stock and reducing losses due to predation and disease. Somatic growth is increased by controlling the diet, provision of near optimum environmental condition, and reducing investment in reproduction.

5.2a Diet

The diet of the cultured species determines the method of feeding adopted. Filter feeders are usually supplied with food from the natural environment – either by being placed 'out' in suitable areas (e.g. oysters and mussels) or by raising in ponds or channels fed from natural waters by sluices and channels. Grazers are usually raised in relatively shallow ponds, in which natural light incident on the pond stimulates plant growth; this may be enhanced by the addition of artificial fertiliser or manure (see **Box 5.2**). Species of higher trophic position are normally cultured on artificial protein rich diets, manufactured from a wide variety of materials, including agricultural and fisheries wastes and industrial fisheries catch.

Fish culture often shows food conversion efficiencies of wet weight gain of stock to dry weight of food intake of 1:1 or 1:1.25. The utilisation of protein is often higher than that in poultry or livestock raising systems, in which as much as half the amino acids consumed may be broken down and excreted – and therefore lost from protein production. In general the economic costs of raising species of low trophic level, such as filter feeders and grazing fish, are less than for species of higher trophic level – predatory fish. To a large extent this is due to the cost of feed and the lower stocking densities required by the latter. However, the market price of predatory species is usually higher and so the higher costs are potentially offset by higher returns.

5.2b Environmental conditions

In extensive culture systems and open intensive systems, environmental conditions are to a large extent beyond the direct control of the aquaculturist, and the initial siting of the farm in an area where conditions are appropriate is a key step. In closed systems there is scope for direct manipulation of water quality, temperature and the light regime to provide optimum conditions for growth. These may differ for the various life stages of the cultured organism.

5.2c Reproductive losses

Investment by the stock in reproductive tissue generally equates to a loss from the marketable product, as resources are diverted from somatic growth and therefore final weight. Exceptions to this are cases where the reproductive tissues form part of the saleable product or are the principal product, such as caviar from sturgeon (*Accipenser* spp.). In subsistence-based systems, harvesting prior to the onset of reproduction is the main means of preventing such losses. In more developed systems additional or alternative measures may include manipulating the environmental conditions that would otherwise stimulate reproduction, such as temperature and photo-period. An alternative is to use selectively bred stock which have a late onset of reproductive development, or even to raise single sex stocks.

5.2d Predation and disease

The high density of the stock and its constrained state makes aquaculture operations attractive foraging sites for natural predators. Farmers may accept some losses or invest in specific anti-predator devices. Enclosed systems are, by their nature, relatively secure from such losses but any outdoor facility will attract predators. The most straightforward control is to put in a physical barrier to predator entry – a net or a fence – but in some situations more technologically advanced and hence expensive controls are required, such as sonic scarers for seals (see **Box 5.3**).

To some extent maintaining the stock in good condition for growth ensures that it remains healthy. This may, however, be offset by the increased chances of disease or parasite outbreaks due to the high stocking densities. In closed

Table 5.1 *Extent of the usage of chemicals in 148 UK fish farms*

Chemical	No. of farms	Chemical	No. of farms
Malachite green	89	Buffodine	5
Formalin	56	Methylene blue	4
Hyamine	23	MS 222	4
Oxytetracycline/Terramycin	16	Potassium permanganate	3
Chloramine T	12	Acinitrasole	2
Sodium chloride	12	Acriflavine/Pioflavine-hemisulphate	2
Copper sulphate	11	Dipterex	2
Tribrissen/Sulphamerazine/Vesadin	10	Enheptine	2
Betacide/Marinol/Roccal	9	Slaked/quick lime	2
Iodophores/Wescodyne/Pevedine/ Vanadine/Polyvinylpyrrolidone– iodine complex	9	Sodium hydroxide	1
Furazolidone/Neftin	8	Sodium hypochlorite	1

Source: Solbe (1982).

systems, the water can be routinely treated to reduce pathogen levels, while in intensive open systems the food may be dosed with antibiotics and other 'health' promoting chemicals. When the disease/parasite outbreaks are noted the farmer usually responds with targeted treatments, often involving dosing the entire farm with an appropriate chemical preparation. A wide range of chemicals are used in this way (Tables 5.1 and 5.2) and many may potentially have impacts beyond the farm (see **Box 5.3** and **Section 5.8**).

5.3 Aquaculture of fish

In this section we briefly discuss the fish species cultured and most common types of aquaculture operations. An exhaustive review of global aquaculture systems is beyond the scope of this work and we direct the reader to Pillay (1993) and Michael (1986). **Boxes 5.1** and **5.2** present case studies of two very different types of aquaculture.

5.3a Carp

Carp culture has the longest recorded history, and to some extent some of the cultured species of carp (there are at least 11 species and many more 'strains' or races in culture) may be regarded as domesticated. Carp are cultured all over Asia, in central Europe and the former Soviet Union, in parts of Africa and Latin America and in some areas of North America and Australia.

Types of carp

Three species groups can be identified:

- common carp (*Cyprinus carpio*), races of which are cultured all over the world; this includes mirror carp;

- Chinese carps (including the grass carp, *Ctenopharyngodon idella* and silver carp, *Hypophthalmichthys molitrix*);

- Indian carps (including catla, *Catla catla*, rohu, *Labeo rohita* and mrigal, *Cirrhina mrigala*).

Combinations of species in polyculture generally give higher yields than monoculture systems, as a result of interactions between species with differing requirements. For example, in India, catla feeds in the water column – the adult diet includes algae, planktonic protozoa, crustaceans and molluscs; adult rohu feed on various types of vegetation, including decaying plants; while the mrigal is a benthic feeder (Pillay, 1993). There is anecdotal evidence that some species feed on the excreta of others, while the activities of bottom feeders may release nutrients into the water column and so stimulate algae production for grazing species.

Carp culture

The dominant culture technique for carp is in ponds, although some cage culture and stocking of natural waterways (ranching) occurs. In some areas the ponds are stream fed with running water, but in most cases they are stagnant or semi-stagnant. This and the ease of obtaining a suitable feed (see below) means that carp are generally seen as a low-value crop suited to large scale production in feed-poor, rural areas. Depending on the species cultured, feed can be generated by fertilising the ponds with commercial fertiliser, manure, farm waste or even sewage (**Box 5.2**). Species such as grass carp, which feed on macro vegetation, can be fed on cut terrestrial vegetation in a fresh or decayed state.

In parts of Indonesia, bottom feeding carp are reared in bamboo cages set on the bottom of waterways carrying high loads of domestic waste. The fish feed on the benthos that flourishes in these enriched conditions. This use of bottom set cages contrasts with the use of floating cages to culture carp in Chinese lakes and salmon in fjords (**Box 5.1**).

While the common carp readily spawns in ponds, the Chinese and Indian groups have traditionally been taken as eggs or fry from the wild, resulting in the polyculture of a group of species that normally occurs together. Wild collection has now disappeared in China and is much less common in India than formerly, as technological advances have overcome the difficulties with pond spawning that formerly occurred. Induced spawning and hatchery rearing is now widespread for all cultured species and allows greater selection of brood fish characteristics and greater survival of offspring.

Common carp are often cultured in monoculture with stocking densities of 4000–5000 fingerlings (2.5–5 cm) per ha or 2000–3000 fish (5–10 cm) per ha being typical. Growth rate is determined by temperature, so fish reach marketable size (1–2 kg) after three years in central Europe, two years in southern Europe, and one year in tropical and sub-tropical regions.

In general, disease and parasite problems are less in extensive pond culture than in intensive systems such as cage culture. Anoxia (lack of oxygen) is a major source of mortality in ponds in tropical and sub-tropical areas. One of the possible benefits of polyculture in these areas is the reduced likelihood of anoxia through removal of detritus by bottom feeders and the control of algal blooms in the water column by species such as silver carp. Aerators and the replacement of a proportion of pond water with water from an external source are now frequently used to combat anoxia and allow higher stocking densities.

5.3b Trout and salmon

Salmonid culture does not have the long history of carp aquaculture, but it has been the subject of intensive scientific research. Initially, during the 18th and 19th centuries, this focused on hatchery production of young, which were used to restock or enhance native populations or to expand the range of desirable species for recreational angling (**Section 8.3a**). The last quarter of the 20th century saw major expansion of salmonid farming, with large-scale development of ponded trout farms and use of cages for growing on trout and salmon, particularly in the sheltered coastal waters of Scotland, Norway, Ireland, Canada and New Zealand. In Scotland production increased from 2500 tonnes (fresh weight) in 1984 to 19 500 tonnes in 1988, and generally exceeded 50 000 tonnes per annum during the 1990s.

One of the major constraints on salmonid farming is the availability of suitable sites with appropriately high water quality. Freshwater rainbow trout farms and salmonid hatcheries require springs, boreholes or clean, fast flowing streams. While recirculating water systems are technically feasible, they are not generally (because of the high capital costs) economic for farms but may be so for hatcheries supplying a number of farms.

Rainbow trout

The principal trout species in culture is the rainbow trout (*Oncorhynchus mykiss*, formerly known as *Salmo gairdneri*). A native of the Pacific coast of North America, it has been introduced to all continents except Antarctica, and although a temperate species, it has been cultured at high altitudes in the tropics and sub-tropics. Rainbow trout were first cultured in pond systems with a through flow of water and, while this remains a major technique, it has been superseded by the use of raceways, particularly in the USA. These consist of long, continuous channels – earth, brick or concrete, dug into the ground or above the surface – fed with a large volume of high quality water. In Europe, farms also use circular tanks (4–10 m diameter) with high water turnover rates. Floating cages as used for salmon (**Box 5.1**) are also used for trout in some sheltered waters, such as lakes and reservoirs. Stocking densities are generally limited by the availability of water to feed the raceways or tanks. If water temperatures fall below 5°C, fish stop feeding, so in high latitudes bore-hole or spring feed water sources with a constant temperature are preferred for on-growing. Harvestable size is typically achieved after 1–1.5 years under good conditions.

Salmon

While the Pacific salmon (*Oncorhynchus* spp.) has been the focus of much effort with regard to rearing for stocking and ranching programmes (see below), it is Atlantic salmon (*Salmo salar*) that is the principal species in cultivation. Smoults are transferred from a hatchery for on-growing in sea cages (**Box 5.1**). Cage culture systems were developed in the late 1960s in Scotland and Norway; this method of culture is now widespread in sheltered sea areas, such as the Norwegian fjords and Scottish sea lochs and is common in Sweden, Iceland, Ireland, North America, Japan, New Zealand and Chile. Stocking densities vary between 1 and 35 kg m^{-3} cage at harvest, which, depending on conditions, is 2–3 years after stocking.

5.3c Tilapia

Tilapia (Family Cichlidae) are native of Africa and are believed to have been cultured in ancient Egypt. Tilapia taxonomy is complex and, while a number of species (>22) are in culture, it is generally believed that few of the cultured forms are pure species any more. They have been introduced for culture into a large number of tropical and sub-tropical areas, often as a component of aid programmes for rural development. Tilapias breed readily, are herbivorous or omnivorous, and tolerate brackish and even saline waters, and poor water quality; they can easily be reared in simple earthen ponds. Their adaptability led to problems for early development projects based upon their culture: tilapia bred so rapidly that they overpopulated many ponds, escaped into the wild and caused ecological impacts on native faunas. To overcome this problem, they are frequently reared in association with predators, to limit their numbers when small. Other problems also centred on their ease of breeding, with fish switching to reproduction at sizes well below marketable size and growth being slowed as resources were invested in reproduction. Modern culture techniques now involve measures such as special feeding regimes and stock management to raise fish of marketable size prior to the onset of reproduction, including only stocking males.

Cage culture has been tried experimentally in a number of lakes and lagoons. It benefits from the fact that any breeding does not lead to increased stock densities in the cage. To date it has not been widely adopted, primarily for economic reasons.

5.3d Other fish

A large number of other species are cultured worldwide, but often only regionally. The principal groups cultured are the catfish, eels, mullets and sea bass (see Brown *et al.*, 1969; Usui, 1984; D'Ancona, 1954; and Pillay, 1993, respectively, for more detailed accounts).

Catfish

Catfish (families Claridae, Ictaluridae, Pangasidae and Siluridae) have been traditionally cultured in south and southeast Asia, but recent developments in large scale culture in the southern USA have increased attention on this group. Pond culture is the dominant culture method for catfish but the species exploited vary. In the USA the Channel catfish (*Ictalurus punctatus*) is the most important, while in Africa and Asia it is species of the genus *Clarias*, particularly *C. batrachus* in Asia and *C. lazera* in Africa. Generally spawning of brood fish occurs in the ponds or in specific spawning ponds, suitable nest materials being a prerequisite. Eggs may be hatched *in situ* or transferred to indoor hatcheries, where fry are reared in troughs before transfer to ponds. In the growing ponds, the fish feed on naturally available prey, but this is often insufficient to provide for rapid growth and is usually supplemented by feed based on offal and cheap animal products. Marketable size is generally reached two years after hatching under optimal conditions.

Eels

Eels (Family Anguillidae) have a complex life cycle involving migration from fresh water to the sea to breed, with the larval stages being passed in the sea. The principal species used in aquaculture are the European eel (*Anguilla anguilla*) and the Japanese eel (*A. japonica*). Adult Japanese eels spawn near the Japanese coast and elvers arrive in the rivers at age one year. The European eel migrates to the Sargasso Sea in the mid-Atlantic to spawn; it takes three years for the larvae to reach the estuaries of Europe as elvers.

As early as Roman times, wild caught adult eels were held in ponds prior to final harvest. While considerable research effort has been invested in trying to achieve artificial fertilisation and larval rearing, it has had limited success; the basis of modern eel culture continues to be capture of wild fry (elvers) as they return from the sea. The most widespread culture method is in pond farms, but culture in concrete tanks ('tunnels') and net cages is also used. Elvers are collected using scoop nets, sieves or plankton nets from the wild during their natural migration up river, then transferred to nursery ponds. As the elvers grow they are thinned out for sale or on-growing. Where feasible, warm water from springs or industrial sources is used to increase growth rates. All eel ponds have vertical or near vertical walls, usually supplemented at the top with other anti-escape devices such as net overhangs.

On the Mediterranean coast, brackish lagoons are widely employed. Elvers are captured by allowing them to enter the lagoons with incoming tidal flows. Sluices are then closed, impounding the stock. The elvers were traditionally fed on the natural food available in the lagoons and were harvested from the lagoons into which they had been carried by the tidal flow. In the more modern and intensive lagoon systems and in pond farms, artificial feed is supplied. This is usually a powder that is mixed with water to form a stiff paste, which is placed into feeding troughs at the pond/lagoon margin. Aeration is often supplied to intensive pond systems.

In Japan consumers prefer smaller eels and marketable size is achieved after about one year; in Europe large eels are preferred and two years' growth is required. As individual growth rates vary considerably, farmers must continually sort and restock ponds to achieve optimum growth.

Grey mullet

Grey mullets (Family Mugilidae) have been traditionally cultured in Mediterranean lagoons and in northern China, India, Java (Indonesia) and Hawaii. They are herbivorous and detritivorous feeders and while prized in some areas are shunned by consumers in others.

The majority of species cultured are from the genus *Mugil*, but in India *Rhinomugil corsula* is also cultured. They are euryhaline and the dominant, and traditional, culture method is rearing in embanked brackish ponds/lagoons, often in polyculture systems with carp (Asia), eels (Mediterranean), milkfish (India) or tilapia and carp (Israel). In traditional lagoon systems, wild fry may simply be obtained by allowing the flood tide to carry them in through open sluices. In more intensive systems this is supplemented or replaced by estuary caught fry. While induced breeding of many species has been achieved, it remains difficult and costly and so has not made a big impact on aquaculture practice.

Mullet are not fed; fertilisation of ponds and sluicing of estuarine water into and out of ponds (through meshes to avoid stock loss) supplies nutrients for plant growth and plankton. The mullet graze on benthic algal growth and plankton.

Sea bass

Sea bass (Families Serranidae and Centroponidae) are also euryhaline, but are predatory in behaviour. They are traditionally part of the multi-species mix cultured in the Mediterranean lagoons, and in brackish ponds in Asia. Their high value and declines in the availability of wild caught fish have lead to increased demand for cultured fish in Europe and the USA.

Traditionally sea bass culture in Europe and Asia relied on natural supplies of fingerlings. This supply was highly variable and the main constraint to aquaculture. In the latter part of the 20th century, induced spawning and hatchery rearing became viable; increasingly farms are being stocked with cultured fry raised in hatcheries. Transfer to growing-on cages occurs at around 70–80 days post-hatching in European sea bass (*Dicentrarchus labrax*). Marketable size is reached most rapidly, about 2–3 years in temperate areas, in floating sea cages supplied with pellet feeds.

5.4 Aquaculture of invertebrates

5.4a Shrimps and prawns

A variety of crustaceans are commonly referred to as 'prawns' and 'shrimps', including freshwater, estuarine and marine species. Shrimps have traditionally been part of the mixed species compliment of culture ponds in Asia; in India they were extensively cultured in rice fields. The rapid growth of market demand in developed countries and falling catches in the capture fisheries has led to the rapid development of intensive cultivation systems, particularly in tropical countries looking for export earnings. Most cultured species come from the families Palaemonidae and Penaeidae. Many species are cultured (see McVey, 1983) including the giant freshwater prawn, *Macrobrunchium rosenbergii* and a large number of *Penaeus* species, including the tiger shrimp/prawn, *P. monodon*, the banana shrimp/prawn, *P. merguiensis* and the green tiger, *P. semisulcatus*.

Traditional and modern culture is based on pond systems (see Hanson & Goodwin, 1977). In traditional systems, stocking is achieved by the intake of tidal waters containing fry into ponds, where they remain until harvest. In modern systems, separate nursery ponds are used and in the more intensive systems, hatcheries are used to source the fry. In all systems stocked from the wild, culture is essentially a polyculture system as there is no means of separating the larvae of different species. This causes problems as growth rates vary and unlike many fish species it is difficult to sort the growing shrimps into different ponds as handling usually results in high levels of injury and mortality.

While hatchery techniques are now available for many species, the demand is so great that it far outstrips supply, so wild caught seed-stock still dominates globally. A variety of net techniques are used and often there is social conflict between the aquaculture sectors fishery of juvenile shrimp and the traditional capture fishing sector who blame poor catches of adults on the removal of the juveniles for aquaculture.

Extensive pond culture systems use low stocking densities (3000–5000 fry ha^{-1}), while more intensive systems (with aeration, fertilisation and supplementary feeding) will stock at up to 50 000 fry ha^{-1}. Depending on conditions, shrimps may reach marketable size in two months.

5.4b Oysters and mussels

Culture of bivalve molluscs is one of the oldest forms of aquaculture, dating from at least Roman times. It is also a relatively low cost and ecologically efficient system. Molluscs account for over 30% of aquaculture production, although the amount of culture has declined recently, due to problems of pollution at the growing-on sites affecting growth rates, impacts of pollution and harvesting on natural populations causing reduced spat (larval) supply, and food/health concerns. As filter feeders, they ingest microbial contaminants from the water, including those derived from incomplete sewage treatment. The increased

occurrence of toxic phytoplankton blooms ('red-tides') (see **Section 6.9**) is also a problem, as these can cause mortality of stock or lead to contamination with toxins, which then makes them unmarketable.

Oysters

Oysters (Family Ostreidae) are the dominant cultured molluscs worldwide, with a variety of species of *Ostrea* (flat oysters) and *Crassostrea* (cup oysters) being grown. In *Crassostrea* fertilisation is external, while in *Ostrea* the female draws in sperm which have been expelled by a male; fertilisation and initial larval development occur within the shells. Larvae are then expelled when they reach the free-swimming veliger stage. From hatching to the late larval stage takes up to three weeks, at which time the shelled larva moves to the sea floor and searches for a suitable attachment site. While any hard surface will do, *Ostrea* shows a definite preference for calcareous materials. Naturally spat settle under low flow conditions. Such areas are often plankton poor areas, hence the requirement to move settled spat to areas of higher flow to achieve enhanced growth.

On-growing of spat, from whatever source, essentially consists of placing the oysters in a location with suitable salinity, temperature, water quality and food supply. Where necessary, protection from predators – crabs on post-spat, starfish and gastropod molluscs (e.g. the oyster-drill *Urosalpinx*) – is provided. This usually consists of mesh fencing or other barriers to predator access.

The earliest culture method, still widely used, is 'bottom culture'. Settled spat are collected from areas of high natural settlement and transferred to other areas of the seabed for on-growing. The provision of old mollusc shells to encourage spat fall may be the only intervention by the farmer, although some pest or predator control may also be practised.

Off-bottom culture systems employ solid structures to encourage direct settlement by spat. A wide variety of devices are used for spat collection, ranging from lime coated stakes and tiles to ropes, wires, tree branches and wire/nylon mesh bags. Spat collection and sale is a profitable industry in its own right, but settlement may be combined with on-growth methods. Of these, the oldest is stick culture, in which spat are allowed to settle on stakes driven vertically into, or mounted horizontally above, the seafloor; they are often coated in lime to encourage settlement. After spat fall they are transferred to on-growing areas. Rack culture is similar with racks supporting tiles, ropes or bags (containing old shells) on to which spat have settled. In suspended culture, bags or ropes with spat are hung from floating rafts. Predation is very low on suspended culture; this often offsets the higher capital costs.

Current production of oysters is still largely dependent on natural spat fall; while some producers have achieved hatchery production, economic constraints have so far prevented a wide take up of this technology. However, hatchery production of seed oysters occurs on a fairly large scale in the USA, UK and France. Spawning stock are maintained in flow through flumes with high quality saline water; algae culture is added to boost the food supply and the temperature regime is manipulated to trigger spawning. The larvae are reared on algae in hatchery tanks in water that is filtered and ultra-violet sterilised, and into which algae cultures are added. *Crassostrea* settles after three weeks' and *Ostrea* after two weeks' rearing. Most hatcheries settle spat on to old oyster shells in plastic mesh bags or trays to ease transport.

Ostrea is generally marketed after three winters; *Crassostrea* can take as little as 8–9 months of on-growing to reach marketable size. *Ostrea* remains the higher value product, but the faster growth of *Crassostrea* gives a more rapid return.

In many regions native oysters have been replaced by introduced species – for either faster growth or disease resistance – which now produce spat to sustain the industry based on the original introduction.

Mussels

Harvesting wild mussels has a very long history in Europe, but mussel culture has only received attention in the past 30 years or so, as rapid growth on experimental farms raised expectations of good economic returns. In Europe the principal species farmed are *Mytilus edulis* and *M. galloprovincialis*, while in New Zealand and the Philippines the green mussel *Perna viridis* is cultured.

Mussel culture techniques are very similar to those of oysters. The traditional 'relaying' of settled spat into estuaries/embayments is analogous to oyster bottom culture, while stick systems are widely used in France and around the Mediterranean. The most intensive production systems are based on suspended rope systems, which have proved highly efficient in sea lochs and fjords. Ropes are laid out on rocky shores or estuarine flats close to natural mussel beds, and after spat fall are lifted and transferred to deeper water where they are suspended from buoys/floats with ropes running vertically and horizontally. Such systems are amenable to mechanical harvesting (Lutz, 1985).

Starfish, birds (intertidal beds) and gastropod molluscs are the main predators, and the need for anti-predation measures varies greatly between sites; again suspended rope culture gains by the low predator losses and the rapid growth obtained by continual submergence.

5.4c Other shellfish

While oysters and mussels dominate shellfish aquaculture, a variety of other species are cultivated including crabs, lobsters/crayfish, clams, scallops and abalone. Considerable research effort has gone into culture techniques for lobsters, crayfish and crabs, but to date there has been little commercial culture although some rearing for release (stocking programmes) does occur (see **Section 5.7**).

Crayfish

At least four species of freshwater crayfish are cultured in the USA, and on a small scale in France the signal crayfish (*Pacifastaeus leviusculus*), a species introduced into Europe from the USA, is reared. The red swamp crayfish (*Procambarus clarkii*) is the only species raised on a large scale globally, either as a rotation crop in rice fields or in dedicated ponds. Brood animals are stocked into the ponds which are then drained. As water level falls the crayfish mate and the females burrow into the banks. The eggs hatch while the female is in the humid burrow and when the pond is re-flooded the female expels the larvae from the burrow. The crayfish are omnivorous and feed on vegetable matter and zooplankton in the ponds; a crop of rice or millet planted while the pond was drained provides a ready food source for the crayfish when the pond is flooded again.

Crabs

The mud crab, *Scylla serrata*, is the only species of crab cultured commercially; it is one of the species which develops in the lagoon/impoundment polyculture systems of southeast Asia. The young enter with the tidal water and develop in the fish ponds, reaching marketable size in about six months.

Clams

Techniques for rearing clams, a term used to cover a variety of soft sediment dwelling filter feeders, are similar to those for oysters and mussels. Natural spat fall is collected and transferred to more productive sites for on-growing.

Scallops

Scallops are highly prized by the seafood industry and wild catches are declining. To date commercial aquaculture is limited to China and Japan. Thermal shock of adults in hatcheries induces spawning. The larvae can then be hatchery reared prior to placing at a suitable site for on-growing. This may involve bottom culture or growing on suspended ropes or in mesh bags.

Abalones

Abalones (*Haliotis* spp.) are highly prized, both for their meat and for the shell. Their high value and declining wild stocks have focused attention on aquaculture. To date hatchery fertilisation and rearing to spat fall has been achieved. This hatchery reared spat has been used to restock depleted natural stocks which can be subsequently harvested. In China the abalones are cultured during growing on. Here the abalones are enclosed in plastic cylinders with mesh end covers suspended from long lines; marketable size is reached in about two years.

5.5 Aquaculture of plants and algae

Cultivation of water cress, water spinach and water chestnut has been going on for many centuries, while rice (*Oryza sativa*) is one of the world's most important crops and is now domesticated. Large scale aquaculture of marine algae is, however, a relatively new phenomenon. Output of seaweeds, in terms of tonnage, is almost equal to mollusc aquaculture (Pillay, 1983). The principal producing countries are Japan, China and Korea, with smaller scale production in the Philippines, Taiwan, the USA, Canada, the UK and several other countries. The bulk of the Japanese and Chinese production goes for human consumption; it is also used in fodder, manufacture of agar, algenates, carrageenan, mannitol and iodine.

Seaweed culture seems to have originated in Japan in the 17th century with the culture of 'nori' (lavar, *Porphyra* spp.). This remains the most important species cultivated for human consumption, along with kelp (*Laminaria* spp., *Undaria pinnatifida*) and the green seaweeds, *Enteromorpha compressa* and *Monostroma* spp.

Figure 5.3 *Rope culture of seaweeds.*

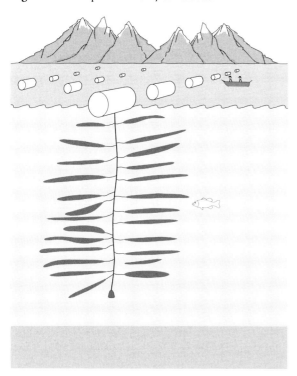

The traditional nori culture system involved placing bamboo bundles or rocks on the seafloor for spores to settle onto. These were then transferred to inshore nursery areas for growing on. These techniques have now largely been replaced by settlement on to nylon nets with split bamboo 'blinds'. Kelp culture can involve seeding the seafloor with stones or concrete blocks to provide suitable settlement, but nowadays the most common technique involves suspended rope culture (Figure 5.3). In the Philippines and Taiwan seaweeds are grown in polyculture in coastal fish ponds.

5.6 Polyculture

Polyculture refers to any system in which several harvestable species are cultured together. Various examples of this have been highlighted in **Sections** 5.3–5.5, including eels, mullet and sea bass in the lagoons of the Mediterranean, and shrimps, crabs and fish in the ponds of southeast Asian marshes. Multi-species aquaculture is common in such extensive systems where the 'seed stock' is obtained from the wild – by tidal flows into ponds or mixed net catches of larvae. Polyculture is also employed in some modern, intensive systems. In China longline suspended rope systems support scallops in bamboo lanterns, abalone in plastic tubes and *Laminaria* growing on the ropes.

5.6a Ducks, terrestrial animals and fish

In Europe ducks have traditionally been raised on fish ponds. They fertilise the ponds, remove weed and stir the bottom, boosting productivity but also providing an additional crop. In southeast Asia there is a long tradition of integrated polyculture, incorporating aquatic and terrestrial species. The most common variant combines intensive chicken or pig rearing with pond polyaquaculture (Eng *et al.*, 1989), the pig sties and chicken sheds being constructed on the pond embankments or even on stilts over the water. The animal wastes stimulate plant growth in the pond, providing grazing for many cultured species: shrimps, grey mullets, milkfish, tilapia, carp, etc.; supplemental feeding may allow an additional crop of predatory fish. In some areas, human excrement has been also used in this way but public health concerns have reduced this practice considerably. India has continued to investigate the possibility of using polyaquaculture as a way of treating human wastes and producing much needed high grade protein (Chandra-Prakesh *et al.*, 1990), a procedure also adopted in Egypt (**Box 5.2**).

5.6b Rice fields

Rice fields provide another opportunity for aquaculture, as they are deliberately inundated for extended periods. The inundation process carries fish and shrimps into rice paddies from the water source; harvesting was probably initially based upon these, although deliberate stocking with fish, shrimp and crayfish later developed. Recent increases in rice cultivation 'efficiency' have lead to a decline in this type of polyculture. However, it is now being recognised as a potentially valuable means of generating high quality protein in several underdeveloped areas. Three models are employed, increasing in complexity and cost. The simplest is to raise a crop of fish or shrimps in the flooded rice field after harvest but before replanting. The second is to continue culture during the rice growing season and to harvest rice and fish together. The most complex involves transferring the fish to other ponds or marginal ditches at the rice harvest, and then stocking them back into the replanted rice field, giving two or more years of fish growth to harvest and so larger market size and greater price per fish. This has to offset the greater costs of moving stock in and out of ditches/ponds at rice harvest time.

5.6c Salmon and mussels

Suspended rope culture of mussels among salmon sea pens has been proposed (Folke & Kautsky, 1989) to remove wastes produced by the salmon (see **Box 5.3**) and to provide an alternative crop and a salmon food source, thus obviating the need for an industrial capture fishery (Figure 5.4). Mussels suspended adjacent to salmon farms have enhanced growth rates as they feed on waste food particles from the farm and phytoplankton growing on soluble nutrients. In turn, mussel flesh can be used to supplement the diet of the salmon, thus: (i) leading to the release of fewer waste products; (ii) cutting down on the required level of fishing; and (iii) enhancing energetic efficiency, as phytoplankton are utilised with fewer intermediate consumers.

Figure 5.4 *The possible use of mussel–salmon polyculture to reduce the environmental impacts of caged salmon farming. N, nitrogen; P, phosphorus (from Folke & Kautsky, 1992).*

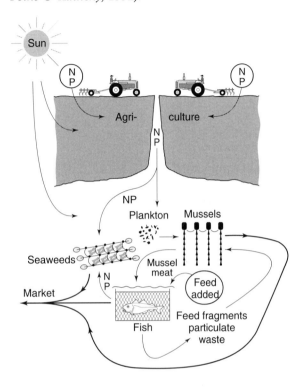

5.7 Ranching

Ranching is the term used to describe the practice of raising stock under controlled conditions and then releasing them. This usually allows the stock to pass the initial high natural mortality period in benign conditions. The release may be to enhance existing populations for 'traditional capture fisheries' or for recreational angling (**Section 8.3a**), or it may be made in an attempt to reduce the costs of 'full culture' by obviating the need for growing-on in controlled conditions. If an anadromous species is the target, its natural homing behaviour can be used to ensure a return of the ranched fish for harvesting.

5.7a Stocking in fresh waters

Some of the most clear cut successes of fish enhancements occur in reservoirs. The damming of rivers to produce reservoirs (**Section 2.3b**) is widespread and has dramatic effects on the aquatic environment, often including declines in the natural fish fauna and increases in opportunistic species (so called 'trash fish'). In many cases stocking with 'economically valuable' species has been successful. Species used include 'food fish' such as trout, but also 'game' angling fish such as pike. Dam construction may also lead to the interruption of migration routes of anadromous fish, so release of hatchery reared post-larval fish may take place downstream of the dam to offset these losses. For example, the River Tyne in northeast England has the largest run of returning salmon of any English river, yet the headwaters of one of its two major tributary rivers are dammed to form Kielder Water. A salmon hatchery was established at the reservoir and returning salmon are stripped of eggs and milt; the larvae raised from this are then released back into the river.

The strong homing behaviour of anadromous species such as salmon and sturgeon – along with their high economic value – has made these species the best candidates for ranching. The various species of Pacific salmon (*Oncorhynchus* spp.) have received the most attention with regard to establishing fisheries in areas beyond their natural range (McNeil, 1979). In northwest USA and Alaska, a combination of hatchery production and the provision of artificial spawning channels is used to enhance juvenile stocks.

In the Caspian Sea region, from Russia to Iran, sturgeons are of considerable economic importance, particularly as the source of caviare. While spawning channels have been tried with limited success, a number of hatcheries are successfully stripping river caught fish of eggs and milt, combining them and raising larvae. The larvae are usually grown on for up to six weeks, to the fry stage, in shallow concrete tanks. The fingerlings (15–35 g) are then released into brackish water areas.

Many reservoir stocking programmes use non-native species and the ecological consequences are unclear. However, as the whole reservoir is an unnatural system, these may be outweighed by the economic benefits. The consequences of artificial stocking are considered further in **Section 8.3a** and **Box 8.1**.

5.7b Marine stocking

Stocking of marine areas is rare due to the scale of the environment and the mobility of the fauna. Cod (*Gadus morhua*) larvae were stocked in Oslo Fjord, Norway, from 1950 to 1971, when the programme was terminated due to a lack of any indication of a benefit. The UK, France and Norway have raised lobsters (*Homarus gammarus*) in hatcheries and released them to enhance stocks, but again these experimental scale releases have not yet yielded strong evidence of any impact on the fishery – probably due to the high natural mortality of small lobsters.

The most ambitious stocking programme operating today is in the Seto Island Sea area of Japan. A number of species are stocked in an area which itself has been altered by provision of wave breaks, artificial nursery areas and additional 'shells' to promote settlement. The main species stocked is the Kuruma shrimp (*Penaeus japonicus*), millions of 1 cm fry being released each year. These reach the fishery at 11–12 cm, 4–5 months later.

5.7c Regulation

It is recognised that the biggest constraint on ranching is regulating the fishing on the released stock; there is little incentive to invest in rearing fry in a hatchery if somebody else captures and sells the subsequently produced adults. To date, most stocking schemes have been operated by government agencies. Exceptions include some schemes on lakes/reservoirs operated by angling organisations or the local land owner, where this includes ownership of the lake's fishing rights. While it might be technically possible to use 'micro-tags' in released fry to show ownership, the practicalities of introducing such schemes with appropriate policing of the fishery make the

concept of ownership of fish in 'common property waters', such as the high seas, nonsensical. Homing andromous fish remain the best candidates for ranching, but the development of wide scale net-pen culture in sheltered marine waters has to a large extent removed the economic incentive for ranching salmonids, potentially the most valuable species.

BOX 5.3   The environmental impacts of caged salmon farming

Water movement

A structure placed into a moving body of water will change the pattern of flow. The degree of effect exerted by a fish cage will be dependent on the mesh size, the type of mesh (knotted versus knotless, diamond versus square), and the diameter of the line (Milne, 1970; Weston, 1986). The extent and type of fouling and the stocking density will also have an effect. There have been a number of empirical studies on the effect of nets on flow regimes; Inoue (1972) recorded flows inside a $3 \times 3 \times 3$ m pen, with a 2.4 cm mesh, as being 70–80% of the initial velocity. If three such cages were moored together, in a raft, flows in the downstream cage were reduced to 10–25% of the initial rate.

In addition to changing the flow within the cage, cages also alter the flow regime around them. Weston (1986) calculated the effect for a cage $12 \times 12 \times 4$ m, showing that current velocities are altered up to 12 m upstream, 24 m either side of and 240 m downstream of the cages. Multiple pens moored together will act as a single, large, structure. A heavy fouling load on the cages, use of nets with a large twine, etc. will further increase these values. The cages will also affect the flow below them, an influence which may extend to a depth three times the depth of the mesh.

Changes in the flow regime will influence the supply of oxygenated water to the stock and affect the removal of waste products, the dispersal of chemicals used to treat the stock, and the degree of sedimentation occurring below the cages.

Water quality

Cages used for fish culture may affect water quality in two ways: a direct effect as fish remove oxygen from

The flux of carbon and nitrogen through a typical 500 tonne per annum production salmon farm (redrawn from Gowen & Bradbury, 1987).

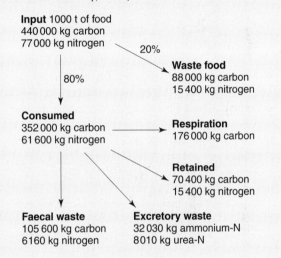

the water column while food and metabolic wastes are added, and an indirect effect as the structures may restrict the exchange and flushing properties of the system. Fish culture will reduce the oxygen content of the surrounding waters as the fish respire, and via the BOD (Section 6.2a) of waste products. In waters with a reasonable tidal exchange the supply of oxygen is, however, unlikely to be limiting.

As much as 20% of the food placed into the cages is unconsumed (see above figure). The metabolic and food wastes released will contain phosphate and nitrogenous compounds (see table). Ervik et al. (1985) recorded an 8–9-fold increase in ammonia concentrations around cages in Norway. In a tidally well mixed estuary,

Box 5.3 continued

Some of the waste produced by a 600 t per annum salmon farm expressed in terms of the effluent produced daily by an equivalent number of people

Parameter	Population equivalent per day
Biological oxygen demand	423 000
Ammoniacal nitrogen (NH_3–N)	114 000
Nitrate (NO_3–N)	300 000
Suspended solids	345 000

with a flushing time of 3–4 days and high inputs from municipal sewage, fish farm developments could potentially lead to a 30% increase in dissolved nitrogen in the waters of the upper estuary (Frid & Mercer, 1989). Gowen *et al.* (1988) review the possible effects of nutrients released from sea cages located in Scottish sea lochs. Their data indicated raised nutrient levels in the vicinity of the cultures, but they did not record any significant effect on local phytoplankton populations.

In general, the time required for phytoplankton bloom initiation is of the order of 2–3 days (Strickland, 1983); during this time phytoplankton will be moved away from culture operations and mixing with nutrient poor waters will have occurred. It therefore seems likely that stimulation of phytoplankton will only occur in waters with low flushing times and high inputs of nutri- ents. Aure & Stigebrandt (1990), in a review of the eutrophication and oxygen demand of fish farming in Norwegian fjords, concluded that in most fjords eutrophication and deoxygenation effects are unlikely, but concede that in areas with poor exchange they could be a problem.

Sedimentation and benthic enrichment

In the sheltered low tidal flow conditions of most aqua- culture operations, benthic enrichment is the domin- ant immediate environmental impact. The reduced flow rates associated with cages may lead to increased natural siltation, while the additional steady input of nutrient-rich particles in the form of unconsumed food

particles and fish excreta falling from the cages increases the supply of fine particles to the sediment in the vicinity of the cages. Weston (1986) and Gowen *et al.* (1988) provide extensive reviews of the sedimenta- tion rates in the vicinity of aquaculture operations. Sedimentation rates beneath cage cultures are typically 2–10 times those in reference areas. As a large propor- tion of the settling material is fine and organically rich, it can have profound effects on the benthic commun- ities impinged upon (Weston, 1986; Brown *et al.*, 1987; Gowen *et al.*, 1988).

Brown *et al.* (1987) characterized an essentially azoic zone directly beneath fish cages in Scottish sea lochs, with a highly enriched, species poor zone, extending from the cage edge out to 8 m. A slightly enriched, transitional zone extended from 8 to 25 m from the cages, at which point essentially normal com- munities were recorded (see figure). In the azoic zone the sea bed is black, partly fragmented food pellets lit- ter the sea bed, and a mat of white bacteria (*Beggiota*) covers much of the mud. The smell and taste of hydro- gen sulphide can even be detected inside diving gear.

Introduction of exotic species

The demography of disease agents in the marine envir- onment is poorly understood, but the accidental intro- duction of a pathogen or parasite could be ecologically disastrous. The specific nature of most disease agents would limit the effect to wild stocks of the cultured fish; however, the economic importance of such wild popu- lations can be considerable. The mesh of the cages will not act as a barrier to the spread of disease agents or parasites, while the tidal flows may disperse them over considerable distances, inoculating neighbouring farms and wild stocks alike. During the 1980s, Nor- wegian farmers suffered from an outbreak of the fluke *Gyrodactylus*. It is widely believed that this outbreak was initiated from stock imported from Denmark, as *Gyrodactylus* does not occur naturally in Norway but is endemic in Denmark.

Genetic effects

There exists the potential for fish escaping from fish farms to join wild stocks on the breeding grounds. This

Box 5.3 continued

(a) Benthic total abundance, biomass and species richness on a transect running down current from a Scottish salmon farm (redrawn from Brown et al., *1987). (b) Scavenging crab amidst the organically rich silt adjacent to a salmon sea cage (photo by C. Frid).*

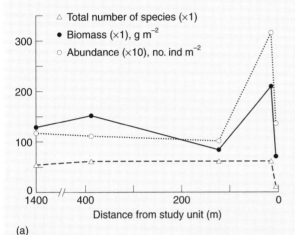

(a)

(b)

could lead to a dilution of the genetic identity of the wild stock and a loss of fitness of that population. As each spawning river is thought to contain at least one genetically distinct race, dilution of the genetic identity of wild stocks is of serious concern. Weston (1986) considers the potential for reduced fitness to be small, because in the USA large numbers of reared smoults are released to supplement wild stocks anyway, and because the highly selected, for cage conditions,

nature of the escaping fish would place them at a competitive disadvantage in the wild.

In Europe wild stocks are not supplemented to the same extent by hatchery reared fish. Furthermore, it is likely that the large size of the escaping fish may give them a competitive advantage at the reds, ensuring that they sire a greater proportion of the young. Given the stresses already exerted on wild populations of salmonids in most parts of their natural range, the genetic effects of escaping fish joining a breeding migration could be considerable.

The Scottish Salmon Growers Association consider the number of escaping fish to be negligible. However, in America it is estimated that 15% of farmed fish escape, while a survey in 1986 showed that 3% of the fish caught in the high seas fishery around the Faeroe Islands were escapees (identified by the presence of the colourant canthaxanthin in their tissues).

Release of toxic chemicals

Fouling loads on the nets of aquaculture operations were initially controlled by the application of paints containing tri-butyltin (TBT). Due to the ecological damage resulting from the build up of TBT in estuaries (mainly from anti-fouling on boats), its use has been restricted by government legislation (see **Section 6.3c**). However, chlorine, sodium hydroxide, calcium oxide and iodine solutions are all still in regular use for this purpose; fish diseases are treated by applications of broad spectrum antibiotics, formaldehyde, malachite green and dichlorvos.

The release of dichlorvos, used as means of controlling sea lice, *Lepeophtheirus salmonis*, from fish farms in Scottish sea lochs has been criticised due to its effects on other organisms within the community (Ross & Horsman, 1988). Dichlorvos is particularly toxic to crustacea, including calanoid copepods and economically important crabs, lobsters, shrimps/prawns and *Nephrops*. It also causes cataracts in the farmed stock and possibly in wild fish.

By 2000 many Scottish farmers were reporting increasing resistance in sea lice to dichlorvos. This has led to the Scottish Environmental Protection Agency licensing use of cypermethrin, a synthetic pyrethroid,

Box 5.3   continued

and azamethiphos, an organophosphate. Cypermethrin is extremely toxic to crustaceans and its use in sheep dip has been widely blamed for the loss of invertebrates in many upland streams. Azamethiphos is ten times as toxic as dichlorvos and also causes mortality in lobsters and fish larvae (Edwards, 1997).

The dispersive nature of the marine environment will ensure that chemicals used at fish farms will be moved long distances; as many modern preparations have high toxicity and long life, there is thus great

potential for direct environmental damage. It would, therefore, seem reasonable to adopt a 'precautionary approach' to the application of chemical preparations to the fish or the fish farm structures.

Effects on populations of natural predators

The high concentrations of fish found within aquaculture cages act as a lure to natural predators. In Scotland seals are the most prevalent predator. Attacks occur all year around, but are more common in winter,

Various modes of attack by predators attracted to salmon farm sea cages.

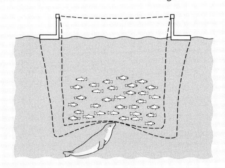

Seals

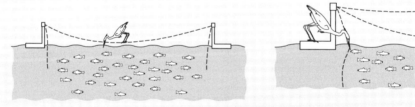

Herons

Cormorants and shags

Box 5.3 continued

after the breeding season, and during bad weather. Ross (1988) found that 80% of Scottish fish farms had suffered losses through seal attacks. Seals attack, predominantly at night, from under water, grabbing the fish through the mesh and sucking or chewing it through the net, often leaving the tail and head in the cage. While seal losses may be the most important economically, often because of the associated damage to nets, the greatest day to day loss is usually attributed to bird predators. Herons stab fish from the walkways or nets, while cormorants and shags dive down and stab fish through the side netting. As they are then unable to retrieve the fish through the mesh, they may continue the attack over a considerable period, damaging large numbers of fish. Otters and mink may also be important at some localities, the 'kill for fun' behaviour of the mink being particularly damaging, as a single animal may leave hundreds of fish dead in the bottom of a cage after an attack.

Various mechanisms have been developed to reduce seal predation. Anti-seal nets, used on some farms, provide a second layer of defence but greatly increase maintenance, and are ineffective in tidal streams as the currents push them close to the main net. On some farms the nets are left slack and act as entangling nets, drowning the entrapped predator and hence preventing further attacks.

There is increasing opinion that underwater predator nets are not cost effective and their use is being abandoned. Instead, sonic seal scarers are now being marketed. Some emit random pulses of sound to scare off seals and the reported effectiveness of these varies considerably. More recently a device has been marketed that emits a pulse of sound only at the approach of a large object underwater.

A common response of fish farmers (64% of those interviewed in 1987), although one that is often illegal, is to cull the offending seals. On some farms this is done as a last resort, while on others it is carried out as a matter of course. Culling can, however, impose a severe additional pressure on species already suffering reduced population sizes due to anthropogenic effects.

5.8 Interactions between aquaculture and the environment

Aquaculture, just like agriculture on land, can have profound effects on natural systems. Construction of any aquaculture facility alters the environment, although the scale of such changes can range from trivial (a handful of floats supporting ropes on which algae grow) to large scale (such as removal of mangrove forests to allow development of shrimp culture ponds). In general we can characterise the environmental effects of aquaculture into:

- modifications to the environment;

- waste products;

- interactions with wildlife;

- use of resources.

In **Box 5.3** we consider in detail the environment–aquaculture interactions for a Scottish salmon farm. As the scale and significance of such effects are so site and type of operation specific, the remainder of this section is generic in nature.

5.8a Modifications to the environment

The most profound environmental changes are those associated with direct habitat modification. This can range from removal of natural habitat to its replacement with aquaculture facilities (e.g. building intensive systems on formerly natural habitat, draining marshes, 'improving' lagoons, removing mangrove forests, and addition of substrates such as rubble or rock for seaweed culture).

Any construction in or adjacent to the aquatic environment will have an influence upon it. Structures placed within a water body alter the flow regime and hydrography, which in turn can alter sedimentation regimes. Construction of buildings, roads, or hard standing next to a watercourse will alter drainage patterns, while major drainage of marshes may be necessary to 'reclaim' them for development. Many of these modifications of the environment are unavoidable, while for others careful farm design can significantly reduce the potential impacts.

5.8b Waste products

As an absolute minimum a farm will generate metabolic wastes from the stock. In addition, however, an intensive system is likely to generate wastes from chemicals; husbandry often involves disinfection of farm equipment, application of pesticides and treatment of diseases. A wide range of chemicals is in common use (Tables 5.1 and 5.2). Excess food and organic wastes produced by the stock place a biological oxygen demand on the environment and release nutrients on breakdown, possibly contributing to eutrophication (see **Section 6.2b**). Food pellets also contain additives, including up to 15 vitamins, 11 minerals and, for salmon, synthetic colouring agents, which are released into the environment in this way. Artificial colourings are used to give the fish's flesh a pink colour similar to that of wild salmon, rather than 'hatchery grey'. Canthaxanthin (E161g) was initially widely used and, although banned in the USA in the 1980s due to health concerns, was still used in Europe until 1990. The more expensive astaxanthin has now replaced it.

5.8c Interactions with wildlife

The high densities of fish in ponds or cages are attractive to natural predators – especially birds and carnivorous mammals. Loss of stock to such predators is potentially a major cost to the fish farmer, encouraging use of antipredator strategies, from culling to antipredator barriers. If effective, barriers do no harm to the predator but are often costly and of only limited use, as predators often find ways of overcoming them. A cheap and attractive alternative is to reduce the predator population by culling, which clearly has a negative and potentially very severe impact on predator populations.

Other interactions with wild populations occur through the release of pathogens and parasites from the farm to the detriment of wild populations. In systems with highly selected stock, any escapees that interbreed with wild stocks could cause disruption to the natural genetic health of the species.

5.8d Use of resources

All intensive systems use energy (generation of which has its own suite of ecological effects) and foodstuffs. In many systems the foodstuffs are based on fish meal, a product of industrial fisheries. The economic value of using this low value resource and converting it to a high value product must be considered against the ecological cost of taking and using a large amount of natural production for a relatively small volume of utilisable protein for human consumption.

Environmental effects are not just restricted to the site of aquaculture operations, especially in systems in which artificial food is supplied. In the early 1990s the Norwegian fish farming industry alone consumed the equivalent of 430 000 tonnes of sand eels and the Scottish salmon industry a further 184 000 tonnes. This represents approximately 25% of the primary production of the entire North Sea. To produce 1 tonne of harvested salmon requires about 5.3 tonnes of fish, which in turn need about 1 km^2 surface area of primary production in a productive sea like the North Sea or Baltic. Therefore the area of primary production is about 40 000–50 000 times the surface area of the cages.

Table 5.2 *Chemicals registered or approved by the US Food and Drug Administration for use in food fish culture*

Product	Use
Therapeutants	
Acetic acid	Parasiticide
Formalin	Parasiticide and fungicide
Romet 30	Bactericide
(sulphadimethoxine and orthomeprim)	
Salt	Osmoregulatory enhancer
Sulfamerizine	Bactericide
Oxytetracyline (Terramycin)	Bactericide
Anaesthetics	
Carbonic acid	Anaesthetic
MS 222 (tricaine methane-sulphonate)	Anaesthetic and sedative
Sodium bicarbonate	Anaesthetic
Disinfectants	
Calcium hypochlorite	Disinfectant, algicide and bactericide
Water treatment	
Fluorescein sodium	Dye
Lime (calcium hydroxide, oxide or carbonate)	Pond sterilant
Potassium permanganate	Oxidiser and detoxifier
Rhodamine B and WT	Dye
Copper sulphate	Algicide and herbicide
Copper, elemental	Algicide and herbicide
2,4-D	Herbicide
Diquat dibromide	Algicide and herbicide
Endothall	Algicide and herbicide
Simazine	Algicide and herbicide
Clean-Flo (aluminium sulphate, calcium, sulphate and boric acid)	Algicide and herbicide
Glyphosate	Herbicide
Potassium ricinoleate	Algicide
Xylene	Herbicide

Source: Rosenthal *et al.* (1987).

5.9 Summary

- Aquaculture refers to production of harvestable aquatic species in confined conditions. These may be completely separate from the natural environment, or receive inputs of water and food from the surrounding aquatic system.

- Many species of organisms are cultured, typically in ponds (freshwater fish) and cages (estuarine fish), or on fixed structures (shellfish).

- Culture in ponds lends itself to polyculture, allowing several different species, each exploiting a different part of the aquatic system, to be reared together.

- Ranching is the raising of fry through their most vulnerable life stage, followed by release into the environment.

- Aquaculture can have severe impacts on the environment, through addition of pollutants.

Part 3   **Aquatic Environment**

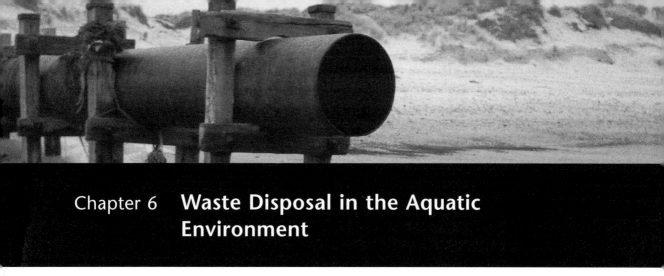

Chapter 6 Waste Disposal in the Aquatic Environment

6.1 Introduction

6.1a The nature of pollution

Aquatic systems have always been a preferred site for disposal of human wastes. Water movement disperses waste materials and removes them from the source, while biological processes are very efficient at neutralising products such as sewage. Indeed, there is an old adage: 'the solution to pollution is dilution'. To a large extent this is true, but as human populations have grown and industrialisation has become more widespread, two problems have arisen with simple disposal of waste products into the environment. Firstly, the concentration of traditional waste products has increased to the extent that the natural environment is having difficulty in dealing with them, and secondly new types of waste product have been developed, of which organisms have had no evolutionary experience. If the environment is degraded by addition of a waste product, it becomes a pollutant. However, a waste product added to a water body is not necessarily a pollutant; if it acts as food or substrate for organisms it benefits them by its presence. Only if added to excess, so that it causes stress to organisms or ecological systems, does it become a pollutant. Even under these circumstances, certain species may benefit from the waste product, for which it is a habitat enhancer. Therefore, the definition of pollution used in **Section 1.1** refers to environmental degradation from a human perspective, because this is the one of most relevance to us.

Types of pollutant

Pollutants are extremely diverse and, by the very nature of the process whereby wastes are added into the environment, a single pollution source may contain many pollutants. It is possible to distinguish several different types of pollutant, although these groups are rather crude, and some waste materials have properties of both.

Biodegradable materials are those which, once in the environment, are subject to microbial attack which leads to their degradation and eventual complete removal. They normally derive from organisms and include perhaps the most widespread type of pollutant: sewage derived from domestic human waste. Also included are wastes from agriculture and from food processing activities. As they provide suitable substrates for microbial activity, biodegradable pollutants remain active until either they are deeply buried in sediments or they are broken down completely.

Non-biodegradable materials are not affected by microbial activity. Inert pollutants, such as solid particles and heavy metals, remain in the environment in essentially an unaltered state,

Figure 6.1 *(a) Discharge plume in a river, showing gradual improvement in water quality. (b) Single discharge curve. (c) Additive effect of sewage discharges without adequate distance between each.*

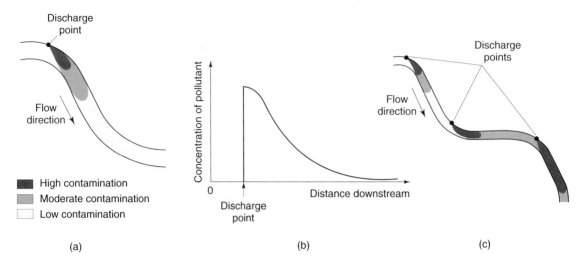

(a)   (b)   (c)

although they will be transported and diluted. Others, such as acids, are degraded by chemical reactions within the water body. If non-toxic, an inert substance will only act as a pollutant during its addition to the aquatic environment and for the length of time thereafter required for organisms to adapt to its physical presence.

A third type of pollutant is energy, particularly in the form of heated water. This will rapidly dissipate, but even its short-term presence can have a major influence on aquatic communities.

Point source and diffuse discharges

Pollutants enter the environment either as point source or diffuse discharges. A point source is a single location that can be identified as the source of the pollutant, examples being a sewage outflow pipe into a river or a designated location for dumping colliery spoil into a coastal sea. In each case, the location at which the pollutant enters the water is clearly identifiable and discrete. A diffuse discharge has no such clearly identifiable location, but enters across a large area, examples being acid rain, falling across an entire catchment, or nutrients derived from fertiliser addition to arable land and leaching into a river along a large stretch of its bank.

Point source pollutants are easier to monitor than diffuse ones, and their effects on the environment are generally clearer. A point source will produce a zone of maximum effect in its immediate vicinity, with decreasing influence as one moves away from this. This is most obvious in rivers, as the pollutant is washed downstream, producing a plume of gradual decline in intensity (Figure 6.1a). If the input is continual, the zones of decreasing concentration will remain fixed in space. A discrete or episodic pollution event will disperse, but will show the same pattern of greatest intensity at the beginning and then decline over time within a single location (Figure 6.1b). In some cases, a single pollution event causes a plume whose progress in space can be monitored; an example of this is an oil spill, whose movements and rates of dispersal can be predicted. In 1986, a fire in a chemical warehouse in Basel (Switzerland) caused the release of 1300 t of highly toxic chemicals, including over 900 t of pesticides, into the River Rhine (Güttinger & Stumm, 1992). It also contained red dye, allowing its progress down the river to be easily monitored

Figure 6.2 *Pesticide plume in the River Rhine, following the spillage on 1 November 1986. The spill occurred at km 159 and the movement and gradual dilution of the plume was tracked over the following few days. (a) Propagation of one of the contaminants – phosphoric acid esters. (b) Different effects of the spillage on organisms in the river (from Güttinger & Stumm, 1992).*

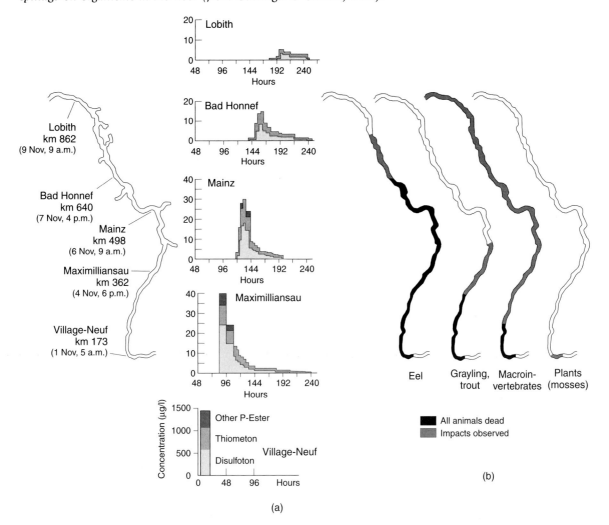

(a)

(b)

(Figure 6.2). The plume of highly polluted water gradually extended in length and diluted as it moved down the river, mimicking over time and space the pattern shown in Figure 6.1a.

6.1b Effects of pollution

Pollution affects the environment in several ways. Some pollutants are toxic, having a direct effect on the metabolism of aquatic organisms (**Box 6.1**). Very few pollutants are, however, toxic to all species. More common are indirect effects – addition of pollutants altering the physical or chemical environment to the detriment of organisms.

Very severe pollution eliminates life comprehensively and therefore destroys ecological communities. Most types and concentrations of pollutant do not, however, have such an extreme

145

Box 6.1 Toxicity

Some substances cause harm to organisms simply by their presence, because they are toxic, having a direct physiological effect upon organisms. We can further distinguish:

- **Acute toxicity** – a large dose of poison of short duration, usually resulting in a rapid, lethal response.
- **Chronic toxicity** – a low dose of poison over a long period, either lethal or sub-lethal in effect.
- **Cumulative toxicity** – toxicity arises from, or increases in severity with, successive additions.

Toxicity is normally quantified by carrying out exposure tests on organisms in the laboratory or in a controlled situation in the field. There are a number of measures of toxicity, including median lethal time and median lethal concentration.

Median lethal time (LT_{50} or LT_m). The time taken at the specified concentration of toxin for 50% of the test organisms to die. The figure below shows the time taken for 50% of test organisms (the copepod *Acartia tonsa* at different temperatures) to die at a particular concentration of toxin. Note the difference in LT_{50} with the time of year the animals were collected.

Median lethal concentration or dose (LD_{50} or LD_m). The concentration of the toxin required to produce mortality of 50% of the test organisms after a specified time period. The time period used varies; it may be as low as two hours for planktonic organisms

(a) Laboratory measures of toxicity. Survival of Acartia tonsa *at various temperatures (numbers on lines °C) at two times of year. Notice how the median lethal time decreases with increasing heat stress and varies with the time (and temperature) from which the animals were collected (data from Reeve & Cosper, 1972). (b) Lethal does (LD_{50}) for spiny lobster exposed to the pesticide 'Dimercon'. Note how the duration of the exposure influences the result (data from Suarez et al., 1972).*

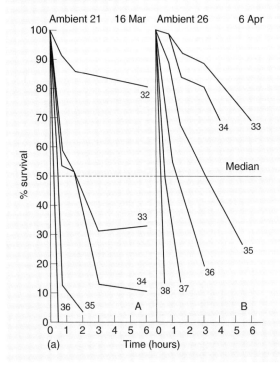

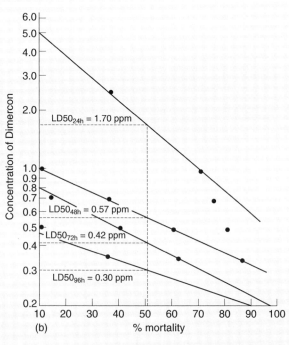

Box 6.1 continued

Sources of variation in bioassays

Source of variation	Ratio for comparing responses	Ratio range
Age of test organism	Old : young	0.01 to 1000
Sex of test organism	Male : female	0.5 to 5 or wider range
Genotypic difference	More resistant genotype : less resistant genotype	1 to 100 or wider range
Acclimation regime	Acclimated : non-acclimated	1 to 10 or wider range
Duration of exposure	Short bioassay test : long bioassay test	1 to 100

Source: Adapted from White & Champ (1983).

but is usually 96 hours. When the end point is some condition other than death, it is termed an effective concentration (EC) or effective dose (ED). Figure (b) on the previous page shows the concentration required to cause mortality of 50% of the test organisms (the spiny lobster, *Panulirus argus*) over a 96 hour period. It also illustrates the difference in values obtained when using different exposure periods.

There are many problems with toxicity testing in this way and in interpreting and extrapolating the results to the natural environment. Different species can vary considerably in their tolerances to toxins, while response of individuals within a species may vary depending on size, age, sex, reproductive condition, nutritional state and history of exposure (see table).

Toxicity tests fail to reveal effects that may only become apparent after a time delay. Furthermore, by using death as a convenient end point they overlook sub-lethal effects, whereas even in concentrations too low to cause death, toxins will reduce fitness, causing loss of breeding potential or foraging ability, for example, which will eventually impact upon populations.

The environment also has an effect. Metabolic processes are simply chemical reactions within an organism's body, whose rate increases with temperature. Therefore, as toxins operate chemically by interfering with metabolic processes, higher temperature increases toxicity. Furthermore, some toxins, metals in particular, can exist in a variety of forms or complexes, each differing in toxicity. Sea water, in particular, is a chemically complex environment in which metal ions may be present in a number of forms and concentrations. Metal ions form complexes with organic molecules, which tend to increase the uptake rate of the metal ion and can be up to 100 times more toxic than inorganic complexes with the same metal ion (Bryan & Langston, 1992; Depledge *et al.*, 1995).

Under natural conditions, organisms are rarely exposed to a single pollutant in the absence of other stresses. Interactions between compounds often result in complex effects. When the ciliated protozoan *Cristigera* was exposed in culture to solutions of salts of zinc (0.25 ppm $ZnSO_4$), mercury (0.005 ppm HgCl) and lead (0.3 ppm $PbNO_3$), they reduced growth rate by 14.2%, 12.1% and 11.8%, respectively, when added separately. If the effects were additive, a mixture of all three would reduce growth rate by 14.2 + 12.1 + 11.8 = 38.1%. The measured reduction was in fact 67.8%, demonstrating that the cocktail was more potent than the compounds in isolation (Gray & Ventilla, 1973). This is known as a **synergistic** effect. Alternatively, there may be an **antagonistic** effect, one pollutant reducing the effect of another. The experiment above was repeated using 0.0025 ppm HgCl (which decreased the growth rate by 9.5%) and 0.15 ppm $PbNO_3$ (8.5%). Combining these would be expected to give an 18% reduction in growth, whereas the measured effect was only 10.7%. Other well known antagonistic pairs are selenium, reducing the effect of mercury, and zinc, reducing the effect of cadmium.

effect – they selectively remove sensitive species, but equally other species may benefit, through addition of food or habitat resources, or elimination of predators or competitors. Therefore, pollution causes an ecological shift. This is commonly observed in areas influenced by continuous pollution, such as the vicinity of a sewage outfall discharging poorly treated sewage (**Section 6.2**), or the heat plume produced by a power station (**Section 6.5**). In these cases, the environment is permanently altered, eliminating sensitive species but allowing tolerant ones to establish and thrive. Discrete or intermittent pollution events, however, will have detrimental effects on the environment, but then not persist long enough for tolerant species to establish, resulting in impoverishment and continually disrupting community processes. It is these, therefore, which may have the greatest effects on the environment, but which are the most difficult to control, as they are often accidental or malicious.

Interactions between pollutants

Various types of pollutants are considered separately below. However, it is important to remember that different types of pollution often occur together, and influence each other. Sewage, for example, is considered mainly in terms of the easily biodegradable faecal waste that makes up the bulk of its pollution load, but it also contains nutrients, heavy metals, organochlorines, inert solids and a range of other contaminants, in varying proportions. Processes such as acidification and heated effluent discharge affect the toxicity of certain metals, as well physical features of the environment. Oil pollution may be exacerbated by treatment techniques that input synthetic toxins into the environment. Radiation emitting substances may also act purely as heavy metals, with the toxicity that this entails. These examples demonstrate that pollution is rarely simple, and emphasise that the distinctions recognised below should be considered to be artificial, and are used

for convenience as much as for their representation of pollution in the aquatic environment.

6.2 Biodegradable wastes

6.2a Sewage

Direct human waste – faeces and urine – is probably the oldest form of pollution, produced wherever people are present. As human settlements began to grow, people initially used the surrounding land to dispose of their wastes, recycling it as fertiliser. An early development for coping with localised populations higher than surrounding agricultural land could absorb was the 'earth closet', a pit adjacent to buildings into which the wastes were deposited. However, as the water supply was drawn from nearby wells, cross contamination often resulted. The introduction of piped water from remote sources (see **Section 2.5**) overcame this, but as towns continued to grow the scale of waste disposal also increased. With the availability of piped water, however, the invention of the 'water closet' or flush toilet offered a solution, using large volumes of water to move wastes away from the site of production. This process dilutes the waste but greatly increases the volume to be treated or disposed of. Introduction of the water closet and the requirement for piped removal of waste led to the creation of the essentially liquid waste we know as **sewage**. Aquatic systems have traditionally been the sewage outlets of towns, wastes being discharged untreated into a convenient large body of water, usually a river or coastal sea, whose natural flow dilutes and transports them away.

Modern sewage is more complex in composition than in the past. In addition to human wastes, it normally also contains large volumes of water derived from cleaning and cooking, and therefore contaminated with cleaning fluids and organic food wastes. Non-organic material present includes sanitary products and barrier contraceptives, while urban sewage can contain water from municipal drains and from industrial discharges. Sewage treatment is considered in **Section 6.7a**.

Direct effects of sewage

Sewage will almost inevitably contain some toxins, but its main effect upon aquatic environments into which it is discharged is not direct toxicity, but its influences on the physical environment and on oxygen concentration. Untreated sewage contains suspended solids at a density high enough to cloud the water, smother benthic species and clog animals' gills. These particles are mainly faecal and provide a very rich energy source for bacteria, which multiply to very large populations on faecal matter even before it leaves the human body. Furthermore, being composed of very tiny particles, sewage has a vast surface area to be colonised, so can support very high densities of bacteria, which consume the organic components, their high numbers depleting oxygen through respiration.

The amount of oxygen consumed by bacteria and therefore required for biological oxidation of waste is termed the Biological Oxygen Demand. Some wastes also have a Chemical Oxygen Demand, as reduced compounds undergo oxygenation. The combined Biological and Chemical Oxygen demand is known as the Biochemical Oxygen Demand or BOD (in older texts BOD may be used to represent the Biological Oxygen Demand alone). Typically, urban sewage has a BOD of 500 mg l^{-1}. This is low compared with some wastes (for example, the waste derived from brewing beer has a BOD of 70 000 mg l^{-1}), but is high compared to natural cleaning processes: oxygen-saturated water has the capacity to deal with a BOD of 8.0–8.5 mg l^{-1}.

Small, widely spaced discharges are not necessarily a problem, as rivers can absorb a certain amount of input with only local effects (Figure 6.1a). Problems arise when towns become larger and more densely populated, the quantity of effluent increases, and they become less widely spaced; one input's effluent has not been completely neutralised before the next is reached (Figure 6.1c).

Ecological effects of sewage discharge

A constant input of untreated sewage into a river creates a distinct plume, as in Figure 6.1a. Similar effects can be identified in the vicinity of marine discharges, although linear patterns of effects do not occur due to the oscillating tidal currents. At the outfall, BOD rises instantly, along with concentration of suspended solids, nutrients and salts. Oxygen concentration declines rapidly, often with a lag (Figure 6.3); in severe cases, anoxia can occur.

Figure 6.3 *The effect on environmental oxygen concentration of the discharge of an oxygen-demanding waste. The horizontal axis represents time since release or distance downstream in a river. Re-oxygenation occurs from the atmosphere while bacterial and chemical oxygen demand lowers levels. This results in an environmental oxygen sag; the critical point for the biota occurs when oxygen levels are at their minima – some distance/time away from the discharge point.*

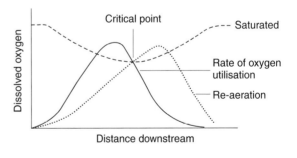

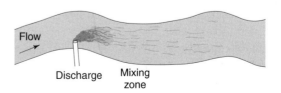

The input of suspended solids alters the structure of the bed sediments, to the detriment of species requiring hard substrates, and the decline in oxygen eliminates sensitive taxa, including fish and most invertebrate species (Figure 6.4a). For those species able to tolerate low oxygen concentrations, however, the fine particulate matter and its associated bacteria provide a rich food source, and the community becomes one dominated almost exclusively by fine particle detritivores and their predators.

In rivers tubificid worms may be the only macroinvertebrates present immediately downstream of a sewage input but, free from predation and with plenty of organic matter to eat, can reach very high densities. Predatory protists, too, become abundant, feeding upon bacteria in the water column. The river bed is altered, not only by particulate deposits but by development of sewage fungus, a complex mix of mainly bacterial and protist species in a slimy biofilm on the river bed or in long floating streamers.

In estuaries and marine sedimentary environments the same pattern of effects occurs; only the taxa involved differ. Close to the outfall the sediments may be anoxic and abiotic, with a covering of the 'sewage fungus' containing *Beggiota*. There is then a zone of low diversity but high abundance, containing opportunistic taxa including oligochaetes such as the tubificids, but also large numbers of small polychaete species such as the capitellids.

Although bacterial respiration appears to be completely detrimental to the benthic fauna, the activity that uses up oxygen is contributing to decomposition and therefore removal of the organic contaminants. Similarly, very high numbers of detritivores rapidly consume and respire the organic inputs. This is therefore the first stage in a self-cleansing process. As bacterial densities decline and oxygen levels start to recover, other detritivores become dominant. In freshwater systems, these include chironomid midge larvae and crustaceans such as hoglouse (*Asellus aquaticus*),

all of which remove organic matter, while in marine sediments it includes more sensitive species of polychaete and deposit feeding bivalve molluscs and echinoderms, especially brittlestars (ophioroids). The gradual self-cleansing of the aquatic environment produces a predictable (linear in rivers) pattern of organism distribution away from the input, each species being found under the conditions to which it is best adapted (Figure 6.4b). Sewage is rich in nutrients such as nitrates and phosphates (see **Section 6.2b**), and these may cause rich algal growth once the water has cleared, further altering the river bed but contributing to the removal of excess nutrients.

The situation described above may be less desirable than having no inputs at all, but the aquatic environment is able to recover. Problems arise when large inputs of untreated sewage relative to the size of the water body result in complete deoxygenation, and therefore loss of all but anaerobic bacteria. Once this occurs, the cleansing process ceases to function and the river is biologically dead.

6.2b Nutrients

The increase in nutrient concentrations in water is frequently referred to as eutrophication. Although it can occur naturally, it is a consequence of many human activities. Sewage is an important source of nutrients: the average person produces 11 g nitrogen and 2 g phosphate per day in faeces. Nutrients also derive from fertiliser runoff in agricultural areas. Nitrate in particular retains its solubility, and up to 50% of all the nitrogen applied to crops ends up in groundwater or surface waters.

Effects of nutrient enrichment

Nutrients added to water enhance productivity by photosynthesisers. Small or moderate levels of enhancement may be advantageous, but too much increase, or very rapid increase, will be

Figure 6.4 *(a) Cumulative number of taxa of major macroinvertebrate groups at various sites on the Esquel River, Chile. The sudden change between sites 8 and 9 is caused by sewage discharge from a small town (from Miserendino & Pizzolón, 2000). (b) Longitudinal sequence of invertebrate distribution downstream of a sewage discharge in the UK.*

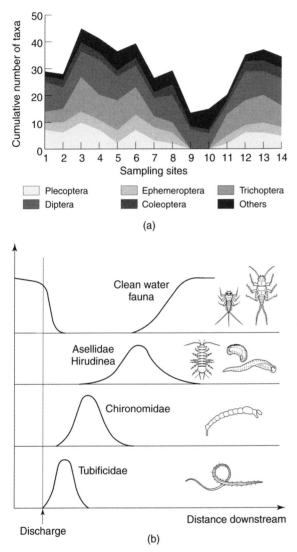

(a)

(b)

algae most. Once algae dominate, they eliminate macrophytes through shading. In freshwater lakes and sluggish rivers receiving large quantities of sewage, the dominance by algae is a well described phenomenon. The transition of the Norfolk Broads (England) from a series of clear lakes, covered in water lilies and other flowering plants with a diverse fish fauna, to a series of murky lakes with no macrophytes and an impoverished fish fauna, is attributed to eutrophication (Moss, 1983).

Algae are able to respond very rapidly to increased nutrient loads, and where these are high they will form dense blooms, which completely shade the water body and produce large numbers of dead cells. The decomposition of this detritus can lead to deoxygenation of a lake below its surface, while algal respiration may cause overnight anoxia on a daily basis. Under warm conditions, Cyanobacteria ('blue-green algae') are favoured by high nutrient loads. Several species are toxic to vertebrates, making toxicity a further effect of eutrophication.

In fresh waters, nutrient inputs have their greatest effects in small or shallow lakes. Rivers are rarely affected by algal blooms, because the flow disperses phytoplankton. However, benthic algae may reach high densities.

Eutrophication in the sea

It is widely believed that eutrophication in the sea is unlikely to occur due to the greater flushing ability and dilution capacity, except in enclosed areas such as bays and estuaries. There is, however, now increasing evidence of nutrient build-up in the Baltic and eastern North Sea (Brockmann *et al.*, 1989; Hickel *et al.*, 1993). In the North Sea it is estimated that nitrogen inputs are now 4 times background, while phosphate inputs are 7 times background. Of these increases, 37% of the additional nitrogen and 68% of the additional phosphate derive from sewage, and 60% and 25% from agricultural runoff respectively (Gerlach, 1988).

detrimental. If macrophytes dominate the system, they may thrive and absorb the extra nutrient load. However, disturbance to macrophytes, or rapid fluctuations in nutrient input, will benefit

151

One of the first effects of enhanced nutrient inputs on marine ecosystems is a change in phytoplankton species composition. Increased availability of nutrients, especially nitrogen, causes a shift in the competitive balance, such that species adapted to the low nutrient conditions are excluded by more aggressive nutrient-hungry species. However, such changes can be subtle and can only be detected if good historic data are available.

A phytoplankton bloom is a clearer demonstration of changes. Marine blooms occur naturally each spring and autumn in temperate areas. However, occasionally blooms of a single species occur outside the normal periods of natural nutrient mixing, often of highly coloured species whose sudden presence may be obvious as a 'red tide' – usually caused by *Gymnodinium* or *Mesodinium* – or a 'brown tide' – caused by *Phaecystis*. While such blooms have been recorded since at least the 17th century and therefore are unlikely to be directly linked to sewage enrichment, the increased availability of nutrients is a prerequisite for any bloom. In May 1988, and again in 1989, a bloom of *Chrysochromulina* was initiated in the eastern North Sea, within the nutrient rich plume from the River Rhine. The bloom swept north into the Skagerrak, and caused large mortalities and considerable economic loss in fish farms in Sweden and Norway (Diaz & Rosenberg, 1995).

6.2c Oil pollution

Crude oil is a mixture of hydrocarbons with different molecular weights, molecular structures and hence properties. The mix and nature of the components varies between oils from different geological reservoirs. Most North Sea crudes, for example, are relatively light, containing a lot of low molecular weight fractions, while Venezuelan crude is very heavy and tar like. Supplies of oil are restricted to locations where the physical and geological conditions have been favourable for its production and accumulation. These are rarely conveniently close to the main users. Hence, oil has to be extracted (see **Chapter 8**) and transported to where it is needed. Oil has to be moved from the production site to a refinery and products moved from refinery to end-user. This is often achieved using pipelines, but about half of the world's annual oil production, which amounts to around 3 billion tonnes per year, is transported by sea. The routes employed are often quite intensely travelled. For example, in 1989 tankers carried 504 million tonnes of oil out of the Persian Gulf, of which 117 mt passed around Cape Horn. In total, 340 mt of oil was carried by sea to Europe and 315 mt was brought to the eastern seaboard of the USA (Clark *et al.*, 1997). During this transport there is the possibility that some oil will be discharged from tankers, either accidentally or deliberately, and therefore contaminate sea water.

Oil spills

Natural seeps from oil reservoirs account for about 10% of inputs of fossil hydrocarbons into the sea, while the total input from domestic waste disposal and sewage exceeds the direct inputs from the oil industry and almost equals the total input from transportation (Figure 6.5). Much input from transportation is non-tanker accidents. Large bulk carriers nowadays carry as much oil in their fuel tanks as was carried by the average

Figure 6.5 *Sources of anthropogenic hydrocarbons in the marine environment.*

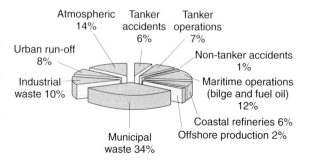

Figure 6.6 *The influence of wind and current on an oil slick. Oil slicks are moved with currents at current velocity and down-wind at 3% of the wind speed. Their track can therefore be predicted using simple vector plots.*

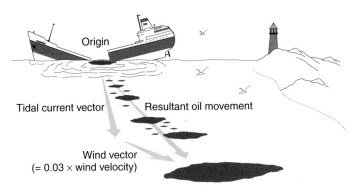

crude oil tanker 40 years ago. For example, Ireland's worst oil spill in recent years resulted from the wreck of the *Kowloon Bridge*, a bulk iron ore carrier that was lost only a few days after loading fuel in November 1986.

When spilt on the sea, oil tends to form a slick on the surface (Figure 6.6), which is subject to mixing by wind and waves. A small component of the oil dissolves into the water, while water is physically mixed into the oil to form an emulsion at the sea surface, often referred to as 'chocolate mousse'. The volume of the mousse slick can therefore be considerably greater than the volume of oil originally lost. Oil weathers as the lighter volatile component is lost and photo-oxidation occurs. Thus crude oil tends to thicken and become more tarry with time. Eventually such deposits become heavier than water and sink. On the seabed they accumulate debris, forming tar balls, which may eventually be cast ashore.

Oil is broken down by bacteria at a rate usually limited by nutrient supply and oxygen. Thus, oil in the water column and on the seabed will usually be degraded but once incorporated into the sediment matrix it is broken down very slowly.

Techniques to reduce oil spillage are considered in **Section 6.7b**, and clean up systems in **Box 6.2**.

Biological effects of oil

Oil contains some toxic components, in particular the aromatic fractions and heavy metals, and its ingestion can therefore lead to a direct toxic effect. It also chemically attacks the lipid in cell membranes, particularly of the gut epithelium. Oil can be tasted at very low concentrations, giving contaminated food a 'taint' and making it unpalatable. Therefore, even if no mortality occurs, fish caught near a spill may be unmarketable.

Sea birds become oiled at the sea surface, lose their insulation and either drown when they become waterlogged, get hypothermia or die from oil ingested while trying to preen. Young seals have suffered mortality when oil has stranded them on pupping beaches, but in general fish and adult marine mammals are able to avoid spilt oil, possibly in response to the taste of the 'tainted' water. Where oil threatens or damages coastal zones, a clean up operation is usually undertaken (**Box 6.2**). Coastal habitats vary in their sensitivity to the ecological effects of oil spills (Table 6.1).

Community effects

At severely impacted sites the smothering effect of the oil results in the complete loss of the fauna and

Table 6.1 *A classification of the sensitivity of the natural environment to oil spills*

Shore type	Sensitivity
Exposed rocky shores	Least sensitive – high natural dispersion
Eroding wave cut platforms	
Fine grained sand beaches	
Coarse grained sand beaches	
Exposed, compacted tidal flats	
Mixed sand and gravel beaches	
Gravel beaches	
Sheltered rocky coasts	
Sheltered tidal flats	
Salt marshes and mangroves	
Coral reefs	Most sensitive – highly sensitive biota

Source: Gundlach & Hayes (1978).

flora; following removal of the oil (by clean-up or naturally by weathering) the system follows an ecological succession. However it is more commonly the case that some species are lost and some survive with only limited mortality, so the balance in the community is shifted. On rocky shores grazing gastropods and barnacles are particularly vulnerable. The latter are prevented from feeding by the covering of oil, while the former ingest oil residues while feeding and suffer toxic effects. Both groups are also sensitive to dispersants used in clean-up.

The gastropod grazers on moderately wave exposed shores are 'key-stone' species in maintaining the shore as a mosaic of patches of barnacles, macro-algae and limpets. Following the *Torrey Canyon* wreck, the rocky shores of Cornwall (UK) were heavily contaminated by oil and many were also cleaned using detergents. This resulted in mass mortality of limpets, periwinkles and barnacles. The following spring the shores exhibited a 'green flush' as the rocks became covered in a diatom dominated algal film, a result of the absence of the normal grazers. As the summer progressed the shores became covered in young macroalgal plants. Once established these persisted for several years, during which time the 'whiplash' effect of their fronds, as they moved

with the waves, swept the adjacent rock, keeping it clear of limpets and barnacle larvae. It was only as these plants died that limpets and barnacles were able to colonise; the ecology of the shore had effectively temporarily switched to an alternative state. It took between seven and ten years for the shores to return to a state approximating the pre-spill one (Hawkins & Southward, 1992).

6.3 Accumulating toxic wastes

The substances that are the focus of this section are resistant to chemical change or degradation. Their concentrations may be diluted as they are mixed away from the discharge point, but they remain in the environment for very long time scales, behaving essentially as conservative properties of the water mass, hence the other term for them, **conservative pollutants**. Substances falling into this group are heavy metals, complex halogenated hydrocarbons, PCBs, many pesticides, radioactive discharges, solid wastes and plastics. Of these, most act as toxins and are considered together in this section. Solid wastes and plastics, however, are metabolically inert unless contaminated with other accumulating wastes, but their

Figure 6.7 *The process of bioaccumulation. In this example, the mercury is a natural input from geothermal emanations into several lakes in North Island, New Zealand (after Kim & Burggraaf, 1999). The increase in concentration between components of the food web is shown. The concentration of mercury in the water is 0.1–0.5 ng l⁻¹.*

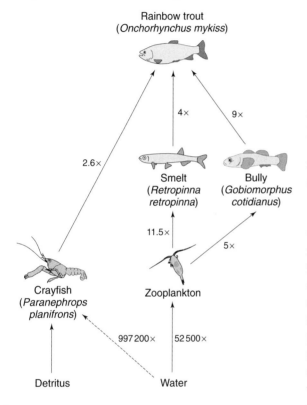

physical presence has a polluting effect; these are considered separately in **Section 6.4**.

Organisms vary widely in their ability to metabolise conservative pollutants. If the organism is unable to metabolise or excrete the pollutant it becomes incorporated permanently into its tissues, with levels being continually enhanced, a process known as **bioaccumulation**. When a predator or grazer consumes food that is enriched with respect to a particular pollutant that it is unable to excrete or metabolise, it too accumulates the pollutant in its tissues (Figure 6.7). However, as its food is enriched with the pollutant, the

rate of accumulation is faster. This is **biomagnification**. Top predators, such as seals, birds, marine mammals and humans, are most at risk from this process.

6.3a Metals

Metals are a natural component of surface waters, making separation of the natural component from an anthropogenic one difficult. They reach aquatic systems from erosion of ore bearing rocks, blown dust, volcanic activity, atmospheric fallout, forest fires, and the reworking, by dredging or natural processes, of metalliferous sediments. Industrial discharges direct to the sea and the contribution from ocean incineration are very small on a global scale, but can be significant locally (Kersten *et al.*, 1989).

Some metals are important as micro-nutrients. Copper, for example, is a key component of the respiratory pigment haemocyanin, which is common in crustaceans. However, it is very toxic at high concentrations for a variety of organisms and is widely used in anti-fouling preparations. Similarly iron is a component of the respiratory pigment haemoglobin, and is also required by many marine algae. Iron is present in quite high concentrations in sea water and many fresh waters, and is only of note as a pollutant at point source discharges. Iron is often associated with acid mine drainage, although the main water quality problem in this case is the acidity of the water, which changes various metals into their more volatile forms. The role of acidity in increasing aluminium toxicity is considered in **Section 6.5**.

Certain metals are highly toxic, even at very low concentrations, including mercury and cadmium, both of which bioaccumulate readily. Lead, too, is a bioaccumulator, although it is of low toxicity to aquatic invertebrates. Its effect on vertebrates can, however, be severe, the most widespread concerns being its effects on humans and on wildfowl ingesting lead shot (see **Section 8.3c**).

Metallic tin is not very toxic. However, organo-tin compounds can be highly toxic and this is reflected in their widespread use in anti-fouling paints. The development of the organo-tin tributyl-tin (TBT) as an antifoulant is discussed more fully in **Section 6.3c**.

Ecological effects of metals

Metals act mainly as bioaccumulators, and their effects are therefore selective, being greatest on predators. Whole community effects are less common. Iron rich discharges may have a physical effect on the environment. Europe currently discharges approximately 5.6 million tonnes of acid iron waste to the sea each year. At the discharge point, heavy crusts of iron oxides are deposited on hard substrata, including the shells of molluscs and crustaceans, while the hydrated oxides of iron form flakes in the water column. These can clog feeding structures and gills. Given the localised nature of these effects it is difficult to quantify the ecological consequences of the extra metabolic cost of dealing with the flakes on the biota (Newell *et al.*, 1991).

Where metal ion concentrations are consistently high, sensitive species may develop tolerance to them. This phenomenon is widespread in rivers and estuaries impacted by runoff from mine workings, which often contain high concentrations of metals and may also be very acid. The Fal Estuary in Cornwall, southern England, has naturally high levels of copper from weathering ores, and a long history of copper discharges from mineral workings. The fauna of the estuary is notable for the absence of almost all species of molluscs, while the polychaete *Nereis diversicolor* and the fucoids appear to be tolerant strains (Grant *et al.*, 1989; Bryan & Langston, 1992).

6.3b Pesticides, PCBs and CHCs

Organic compounds containing halogens such as chlorine are not amenable to biological degradation in the way that oil based hydrocarbons are. They are essentially permanent additions to the ocean and are entirely artificial. Many are heavily bioaccumulated, particularly in fatty tissues, and are very toxic, hence the use of many as pesticides.

DDT (dichloro-diphenyl-trichloroethane) came into widespread use after 1939. In recent years it has been superseded in the developed world, although it is still used in developing countries. As a pesticide it is extremely effective, being highly toxic to insects, but having a low toxicity to other groups; it is cheap and easy to handle and a single application remains effective for a long time (its half life in the soil is about ten years). It is also highly bioaccumulating. DDT breaks down into the less toxic DDE (dichloro-diphenyl-ethane). Most of the chlorinated hydrocarbons in the sea, and 80% of those in marine organisms, are DDE, mostly derived from DDT. Lindane (gamma hexachlorocyclohexane, gamma-HCH) was introduced at about the same time as DDT, but its use is now much reduced. Although heavily regulated it is still widely used in specialist applications, especially the veterinary treatment of sheep scab, and in seed treatments.

Aldrin, dieldrin, endrin and heptachlor are all extremely persistent organo-chlorine pesticides whose breakdown products are also highly toxic. Many uses of these chemicals were phased out in the 1970s, although some specialist uses continue (and large quantities were sold off to developing nations). Their persistence ensures that they are now widespread in the environment. PCBs (poly-chlorinated biphenyls) are used in a wide range of electrical apparatus and as flame-retardants. They are extremely persistent in the environment and attempts have been made since the 1970s to reduce PCB accumulation, culminating in the 1990 undertaking by the EU to phase them out by early in the 21st century. The only effective disposal option for PCBs is high temperature incineration.

Effects of chlorinated hydrocarbons

Chlorinated hydrocarbons (CHCs) are not very soluble in water, but are highly soluble in fats. Sewage disposal, for example, often introduces these components into the system bound to lipids. Once in the environment they bind to organic and inorganic particles, from where they can be ingested by filter feeding organisms. Given their affinity for lipid and their low excretability, they tend to be bioaccumulated and biomagnified and may become biologically active at times of lipid metabolism, such as starvation periods while brooding eggs or rearing young.

In the Baltic Sea, ringed seal (*Phoca hispida*) populations have only recovered slowly since the cessation of hunting, pregnancy rates being typically 28%, as opposed to an 80% rate in a healthy population. Non-pregnant females contain an average of 77 ppm PCBs in the blubber, pregnant females only 56 ppm. About 40% of females autopsied have one or both uterine horns blocked by occlusions, making them sterile. Grey seals (*Halichoerus grypus*) from the area show the same pathological symptoms, with PCB levels averaging 110 ppm in the non-pregnant females and 73 ppm in the pregnant females.

PCBs interfere with ovulation and development in some mammals, but this has never been tested on seals and there is no direct evidence linking PCBs to the uterine occlusions. So while the difference in PCB levels between pregnant and non-pregnant females is interesting, PCBs cannot be definitely implicated as the causal factor. Grey seal populations on the Farne Islands (North Sea) have risen so high as to require their culling in spite of average tissue PCB levels of 122 ppm.

6.3c Endocrine disrupters

The endocrine or hormone system plays a central role in the regulation of biochemical, physiological, developmental and reproductive activities in metazoans. The hormones which are the chemical messengers of this system are often highly specific in their function, but also are frequently conserved across taxa; for example, the female sex hormone oestrogen is chemically similar in invertebrates, fish and mammals. While it was recognised in the 1950s that certain pollutants (e.g. DDT) acted by disrupting hormone mediated processes, it was the observation of altered sex functions in a range of vertebrate species in the 1980s that lead to the recognition of a new category of pollutant – endocrine disrupters (EDs).

The effects most studied so far are mimics or antagonists of steroid proteins, including oestrogen. Oestrogen is responsible for regulating the development of female sex organs and control of the reproductive cycle. While sex is genetically determined in mammals, it is reinforced during development by the action of the sex hormones; in other vertebrate groups exposure to oestrogen during development may cause development as a female irrespective of the genetic make-up. There is evidence from a range of organisms of 'feminisation' of males when exposed to oestrogen mimics in the environment. As EDs act by stimulating the receptor in the organism, there is scope for substances present individually at extremely low concentrations to act additively as receptors. An observed response may, therefore, be the effect of many substances, while a similar response at different locations may be due to different combinations of EDs. This raises very serious concerns about the ability of current, chemical-specific, environmental risk assessments quantitatively to predict the full environmental consequences of novel chemicals.

In the early 1990s it was observed that male trout living in the lagoons used as the final purification stage of sewage treatment works were showing signs of abnormal sexual conditions (Harries *et al.*, 1997). Many of the observed changes – egg yolk protein vitellogenin in the plasma, testes degenerating or showing signs of egg production (ovo-testes) – were similar to effects seen in males exposed in the laboratory to oestrogen. A wide range of compounds have now been shown to have hormone activity. For example, a spillage of dicofoil (a DDT based insecticide) into Lake Apopka (Florida) occurred in 1980, causing game fish and alligator populations to crash (Colborn *et al.*, 1993). Even by 1987 75% of alligator eggs failed to hatch (compared with 5% at control sites); those that did hatch showed high levels of sexual abnormalities, including 90% of the male offspring failing to produce testosterone

(a male sex hormone). The alligators contained DDE at around 0.1 ppm, too low to cause toxicity but high enough to interfere with development.

In the marine environment the clearest examples of ED impacts are on reproductive health in flat fish from urbanised estuaries and embayments. Effects have been seen in English sole (*Pleuronectes vetulus*) in the USA and European flounder (*Platichthys flesus*) in UK estuaries (Figure 6.8). In the Tyne Estuary up to 94% of

Figure 6.8 *The level of reproductive abnormalities (a) from the industrialised Tyne Estuary (sites T1–T3) and the relatively clean Solway Firth (site S) in northern England. The percentage of male fish with egg yolk protein (vitellogenin, VTG) in their serum is given under each bar; the bars show the mean concentration of VTG recorded. (b) The abnormalities in European flounder* (Platichthys flesus) *(photo by M.E. Gill).*

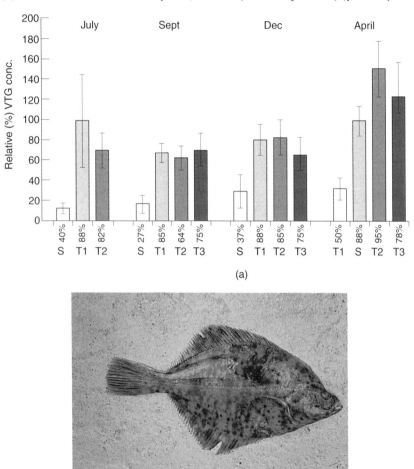

(a)

(b)

male flounder had vitellogenin in their serum (Lye *et al.*, 1997, 1998). Testicular morphology and sperm production were also altered compared to fish from an area receiving only limited sewage and industrial waste discharges. Other indications of ED effects in the marine environment include observations of increased intersexuality in harpactoid copepods from near sewage discharges, impaired reproduction in gobies exposed to suspensions of sewage sludge, and decreased levels of pregnancy in common seals (*Phoca vitulina*) fed on PCB-contaminated fish (Matthiessen *et al.*, 1998).

TBT

Since the interest in EDs has developed, the results of many earlier studies have been reinterpreted as possible ED effects. The impacts of the antifouling agent TBT were originally seen as an 'organo-heavy metal' effect, but would now be recognised as an ED effect. **Fouling** is defined as the growth of marine plants and animals on ships' hulls, fish farm equipment and pier/rig legs, and is a major cost to maritime industries. The growths impose extra stress on the structures and increase drag, resulting in greater energy cost for ships and more wave damage to static structures. Antifouling technology seeks to limit this growth. In the 1970s a new generation of anti-fouling paints containing TBT were introduced. These were long lasting and very effective. However, by the 1980s there were reports of effects on mollusc populations from many areas, including altered shell growth in cultured oysters and development of abnormal sexual conditions in the dog whelk (*Nucella lapillus*), a predatory snail found on European rocky shores (Gibbs *et al.*, 1991). In *Nucella*, exposure to TBT causes the females to develop a vas deferens and penis in addition to the normal female sexual organs. This condition is known as imposex and in severe cases the abnormal penis occludes the female genital opening, causing the female to be sterile. These dramatic effects, in particular the impact on oyster farms,

stimulated a ban in the mid-1980s of TBT-based antifoulants on small (<25 m vessels) and static structures. In spite of evidence for widespread recovery under the current regulation regime (Evans *et al.*, 1996; Evans, 1999), the International Maritime Organisation (IMO) has introduced a global ban on TBT use from 2003 and a phase out of existing coatings by 2008.

6.3d Radioactive discharges

Radioactive substances consist of radioisotopes or radionucleotides. These are unstable atoms that decay into a more stable form by emitting subatomic particles or energy, known as radioactivity.

We can distinguish four types of radioactivity. **Alpha (α) particles** consist of two neutrons and two protons. They have low penetrating power – they are stopped by 40 μm of tissue – but are potentially very damaging to biological systems if an α emitter is ingested or inhaled. **Beta (β) particles** are either electrons or positrons, with moderate penetrating power (40 mm tissue); again their main biological effects arise from ingestion or inhalation of a β emitting substance. **Gamma (γ) rays** are similar to x-rays and are deeply penetrating. The fourth category of radioactivity is **neutron radiation**, which arises from nuclear fission – when a large and heavy atom fragments into two atoms of a lighter substance and emits excess neutrons. These are highly penetrating and can cause damage to molecules with which they collide, causing the emission of γ rays. Some elements undergo spontaneous fission, but generally neutron radiation is restricted to nuclear reactors.

As a radioactive material emits radiation, in whichever form, it moves into a more stable state. The rate of radioactive emissions therefore decreases with time – hence the term radioactive decay. The time taken for the rate of radioactive emission to decrease by one half is known as the **half-life** of the substance. Each radionucleotide has its own characteristic half-life, which is constant and not influenced by the external environment.

Fresh waters will naturally pick up a radio-nucleotide signature reflecting the nature of the geological areas they flow through. Sea water naturally contains radionucleotides, in particular potassium-40, rubidium-87, uranium and radium. In addition to these there are 'artificial' radionu-cleotides deposited from the atmospheric atomic bomb tests of the early 1960s and from discharges from nuclear power stations and reprocessing plants. The latter can be easily observed via the radionucleotide caesium-137, which has a half-life of 30 years and does not occur naturally; thus its dis-tribution has allowed the hydrography of the North Sea to be mapped (Figure 6.9). Also discharged from the Sellafield and Le Hague nuclear waste repro-cessing plants are caesium-134, strontium-90, ruthenium-106, antimony-125, armerian-241 and tritium (^{3}H). Low numbers of nuclear submarines and nuclear weapons have been lost at sea, but in addition to this, dumping of nuclear waste at sea, typically in depths of 3–4 km, began in 1947. This continued until 1972 for high-level radioactive waste and 1982 for intermediate and low-level waste.

In the sea the fate of the radionucleotides depends on their physical/chemical nature. Cae-sium tends to stay in solution, while plutonium and ruthenium tend to bind to particulate matter and become deposited along with fine materials on the seabed.

Effects of radiation

The biological impacts of radioactivity depend on the tissue affected. High concentrations of radio-nucleotides damage chromosomes and prevent their normal functioning. At lower levels of expos-ure there are **somatic effects** – impacts on grow-ing tissues, giving rise to various forms of cancer – and **genetic effects** where the damage is to the gametes and is manifested in the subsequent gen-eration as defects or abnormalities.

Ecological effects of radionucleotides in the aquatic environment are difficult to assess. Most concern has focused on pathways, referred to as **critical pathways**, that may bring the radioactive material into the human food chain. It is to be expected that any exposure to radiation will cause an increase in the frequency of cancers, genetic defects and mortality, irrespective of whether the radiation is from a natural or man-made source. However, given the very high natural mortality in most aquatic species it is very difficult to detect any effect at the population level in the field.

On 26 April 1986 an explosion and fire at the Chernobyl Nuclear Power Station in the Ukraine resulted in the release to the atmosphere of around 3.5% of the nuclear material in the reactor. The radioactive cloud initially moved over Scandinavia and then southward, covering central Europe and northern Britain. Increased levels of radioactivity were subsequently measured around the Mediter-ranean and over much of the world (Savchenko, 1995). Of the emitted material the most problem-atic substance was caesium-137. Levels in fish from Scandinavian lakes increased rapidly after the accident and remained above background lev-els for over five years (although there was much variation between species and between individuals within a species). In central Wales (which was out of the main region of deposition), radioactivity in otter spraints (faeces) in the September after the accident was more than double that in samples from before the accident, but by January 1987 they had returned to normal (Mason, 1990).

Apart from their radioactivity, radionucleotides behave chemically in the same way as the non-radioactive isotopes. As many of the radioisotopes arising from human activities are heavy metals, this means they tend to bind to lipids and bioac-cumulate and biomagnify in the food chain.

6.4 Inert solids

Inert solids are those that remain in particulate form. Their effects, therefore, are mainly physical, although many are contaminated with heavy metals and so include a toxic element.

Figure 6.9 *The distribution of 137 Cs and 134 Cs in the waters around the UK in 1984. The contours allow the underlying hydrographic circulation to be deduced (inset) (from Salomons et al., 1989).*

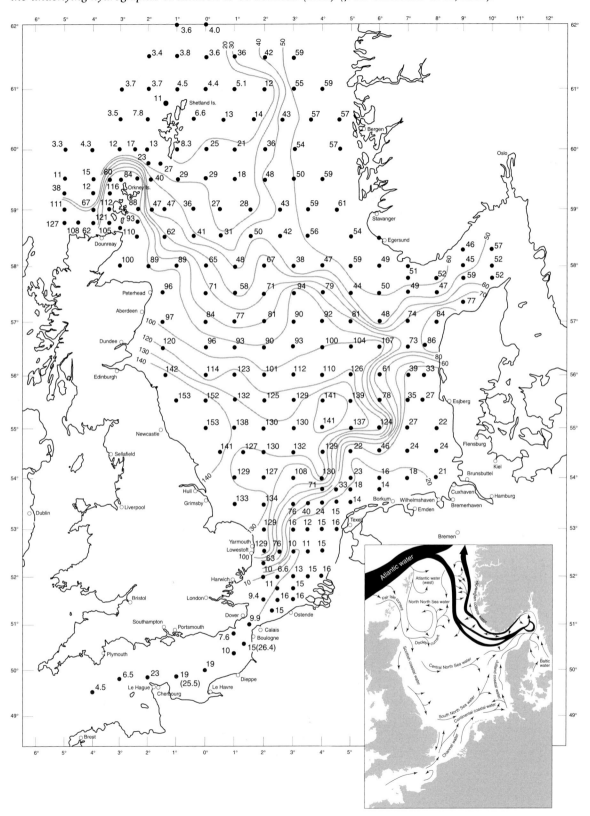

6.4a Types of inert solids

Mining for mineral resources typically involves the removal of considerable quantities of 'mine-stone' or 'spoil', the material that has to be dug out to provide access to the target minerals. If mining is adjacent to the sea, it may be used as a convenient dumping ground for this spoil. Some power stations also dump solid residues. Power generation at coal-fired plants results in considerable quantities of by-products, including gases and solid wastes like fly-ash, clinker and, with appropriate technology, flue gas desulphuration (FGD) residues. Fly-ash or PFA (pulverised fuel ash) is the predominantly silty fuel residue that remains following combustion, and amounts to over 80% of the solid waste fraction. Marine dumping of PFA is unusual, but occurs in certain localities. The Israeli Hadera Power Plant dumps at a deep water site in the eastern Mediterranean Sea (Golik & Krom, 1989; Kress *et al.*, 1993). Between 1956 and 1992, the Blyth Power Station (northeast England) employed a 1200 t barge to tip nearly 14 million tonnes of PFA and clinker – around 1000 t per day – at a 3 km² dumping site, 6 km off the coast.

Macro-litter is defined as identifiable large particles. It has been estimated that between 1.1 and

Figure 6.10 *Common items of marine refuse with an indication of the degradation times. Plastic does not degrade, although it may be physically broken into smaller particles.*

Figure 6.11 *Macro-litter recorded from the River Taff, South Wales. Percentages are of numbers of identifiable particles (after Williams & Simmons, 1999).*

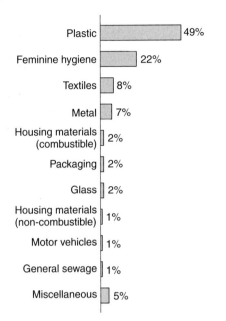

flow level; it can then persist in such places for many weeks.

6.4b Direct effects of inert solids

Dumping mine waste at the coast and into the sea has been widespread due to assumed negligible effects. However, such waste alters sediment stability and porosity, changes sediment particle size and smothers rocky outcrops. The process of addition smothers organisms and clogs gills, while light availability and organic matter are reduced. Changes to sediment structure may alter its suitability for colonisation by juveniles and possible changes in dominant fauna.

Daily dumping of residues from the Blyth Power Station caused almost total abiotic conditions in the impacted sediments throughout the disposal period, and smothering by fly-ash was the main cause of benthic mortality (Bamber, 1984). Only limited recovery had occurred nine months after dumping ceased, due to the continued prevalence of pure PFA at the dumpsite (Herrando-Pérez & Frid, 1998).

In addition to physical effects, spoil from coal mining may contain toxic metals, especially arsenic, and some leaching must be expected over time. Colliery spoil tailings are not chemically inert in marine sediments, and when diluted in sea water, electrostatically bound metal ions are exchanged with calcium and magnesium ions (Hoff *et al.*, 1982). Mine tailings do not release heavy metals into solution when deposited into fresh water, but oxidation by exposure to air significantly increases the availability of such metals (Pedersen, 1983). Indeed faunal species diversity correlates negatively with increasing amounts of metalliferous mine tailings (Ellis and Hoover, 1990). However, the presence of coal particles in the sediment does not necessarily reduce its suitability for infauna as coal itself does not exert a toxic effect – as evidenced by the use of coal

2.6 kg per person per day is thrown overboard from ships (Horsman, 1982). When the input from cargo wrappings and the fishing industry are included, this gives a global estimate of 6.5 million tonnes per year. Macro-litter is increasingly cast up on beaches, beach litter being dominated by packaging materials, predominantly composed of polyethylene (Gabrielides *et al.*, 1991; Lucas, 1992; Haynes, 1997), whose average age is 2.9 years (older materials having been physically degraded into smaller particles). Plastic is essentially not biodegradable (Figure 6.10) and therefore persists in the aquatic environment. Much of the concern about litter arises from its visual impact, and much from the impacts on charismatic groups such as birds, turtles and marine mammals (e.g. Beck & Barros, 1991). Plastic is a major visual pollutant in urban rivers (Figure 6.11), as rising flood waters deposit it on retentive structures, such as vegetation, above the normal particles in the tubes of filter feeding polychaetes (Bamber, 1980).

Apart from its visually intrusive nature, much macro-litter probably has little negative effect on the environment. Eventually physical attrition breaks it down into smaller particles, which are dispersed and incorporated into sediments. Plastic pellets, used in packaging, are now widespread throughout the ocean (Pruter, 1987), their release having occurred during trans-shipment, loading, etc., and accidental losses at sea. Given their small size, low density and inert nature it has been assumed that probably they were neither an aesthetic or ecological problem. They may be ecologically significant in areas where they are abundant enough to displace natural material or be taken by filter feeding organisms in place of zooplankton. Again it is difficult to establish a causal link, as it is only at times of food stress that an effect would become apparent. Chemical analysis of these pellets has now shown them, at least under certain conditions, to adsorb toxic organic molecules including DDE (the breakdown product of the pesticide DDT) and PCBs, and the endocrine disrupter nonylphenol from the sea water (Mato *et al.*, 2001). Concentrations on the pellets are higher than in the surrounding water and increase over time. The contaminants become bioavailable if the pellet is ingested by a marine animal, which may occur if they are mistaken for fish eggs by planktivores, taken by birds, or used as aggregates in the crop.

Some plastic macro-litter is, however, highly impacting. Drift nets made of monofilament nylon are extremely strong, and being almost invisible, are very efficient. Unfortunately they are not selective, and whether still in use, lost or discarded they catch fish, seabirds and marine mammals (see Laist, 1987). It is estimated that the Japanese Pacific salmon fishery took annually 214 500–715 000 birds between 1952 and 1975, and 350 000–450 000 birds between 1975 and 1978. The reported catch of small cetaceans was 13 000 each year, but the actual catch may be much higher (see also **Section 4.3d**).

6.5 Heat

Many industrial processes and thermal power stations (fossil fuel and nuclear) generate excess heat which needs to be disposed of. The use of natural water bodies for cooling in industry is widespread, because not only is such water readily available, it is also an efficient absorber of heat energy. However, in the process of cooling, the water itself becomes warmed. In thermal power stations the fuel is used to heat pure water to steam, which turns the turbines to generate electricity. To gain maximum efficiency from the process, the spent steam has to be cooled back to water to be recycled around the system. This is done using condensers, which in turn are cooled by water drawn from the environment. At coastal sites there is no shortage of water and a continual flow is drawn in, used to cool the process water and then discharged. The volumes used are sufficient to ensure that the discharged water is less than 10°C above ambient. At sites using fresh water, the volume of water available is much lower, and to achieve a return flow only 9–10°C above ambient the water temperature is reduced in cooling towers before being returned to the river.

Natural changes in temperature occur in water, but artificially induced changes – known as calefaction – have three important differences.

- The magnitude of change is normally greater than occurs naturally.

- Natural increases in temperature are normally temporary or seasonal, whereas industrially produced changes are permanent or erratic.

- Natural temperature fluctuations are rarely as sudden or rapid as those produced artificially.

In the sea the effects of the warm water are generally small; the vast volume and the mixing induced by waves and tidal currents rapidly dissipates the warm water plume. In estuaries and rivers, however, the effects are more marked. Once released back into a river a plume of warmer water is produced because, as the temperature drops back to ambient levels, so will the rate at which the temperature drops. Therefore, a plume from a single point source can be measured over long distances. The warm water effluent from power stations in the UK increases river temperatures for several kilometres downstream by as much as 6°C and can delay the onset of autumnal cooling (Langford, 1983).

6.5a Effects of increased temperature

The environmental effects of thermal pollution can be considered at two levels – the environment and the organism.

The effects of increased temperatures on organisms include both direct physiological changes and effects on biological cycles. For plants, invertebrates and poikilothermic (cold-blooded) vertebrates an increase in temperature leads to an increased metabolic rate. This change may be sufficient to be lethal, but even if it is not the increased metabolic rate will impose greater metabolic costs and so may exceed the organism's ability to secure sufficient food or other resources. In tropical regions species may live close to their thermal limits, while in temperate regions some species may be well able to cope with the increased temperature for most of the year, but in summer the additional stress may be too great. This effect may be exacerbated if oxygen levels are low (see below), as the greater metabolic rate will also require additional oxygen to support it.

A rise in temperature increases the rate of chemical reactions. Toxins react chemically with the body. As most chemical reactions increase with temperature, they become more toxic in warm water. Increased temperature also reduces the concentration of dissolved oxygen. This is not normally a problem in clean systems as the turbulence of the discharge restores the oxygen content, but in very enclosed creeks with high BOD,

anoxia can occur. Heated effluent discharge is particularly problematic, therefore, when combined with organic pollution (see **Section 6.2a**).

There may be behavioural effects of heated effluent discharge. Fish may aggregate in the warm water plume, leading to increased pressure on their food supplies.

6.5b Ecological effects of heated effluent

Many species in temperate regions have distinct annual cycles; these may be cued by environmental temperature and so changes in the external temperature regime may impact on the reproductive cycle, for example, leading to a failure to spawn at the correct season.

6.5c Invasion by alien species

Permanent warm water flumes are home to subtropical species in many parts of the world. The tropical fish living in a canal in northwest England are considered in **Section 8.3a**. Also in northern England are fig trees living along the banks of the River Don in Sheffield. For most of the 20th century, the River Don was permanently warmed by discharges from steel works. Human sewage frequently contains large numbers of fig seeds, which are viable and able to germinate. Fig seedlings are, however, intolerant of frosts, which kill them in most parts of England. Along the River Don, however, a combination of sewage (and therefore fig seed) inputs and a narrow band of river bank kept free from frosts by the warm water allowed figs to germinate. The steel works have now closed down and the river runs at more natural temperatures, so no more seedlings will establish, but the mature fig trees already present are now large enough to withstand frosts and will survive until senescence.

In Southampton Water (southern England) the discharge from Marchwood Power Station into the estuary is believed to have allowed a population of the American clam, *Mercenaria mercenaria*, to become established and breed in high enough numbers to support a fishery. When the power station closed, temperatures in the estuary dropped by around 2°C; while the adult clams survived, the temperatures were not high enough for them to breed. This in turn caused the eventual decline of the clam fishery.

6.5d Chlorination

In estuaries and coastal waters, fouling of the cooling water pipes by sessile organisms, and in particular the mussel, *Mytilus*, means that in many power plants the inflowing water is dosed with chlorine to kill the mussel larvae. This means that the discharged effluent is not only warm, but also contains chlorine. The exact form is highly variable – chlorine as hypochlorite, as chloride, as chlorine or reacted with organic material to form an organo-chloride. There is no evidence for any adverse effects of the chlorine residues away from the discharge, but concern has been expressed about the general undesirability of releasing active chlorine into the environment and the possibility of generating organo-chlorides in particular.

6.6 Acidification

Acidification occurs where the atmospheric deposition of strong acid anions (SO_4^{2-} and NO_3^-), accompanied by hydrogen (H^+) ions, exceeds the buffering capacity of the water or the catchment soil. These ions derive from industrial activities, particularly those burning fossil fuels, which release them into the atmosphere. Their normal means of transmission to surface waters is in rainfall; precipitation whose pH has been lowered by the presence of these pollutants is referred to as **acid rain**. Once in the soil or water, acid anions react with bases, which in turn neutralise the acid. The ability of soil to neutralise acid is its **buffering capacity**, and is determined by the con-

centration of bases such as calcium (Ca^{2+}) and magnesium (Mg^{2+}); the greater the concentration of these ions, the higher the buffering capacity and the greater the acidity that can be absorbed without showing any detrimental effects. Sea water is extremely well buffered, as are fresh waters draining limestone. Eutrophic waters are also well buffered, because the nutrients causing the eutrophic status are effective buffers. The interaction between acid and base also neutralises the base, however, so their replacement is required, normally from bedrock weathering.

Acidification of surface waters occurs where acid precipitation falls onto rocks which are hard and therefore weather more slowly than buffering ions are leached. Eventually the buffering capacity will be exhausted, so the soil suffers reduced pH and excess acid water runs into watercourses. The problem is compounded further by the slow rates of erosion of hard rocks; their persistence means that they often form upland areas, which themselves receive more rainfall than adjacent lowlands.

Acidification is most closely associated with areas downwind of industrial activity. It is, therefore, a regional rather than a global problem, but one whose effects are spreading more widely as industrialisation expands. Many soils overlying susceptible geology have buffering capacity to the extent that they can absorb pollutants for many decades without showing any major effects, but then acidification will suddenly occur once this buffering capacity has been exhausted.

6.6a Direct effects of acidification

Lowered pH has direct physiological effects on aquatic organisms, particularly those to do with ion exchange (Rosseland & Staurnes, 1994). Freshwater animals require blood concentrations of ions, such as sodium and chloride, many times higher than occur in the surrounding water, so they employ active ion exchange across their gill membranes to maintain the correct balance. High concentrations of H^+ ions interfere with this process.

Elevated H^+ ion concentration does not occur in isolation, and one of its most serious biological effects relates to its influence on other metallic cations. Aluminium is toxic to many aquatic organisms, and is very abundant in soils that form over granitic rocks, themselves highly susceptible to acidification. Under acid conditions, aluminium ions are released into solution in the Al^{3+} form, which is both the most soluble and the most toxic form of aluminium. Therefore, many of the direct toxic effects of acidification are caused not by the acid itself, but by the aluminium that it releases. The aluminium readily forms complexes with organic compounds, whose presence can reduce its effects (Weatherley *et al.*, 1988). However, it forms similar complexes with phosphates, to the extent that phosphate can aggregate into particles that sink out of the water column. This may cause nutrient depletion, although nutrient levels are normally so low in lakes susceptible to acidification that it is difficult to detect in the field (Schindler *et al.*, 1985).

The effects of acidification differ among groups of organisms. Fish are particularly susceptible and their disappearance is an early indication of acidification. Bivalve and gastropod molluscs have a further complication in that their shells are heavily calcium-based, and therefore dissolve, while many crustaceans have calcium-based exoskeletons. Most mayflies are sensitive to disrupted ion exchange. Some groups, however, have a high tolerance, particularly certain members of the insect orders Plecoptera (stoneflies), Trichoptera (caddis flies) and Diptera (true flies).

6.6b Ecological effects of acidification

Acidification has normally happened before background information was available, so we have little direct evidence for ecological changes. Sediment cores from lakes have, however, revealed changes in phytoplankton species composition at around the time of acidification (Renberg & Battarbee, 1990), while anecdotal evidence for decline in fish abundance is often an early warning of problems (Stoner *et al.*, 1983). Much of our evidence for ecological effects of acidification derives from manipulative studies in which pH has been artificially lowered. In the lakes region of Ontario, Canada, lake pH has been experimentally reduced for up to eight consecutive years, so that interannual effects can be monitored (Schindler *et al.*, 1985). Food chains were altered considerably, as key species were eliminated. Fish, which occupy many of the higher trophic levels, persisted in acidified waters, but stopped reproducing, so that eventually they died out. Crayfish (*Orconectes virilis*) populations were severely affected, first by incomplete exoskeleton hardening after moulting and, after several years, by heavy parasite loads. Small phytoplankton species were replaced by larger ones, while larger zooplankton species were replaced by smaller ones, so grazing efficiency was reduced. Despite these changes, however, primary production, decomposition rates and nutrient levels remained unchanged. This contrasts with findings from lakes that had undergone non-experimental acidification. It is important to emphasise that studies such as this have generally been carried out only with addition of H^+, and not of aluminium, so their application to other situations must be treated with caution.

6.7 Cleaning up pollution

There are two approaches to reducing the effects of pollution. The first and most desirable is reduction, ensuring that the pollutant is not discharged in the first place. Where this is not possible, treatment is required. However, both of these approaches entail costs, and their adoption

Figure 6.12 *(a) The stages in a traditional sewage treatment works. The main objective is to release an effluent with a lowered BOD such that final degradation in the receiving environment does not cause major ecological effects. Biological oxidation may be a one or two stage process and, depending on the degree of treatment required, effluent may be discharged after primary settlement or one or two stages of oxidation. Chemical treatments and disinfection are only used in certain circumstances. (b) Typical primary settlement tanks on a sewage plant discharging into an estuary. (c) Fine trickle filter beds typically form part of the tertiary treatment regime. (Photos by C. Frid)*

(b)

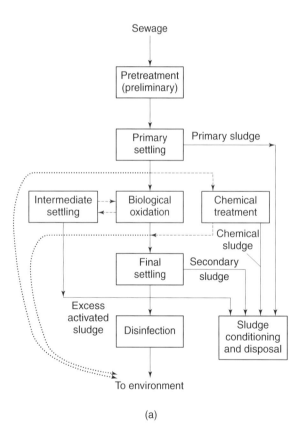
(a)

(c)

usually requires legislation, which is becoming more and more a controlling influence on pollution and its mitigation.

6.7a Reduction

The simplest way in principle to tackle pollution is to reduce waste or, failing this, to ensure that it is disposed of in a way that will cause least damage to the environment. Industrial processes are becoming more efficient at recycling wastes to extract potentially valuable components, reducing discharges of heavy metals into the environment, while the addition of extra components to the industrial process can strip potential contaminants before discharge of the residues. An example of the latter is flue gas desulphurisation, which removes acid-forming sulphates from power station discharges, therefore reducing the problem of acid rain. The problem of macro-litter at sea is decreased considerably if ports provide shore reception facilities for ships' refuse, a process enhanced considerably by recent legislation. The UN MARPOL convention now requires this, along with a ban on marine dumping within 12 miles of land; only readily biodegradable material is permitted to be dumped outside the 12 mile limit.

The example of sewage

In some cases, such as production of sewage, reducing the amount of waste is impossible, but treatment before discharge provides a suitable compromise, by reducing the amount released into the general environment. It is technically feasible to treat sewage to the extent that the water produced at the end is of very high quality and fit to drink, by passing it through several levels of treatment (Figure 6.12). Reducing pollution by means such as pre-disposal treatment requires efficient functioning of the collection and treatment infrastructure. Many sewage systems incorporate combined sewage overflows (CSOs), mechanisms that allow sewage pipes to overflow into rivers, estuaries and the sea in the event of major flooding or heavy rain. These act as safety valves, to ensure that sewage systems are not damaged by being overloaded, but allow raw sewage to enter the environment even where several levels of treatment are normally employed. The most basic form of treatment is removal of large particles – preliminary treatment, a level through which all sewage discharged into fresh waters in the UK should pass under normal circumstances. However, discharges via a CSO do not even have these gross solids removed. Figure 6.11 shows the effect this has on rivers: 22% of all solid effluent particles in the River Taff in South Wales were feminine hygiene products (tampons and sanitary towels), which enter the river only during relatively infrequent flood events (Williams & Simmons, 1999).

Pre-disposal treatment also produces some polluting residues that require disposal. Sewage sludge can be deposited in landfill sites, used as fertiliser and soil conditioner, or incinerated. Until 1998 the UK used a further method of disposal, that of dumping at sea. In the 1990s the UK disposed of 11.5 million tonnes wet weight of sludge a year at 12 marine disposal sites (Figure 6.13). This practice is now discontinued, but its effects on the dumping sites are expected to persist for some years.

Technical solutions

Adopting appropriate technology can considerably reduce inputs, a process well illustrated by reductions that have been achieved in oil discharges from tankers. When a tanker unloads, oil inevitably clings to the sides of the tanks. In a 200 000 t capacity tanker this 'clingage' may be as much as 800 t. On the return voyage the tanker, for safety and efficiency reasons, needs to load ballast. Initially this consisted of pumping sea water into the now empty oil tanks, and discharging it

Figure 6.13 *Marine sites around the UK used for the disposal of primary sewage sludge up until 1998. The main sites were: 1, Thames estuary (5 million t yr $^{-1}$), 2, Liverpool Bay (1.5 million t yr $^{-1}$), 3, Clyde (1.5 million t yr $^{-1}$), 4, Tyne/Tees (0.5 million t yr $^{-1}$).*

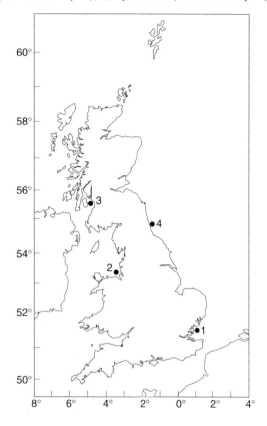

back into the sea, along with some of the oil with which it inevitably became contaminated, as the tanker approached its port of loading. Modern developments in tanker design have reduced this problem considerably – for example, the **load on top** method, in which only uncontaminated ballast water from the bottom of each tank is discharged. Water from the interface is pumped to a slop tank where it further separates, allowing more water to be drawn down. The fresh oil is then loaded on top of the residue from the previous cargo. More modern tankers have segregated ballast tanks, so that oil and water never come into contact.

Biological solutions

For many waste creation activities the technology involves biological treatment prior to discharge of the waste, effectively bringing processes which formerly occurred in the environment after discharge under the operators' control. Much of the sewage treatment process exploits natural processes, such as bacterial activity in filter beds. These require relatively little maintenance, because bacterial biomass is controlled by predators, effectively creating an ecological community within the filter bed. The same principle of exploiting natural biodegradation mechanisms is used in constructed wetlands.

Constructed wetlands take advantage of the fact that many wetland plants are effective at stripping pollutants from water passed through them. Species such as reeds (*Phragmites australis*) are very productive and can absorb nutrients, while the physical structure of the reedbed traps particulate matter, incorporating particles and toxins into sediments and retaining biodegradable matter so that it decomposes *in situ*. Their efficiency and low maintenance makes them attractive, and they are now being widely used to cleanse waste water, from small domestic sewage inputs to industrial plant residues. A further advantage of constructed wetlands is that a nature conservation element can be included (Worrall *et al.*, 1997). Initially conservation value of the wetland itself was considered to be incidental to its major function, but careful design to optimise habitats increases its wildlife value, as well as its treatment efficiency. Increasing habitat diversity can enhance biodiversity within a wetland (Figure 6.14), but for many species, especially birds, the size of the wetland may be the major consideration (Guillemain *et al.*, 2000). The nature of the waste being treated must be taken into consideration; if, for example, a wetland is being used to strip bioaccumulating toxins from water, then these may, in turn, be incorporated into tissues of animals attracted to the wetland. If the wetland is small, it may provide a

Figure 6.14 *A constructed wetland at Slimbridge Wildfowl and Wetland Trust reserve. Although designed primarily for waste water treatment, conservation was an important consideration, stimulating the complexity of the design (from Worrall* et al., *1997).*

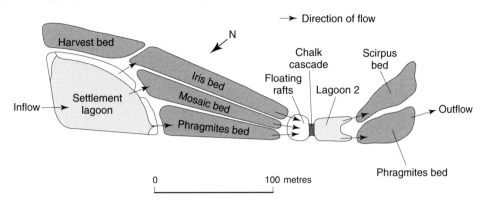

source of food for predators in the wider environment, dispersing bioaccumulators away from the point of treatment. On the other hand, non-toxic pollutants, such as nitrates, may enhance production to the extent that they actually benefit species of conservation importance (Worrall *et al.*, 1997).

Best practical environmental option

However much the production of waste is minimised, ultimately there will always be some wastes to dispose of. There must therefore be a means of assessing the relative merits of alternative disposal and treatment options. For example, sewage sludge may be dumped at sea, placed in landfill or incinerated. Deciding which is the best option is difficult; the marine environment benefits from landfill disposal, but this impacts terrestrial environments. Incineration is good for both, but increases atmospheric CO_2 and so contributes to global warming.

The framework adopted to address such issues is the BPEO (best practical environment option). This seeks to make rational decisions about the total environmental impact of disposal and to seek the combination of treatment and final disposal that minimises adverse environmental effects. In practice the approach used in many countries also includes an explicit economic consideration – the cost of the BPEO may be judged to be too great. Therefore, the procedure is often adopted of seeking the BATNEEC (the best available technology not entailing excessive cost).

Sewage treatment using the processes illustrated in Figure 6.12 is expensive, and a pragmatic approach to treatment is normally adopted, determined by cost and the perceived tolerance of the receiving water body to pollution. In the UK, for example, in 1992 almost all discharges to fresh waters received at least secondary treatment, but 87% of coastal discharges and 58% of estuarine ones were to preliminary or primary levels only and needed upgrading (Table 6.2). This derives

Table 6.2 *The degree of treatment received before discharge by marine outfalls serving large communities (population >10 000) in the UK in 1992 (almost all freshwater discharges now have secondary treatment)*

Treatment	Coastal (%)	Estuarine (%)
None	37	11
Preliminary	50	24
Primary	11	23
Secondary	2	42

from the general perception that the sea is able to deal with sewage more effectively than fresh waters, because its large volume and great mobility facilitates dilution and dispersal.

In the developing world, even treatment for discharge into fresh waters may be prohibitively expensive, and the use of waste stabilisation ponds is widespread. These are simply ponds into which waste water is discharged and then allowed to settle, removing sewage sludge (by settling and bacterial activity), industrial wastes (degradation and precipitation) and nutrients (algal growth). Although they may require considerably more land area than more technologically advanced treatments, they are inexpensive to construct and low in maintenance costs (Pearson, 1996). Their

efficiency is good; for example, the Dandora pond complex serving Nairobi (Kenya) reduces BOD by over 90% (Pearson *et al.*, 1996). An unusual benefit of this site is that waste water discharged from the treatment works into the Dandora River is considerably cleaner than the river water itself, to the extent that downstream of the effluent discharge, BOD is approximately 20% of its level upstream. Effluent treatment in ponds can be combined with other activities, including fish production, either in ponds receiving raw sewage directly, or those receiving algae-rich effluent from the first stage of a multi-pond system. The role of treatment ponds in aquaculture is considered further in Box 5.2.

Box 6.2 Possible responses to oil at sea

There are four basic options for dealing with an oil spill (Tramier *et al.*, 1981).

1 **Monitor only**. This allows for natural dispersion to occur. This option is often the best environmentally but is often seen by the press and public as an attempt by the polluter and the authorities to save money by doing nothing. There is usually considerable socio-political pressure, therefore, to be seen to be acting to combat a spill.

2 **Disperse**. Slicks can be broken up, creating a greater surface area for natural weathering and biodegradation. This also reduces the visual impact of a spill. Dispersed oil is often more biologically available and the greater surface area may increase the speed at which toxic compounds enter the water column. Oil can be dispersed purely by physical means – ships towing 'breaker boards' through a slick to break it up – or more often by chemical means. Chemical dispersants, sprayed on the slick from vessels at sea or by low flying aircraft, aid its break-up when agitated by wave action.

Dispersants should not be used near fish breeding grounds or other areas of high ecological value, as dispersed oil is much more bioavailable and toxic than non-dispersed oil. Modern dispersants are 'low

toxicity' in themselves but early formulations were industrial detergents which were more environmentally damaging than the oil they were supposedly cleaning up.

3 **Contain and recover**. As oil floats on the surface, other floating structures can be used to contain it. The most common devices are oil booms, which float on the surface with a curtain suspended below. In calm conditions booms are quite effective, but in winds or currents they are difficult to handle and oil is splashed over or pulled under the boom. Once the oil is contained it can be recovered. However, containment does not prevent the soluble fraction from leaching out or the volatiles from evaporating, so the oil continues to weather and become more viscous and difficult to handle. Furthermore, emulsification continues to increase the volume of material to be dealt with. Techniques for recovery from the sea surface include pumps, skimmers/ weirs and mops.

Contain and recovery may be a viable option in enclosed water bodies and if the equipment is readily available. However, even under optimum conditions, recovery rates are only 10–15% of the spilt oil.

Box 6.2   continued

Oil spill containment and recovery devices. (a) Various types of floating boom contain oil and prevent it impacting key features. (b) Weir skimmers use the tendency of oil to float on water to separate and recover it from the water surface. (c) Oil mops use the tendency of oil to stick to certain materials (oleophilic) to remove the oil.

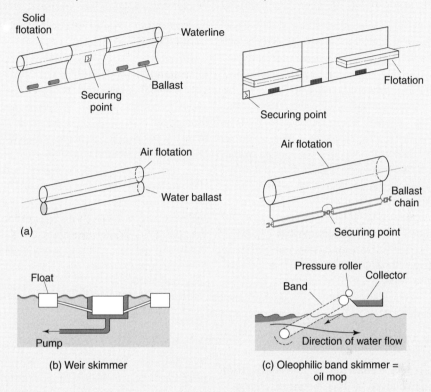

(a)

(b) Weir skimmer

(c) Oleophilic band skimmer = oil mop

4   **Shore clean-up methods**. Once oil is on the shore, some ecological and probably amenity damage has occurred. The critical decisions concern the likelihood of the oil moving off on the next tide to contaminate a further area and the ecological effects of clean-up. In most cases clean-up will do more ecological damage than the initial impact of the oil. However, if the area is of high amenity value or the oil is likely to impact other shores, containment and clean-up are indicated.

The options for clean-up of oil once it is on the shore include:

■ **Removal of oil**. Physical removal of oil by suction pump or oleophilic pads, low or high pressure washing, or steam jetting. If the oil is fresh and fairly fluid it can be pumped from natural collection points in gullies or absorbed onto pads. These are labour-intensive but relatively low impact clean-up techniques. Oil can be washed from hard substrata (e.g. rock, harbour walls), for which weathered oil requires higher temperatures and/or pressures. Low pressure washing with sea water has minimal impact on the biota, but high temperature or pressure washing will effectively denude the shore. The oil/water mix can be collected in pits down shore (on sand or mud shores) to be pumped away, or contained in booms at the low water mark to be picked up from the water surface by skimmers.

Box 6.2   continued

■ **Removal of oiled substrata/biota**. When extensive areas of sandy shore are impacted, the preferred option may be removal, by bulldozer, of the affected sand. This can be cleaned and returned or simply disposed of. On rocky shores macro-algae often trap oil in their fronds. If these are cut off, leaving the holdfast intact, the algae quickly regenerate and the oiled fronds can be sent for disposal.

■ **Dispersant use**. While dispersant spraying on the high seas can be effective, on the shore there is less volume into which to dilute the dispersed oil. In general dispersed oil is more toxic than oil on its own – the dispersant tends to make it more readily assimilated. Dispersant use should only, therefore, be considered for high amenity value shores. There are now formulations available which aim to spray a mix of dispersant, bacteria and nutrients on to the oil – the idea being to disperse the oil into small droplets, to seed these with oil degrading bacteria and to supply them with nutrients. Other formulations aim to disperse the oil off the shore, but then

to rapidly stop working. The oil forms into a new slick, which can be contained by booms and skimmed off the water surface. Such techniques may have application under certain, ideal, circumstances but their general applicability has to be questioned.

In assessing the options to implement a response to a spill, the authorities need to consider ecological and socio-economic factors, including the threat to fishing/fish spawning grounds, shellfish beds, fish farms, water abstraction points, amenity/tourist areas, boat moorings and the natural environment (Table 6.1). The threat from both the oil and the various clean-up options must be balanced. Oil spill contingency plans attempt to do this in advance of a spill, so that such critical decisions do not have to be made in the heat of an incident. Such documents describe the features of the area and the recommended responses, often differentially depending on season (breeding season, tourist season).

6.7b Treatment

Treatment involves removing a pollutant from the environment after it has been discharged. Normally this is neither feasible nor economical. It should never be considered as mitigation for deliberate discharges, as these can be more effectively controlled at source. It is, however, a mechanism that is sometimes practical for certain accidental discharges, or those which occurred before their environmental effects were fully understood.

Oil spills

Despite the reductions achieved in oil discharges, spillage can still occur when oil tanks are accidentally ruptured. Tankers were originally built with a single sheet of steel between the cargo and the sea, so that even a relatively minor collision or

grounding would breach the oil space and result in a spillage. There is now a requirement that all new tankers are built with a double skin system, so breaches are less common. However, they do still occur, so methods of dealing with them need to be available (see **Box 6.2**).

Liming

Liming is a mechanism for combating the effects of acidification in surface waters. It works on the very simple principle that the problem is caused by shortage of buffer in the catchment, so buffer is added, in the form of calcium carbonate.

Restoration

In other areas attempts have been made to restore, by active intervention, areas damaged by pollution. This has opened up a whole new field of 'restoration ecology' (Cairns, 1988; Frid & Clark, 1999). In the marine environment, intervention is not normally required at most open coastal sites because once the damaging activity has ceased recovery begins (Herrando-Pérez & Frid, 1998; Barnes & Frid, 1999; Hawkins *et al.*, 1999). However, in environments that are semi-isolated (Hawkins *et al.*, 1999) or with major structural elements (Clark & Edwards, 1999) recovery requires an active initiation to overcome the isolation or to provide essential physical structure.

6.7c Legislation

When environmental problems ensue from waste disposal, these need to be dealt with, but there will be an economic balance between degree of treatment preferred and the cost. For a small cost, a relatively large reduction in the waste can be achieved (Figure 6.15), but to produce a progressively 'cleaner' waste, increasingly greater amounts of money must be spent. Therefore there will always be, in a free-market system, a powerful economic argument for minimal treatment

Figure 6.15 *The typical relationship between the relative cost of cleaning up an effluent compared with the degree of clean-up achieved by a biological waste treatment facility.*

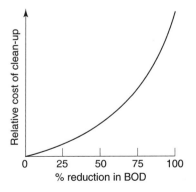

prior to disposal to the environment. Pollution management will always have to work against this background, although some management techniques attempt to redress this by direct intervention in the economics, for example by imposing pollution taxes or charges. Mechanisms such as these have benefited from major changes in society's attitude to the environment. Over the past 30 years, legislation and controls on polluting activities have been enacted in most countries (see **Box 6.3**).

Approaches to pollution regulation

There are two possible approaches to the regulation of pollution discharges.

1   **Uniform emission standards** (UES). The regulator sets a flat limit for each type of installation, which applies wherever it is located, or the use of the water into which it is discharging. The values used for the UES are usually derived on the basis of what is technically achievable at reasonable cost. Most EU states use this approach.

2   **Environmental quality objectives** (EQO). In contrast to the UES approach, the UK and US adopt an EQO approach to the management of water bodies. For each body of water, the discharge limits are set in relation to: the capacity of the water body to disperse or assimilate the pollutant, the uses to which the water body is put, and other inputs to it. Thus different discharge limits will apply to similar installations depending on their location.

The EQO is the water quality objective that the regulatory authority is aiming to achieve. For this to be achievable, tables of environmental quality standards (EQSs) that are commensurate with the categories of water use are required. EQSs may be standards required for drinking water abstraction, bathing, non-

Box 6.3 Pollution and clean-up of the North Sea

The North Sea is a relatively enclosed shallow coastal sea, receiving an inflow of oceanic water from the North Atlantic into its northwest sector and a smaller inflow via the English Channel in the south west. It receives freshwater inflows from the Baltic and all the major rivers draining northeast Europe. It is highly productive and a major site for fishing, supporting around 5% of the annual global fish catch, despite its small area. The North Sea is also a major site for extractive industries – fishing, oil, gas, and sand and gravel dredging, while the Straits of Dover and the southern North Sea are amongst the most intensively trafficked areas of the global ocean. Historically, each of the waterways flowing into the North Sea was used for waste disposal and there were also direct discharges. By the 1980s many countries were expressing concern over the levels of contaminants in the sea and the possible implications for human health, the health of exploited fish stocks and the ecosystem. The principal concerns were eutrophication, conservative pollutants, oil and the effects of fisheries (see Salomons *et al.*, 1989).

Atmospheric and riverine inputs make up a large proportion of the wastes entering the North Sea. These therefore derive from a large area and not simply from the coastal zone. In recognition of the international nature of these problems a series of international initiatives have been developed, including the Oslo Commission (which regulates dumping and discharges

from ships), the Paris Commission (which regulates discharges from land based sources) and initiatives by the European Union. To further this international management regime, the environment ministers of all the states bordering the North Sea have met periodically in what are known as the Ministerial Conferences on the North Sea. While their agreements are not legally binding on the nations, they provide a forum for agreement and intense political pressure to act on implementing agreements.

One of the actions of the first ministerial meeting was to set up a group of international experts to provide a report on the status of the North Sea – the *North Sea Quality Status Report* (North Sea Task Force, 1993). This exercise has been repeated at intervals to provide a series of reports which summarise the current health and provide a check on progress in improving conditions in the North Sea.

Other initiatives of the ministerial conference have seen major reductions in inputs of heavy metals and other highly toxic substances; for example, the 'red list' (which lists the most highly toxic wastes whose removal from discharges is of highest priority), the phasing out of sewage sludge dumping at sea (agreed at the third ministerial conference in 1990 before the EC Urban Wastewater Directive was published), the end to ocean incineration in the North Sea, and initiatives to curb inputs of nutrients from diffuse sources.

---

immersion water sports, immersion water sports, shellfish culture and so on. For each of these there is a table of acceptable limits for pollutants. For example, in the European Union the standards to ensure water is fit for bathing are based on the EC Directive (see **Box 6.4**). A coastal sewage discharge near a bathing beach would need to be treated to ensure that high standards are met on the bathing beach, while one to a stretch of coast used for non-immersion water sports rather than bathing would have to meet less rigorous standards.

Pros and cons of the UES and EQO approaches

There are several important disadvantages of the UES method. Where there are multiple discharges to a single receiving body or when there are diffuse sources, no account is taken of the total input. For example, the River Rhine has many factories discharging effluents into it, each of which meets the UES set by the various states responsible for its catchment, but the total number of discharges means that the river is grossly polluted in places. Furthermore, it may not be the most

Box 6.4   EC Bathing Waters Directive

The 1976 European Directive on the quality of bathing waters required all members of the European Community to designate sites at which bathing traditionally occurred and by 1986 to have introduced measures to ensure the quality of the water at those sites. The UK has over 440 beaches officially designated as bathing beaches under the Directive, all except one being marine sites.

European Community law requires these to meet certain bacteriological, aesthetic and other physico-chemical water quality parameters during the bathing season (May to September in northern Europe). For example, the authorities are required to take 20 samples in the bathing season; total coliforms density should be below 10 000 per 100 ml, and faecal coliforms less than 1000 per 100 ml in 95% of the samples. Limits are also set for contamination with mineral oils, phenols, surface active substances and for the pH, clarity and colour of the water. Viral standards were also set but due to technical problems they have proved to be difficult to enforce.

There is a similar directive governing the quality of waters used to harvest or rear shellfish. The European Community Shellfish Waters Directive requires that each member state designate certain coastal and brackish waters as needing protection or improvement in order to support shellfish life and growth. These areas have to meet certain criteria for physical and chemical water parameters and bacteriological standards for the shellfish flesh.

cost-effective if, for example, it forces a particular technology on the discharger. Finally, it provides no framework in which to regulate discharges from diffuse sources. The method does, however, have two advantages over the EQO approach: it is easy to apply and regulate, and it minimises inputs from point surfaces.

Recently, a new framework for pollution management has been advocated, the **precautionary principle**. This puts the onus on the potential discharger to prove no environmental effect before a discharge is licensed. In reality, this has tended to lead to the simultaneous application of both absolute discharge limits (a UES approach) and a consideration of the effects in the receiving environment (an EQO approach).

The trans-boundary dimension

Much of the pollution entering the larger fresh waters and all of the marine systems of the world has a trans-boundary dimension. Pollution within a single water body may come from sources under many different jurisdictions, while that deriving from one country may have its worst effects in another country. Regulation of trans-boundary effects is particularly problematic, as international agreements are required. There are a number of international bodies that provide frameworks for regulating pollution, including the European Union, the Ministerial North Sea Conference, the UN Environment Programme, and various international conventions on specific pollution issues. As with all international agreements, however, there is a considerable political dimension to the negotiations and these can, at times, override the environmental science.

The European Union has the advantage that its directives are legally binding on member states. UN agreements require ratification by a certain proportion of the membership before they become binding. Work on the UN Convention on the Law of the Sea began in 1958, but it was not until 1994 that it entered into force. Examples of other international agreements on marine pollution include the London Convention on Dumping, the Oslo Commission on North Sea Dumping and Discharges, the Paris Commission (North Sea direct discharges from land), the UN Law of the Sea, and the UN MARPOL agreements. The Oslo and Paris Commissions have now merged to form 'OSPAR Com'.

Within fresh waters, the most effective management of pollution is on a catchment basis, whereas political boundaries are often marked by rivers, separating the catchment into two or more administrative regions. In 1974, water resource management in England and Wales was restructured on a catchment basis, without regard to political boundaries. This strategy is now being widely adopted elsewhere. Most large rivers have an added complication, however, in that their catchments cross international boundaries, requiring co-operation at a diplomatic level if the catchment is to be managed effectively. Some limited trans-boundary agreements are in existence, particularly in Europe, and the United Nations issued a Convention on the Protection and Use of Transboundary Water Courses and International Lakes. The first large scale implementation of cross-border management is, however, taking place only as a result of the European Union's Water Policy Framework Directive (European Commission, 2000), requiring that any river basin which crosses boundaries within the EU is managed as a single unit.

6.8 Summary

- Pollution arises when substances (contaminants) present at higher than natural levels cause an effect in humans, biota or ecosystem functioning.

- Biodegradable wastes – sewage, oil – are eventually broken down in the environment, but the capacity of the environment to assimilate them must not be exceeded.

- Conservative pollutants are diluted and dispersed, but not removed from the environment. Bioaccumulation and biomagnification can mean that concentrations in the biosphere are above those in the discharge.

- Recent years have seen major advances in dealing with many pollutants, but new threats have been recognised, e.g. endocrine disrupters, plastic litter.

- The restoration of degraded environments is using ecological science to unravel some of the damage done by man.

- Cost-minimisation economics leads to environmental degradation as dumping is always cheaper than treating a waste. Therefore pollution control involves regulation and enforcement.

- Marine pollution in particular is trans-boundary and so requires international responses to its management.

Chapter 7   Extracting Mineral Resources from the Aquatic Environment

7.1 Introduction

Extraction of mineral resources impacts many aquatic environments. Some of the most severe impacts are those associated with pollutants derived from mining operations. Acid mine drainage, for example, is a widespread phenomenon whereby tailings from active or abandoned mines, highly acid and with high concentrations of heavy metals, contaminate river systems into which they drain. However, such perturbations are caused by activity not directly connected with the water body, other than as a conduit for the disposal of waste, and are best considered as pollution (**Chapter 6**). In this chapter, in contrast, we consider processes involving the direct exploitation of the aquatic environment itself for mineral resources.

It is possible to distinguish two different types of mineral that are extracted from aquatic environments. The first are those that are created by the water body itself, and are therefore intimately associated with aquatic environments. These include peat, derived from plant remains deposited in wetlands, and gravels and manganese nodules created on the bed of a water body by the sorting and aggregating activities of the water. The second group covers those minerals whose presence is incidental to the water body, but whose extraction involves perturbation of the aquatic environment because they are situated in geological layers beneath its bed. These are offshore reserves of oil and gas, whose extraction requires different technologies to those extracted from land, but if managed correctly should not necessarily lead to detrimental ecological effects beyond localised physical disturbance.

In addition to the resources highlighted here, various other minerals are extracted from aquatic environments. Salt is extracted from sea water by evaporation in many coastal areas, and industrial extraction of other minerals, such as magnesium and bromine, also occurs. Gold panning is an important, if relatively small scale, industry in many rivers, while oyster shells are mined in San Francisco Bay, California, for cement making; the effects of these processes are similar to those of sand and gravel dredging. Various potential mineral resources have been highlighted from the deep ocean bed, including barite nodules, polymetallic sulphides, titanium, gold and platinum (Resources Agency of California, 1997), although none are currently commercially viable.

7.2 Sand and gravel dredging

7.2a Aquatic sediments

Almost all aquatic environments support at least some areas of soft sediment aggregation on their beds. Sediments are discrete, unattached particles, including fragments of bedrock, soil, detrital particles and shell fragments. They are heterogeneous, ranging from fine silts (with a particle diameter less than 63 μm) to cobbles and boulders with particle diameters measurable in centimetres. Each size fraction will respond differently to a given flow regime, so that water moves them differentially. Even very still water is likely to have fine silts in suspension, whereas large particles will be moved only by very rapidly flowing water.

The different physical properties of different particle sizes allow them to be sorted by water movement. A rapidly flowing coastal current will carry all but the largest particles but, as water flows into a sheltered area, progressively smaller particles will settle out as the water loses its momentum. Rivers, too, grade sediments in this way, with larger particle sizes dominating rapidly flowing areas while finer sediments occur in pools and sheltered areas close to the bank. Many rivers in northern Europe and North America in particular run across large terraces of alluvial gravel. These were laid down by glacial ice during the Pleistocene, or even by the river itself at the end of the last glaciation, when it was swollen by glacial meltwater many times its current maximum size and therefore able to deposit sediment over a wider area than today.

Sediment aggregations on the bed of a water body have a large volume of interstitial space, sometimes over 30% by volume. Fine particles in water absorb metal salts and organic solutes from the water column, and then carry them into the bottom sediments, where they settle within these spaces. Sediments therefore act as a sink for contaminants from the water and its catchment.

7.2b Extracting gravels

Gravels are widely used in the construction industry. The natural deposits of sediments in the aquatic environment are an attractive source of aggregates for industry. The fact that they have been sorted by the flow regime means that it is relatively straightforward to win sediments of a particular type. The amount of material dredged for use as aggregates varies, depending on the economies and construction needs of the surrounding nations. For example, during the 1970s around 50 million tonnes of aggregate was extracted from the North Sea each year, but with increased construction work in the 1980s it rose to around 150 million tonnes per year (de Groot, 1996b).

Gravels derived from glacial processes are the most straightforward to extract, as they occur on land, in highly accessible features relatively devoid of mineral or soil contamination, and are well sorted, ensuring little need for sediment grading and washing (Knight et al., 1999). Where such deposits are abundant, there is little need to extract from active water bodies. However, in most parts of the world the best quality aggregates are obtained from active river or coastal water bodies. Extraction from shallow or braided river systems is relatively straightforward, particularly if there is a seasonally predictable period of low flow. Coastal sediment extraction is more complex, because of the permanent nature of the aquatic environment and the depth from which sediments may need to be removed, and therefore the large scale use of marine sediments is a relatively recent development. In the UK, for example, marine sediment extraction in 1950 was confined to a small number of sites in the lower Severn Estuary, whereas now it is widespread off the east and south coasts of England. However, the industry is still relatively small: the extraction from the North Sea represents impacts on around 0.03% of the seafloor, compared with fishing, which impacts around 54% of the seafloor each year, or shipwrecks which cover around 0.05% of the seabed.

River sediments

Gravel is mined from river sediments in three ways. **Dry-pit mining** refers to the digging of pits above the water table, in ephemeral river channels and exposed bars, using conventional mining machinery such as bulldozers, scrapers and loaders. **Wet-pit mining** involves the use of a dragline or hydraulic excavator to remove gravel from below the water table, either in a pit dug from a dry surface or from the active river channel. **Bar skimming** or **scalping** refers to a process whereby the top layer is skimmed off a gravel bar, keeping the excavation above the water table.

Marine sediments

Marine sediments are almost always extracted directly from the active water body, generally from a dredging boat. Sediments may be obtained by bucket or grab dredging (Figure 7.1a) or by suction dredging, in which a vessel may be anchored or the dredger may move, towing the dredge head over the seafloor (Figure 7.1b, c).

7.2c Effect of extraction

Wet-pit mining and marine sediment extraction are of primary concern here, because they involve extraction directly from the water and will therefore have an immediate effect upon its ecology.

Figure 7.1 *Methods used to extract marine sediments. (a) Grab dredging. (b) Static suction dredging. (c) Trailed suction dredging. (d) Low capacity static suction dredging vessel used for maintenance dredging in a small port (photo by C. Frid).*

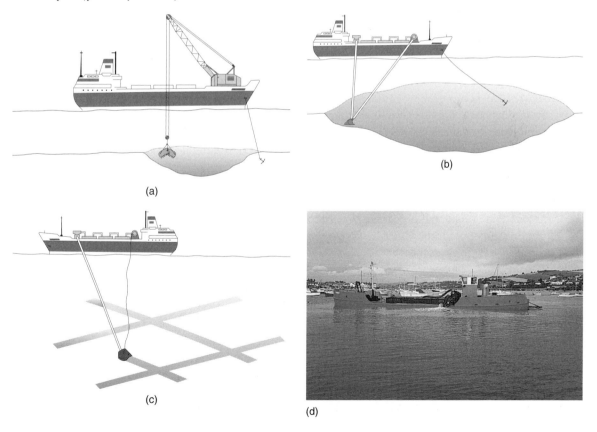

(a)

(b)

(c)

(d)

However, dry-pit mining and bar skimming alter the structure of dry environments which will eventually be rewetted, with probable effects upon their ecology.

Effects of extraction may be divided into two types – the direct effect of the operation itself and the knock on effect of habitat alteration that it causes.

Direct effects

Any dredging operation has an immediate and profound effect upon the area being dredged, as substratum is removed and the bottom topography is changed. Organisms caught by the dredging process will almost certainly be killed, and aggregations of fine sediment or algae will be dispersed, reducing the energy resource for colonists once the disturbance has passed. In addition, however, fine particles and their associated contaminants will be resuspended. Furthermore, in many cases the dredge hopper is fitted with an overflow, so that as it fills with dredged material, the displaced water, laden with fine silts, overflows back to the environment, creating a plume of turbid water at the dredging site.

By these processes, direct extraction from the water body leads to knock on effects down current. The sediment plume will settle, clogging interstitial spaces and gills and shading algae and macrophytes, with a consequent impact on local primary production. The settling silt may alter sediment properties a considerable distance from the site of the dredging operation.

Habitat effects

In principle, gravel extracted from a water body is a renewable resource; the gravel is present because it is being continually moved and replenished by the river or by marine currents. Problems arise, however, when the rate of extraction exceeds the natural replenishment rate of the water body.

Hydrological characteristics will alter in response to the reduced sediment supply, often leading to erosion and restructuring. There is speculation that some coastal extraction is causing erosion of beaches, because the erosive forces still operate, but the depositional forces are starved of sediment and are therefore unable to replace what has been lost. Structural effects of sediment extraction are, however, most clearly understood in river systems.

7.2d Effect on rivers

In rivers, the normal sediment size class in a given area is determined by occasional extreme flows rather than the normal flow regime, so resuspended fine sediment may settle onto areas normally dominated by coarse gravels, and therefore supporting organisms unable to tolerate siltation; it will remain there for weeks or even months until the next high flow re-sorts the sediment. Other short-term effects of extraction include the operation of heavy machinery in the channel, compressing sediment, increasing turbidity and leaking pollutants such as fuel.

Large scale sediment extraction can have even more profound effects than local scale siltation. River channel morphology is determined by the amount of sediment input from the catchment, the river adjusting its form to one that optimises transport of the sediment without major changes in channel morphology. If the sediment load changes appreciably, the river will readjust its channel form. This leads, in turn, to bank erosion, channel incision and channel widening or narrowing as it reconfigures itself (Nicholas et al., 1999). The River Alfios in Greece was a source of instream gravel extraction between 1967 and 1996, during which time in excess of 17×10^6 m^3 of gravel was removed (Figure 7.2). Extraction started around the time that the river was dammed upstream of the gravel extraction reach, cutting off potential inputs of fresh gravel from most of the river's

Figure 7.2 *Volumetric rates of gravel extraction from the River Alfios, Greece, between 1967 and 1995 and the estimated cumulative depth of channel degradation within the 8 km long region from which the gravel was extracted (from Nicholas* et al., *1999).*

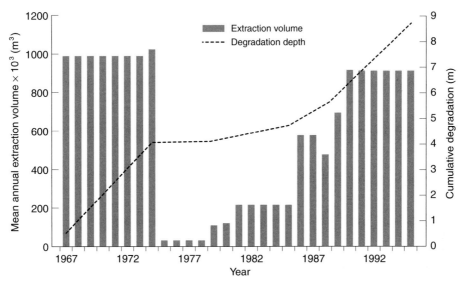

catchment. By 1995, Nicholas *et al.* (1999) estimated that the channel had incised by 9 m, demonstrating that the rate of extraction far exceeded the capacity of the catchment to replenish losses. As a consequence, the river has restructured its channel, with rapid bank erosion, to the extent that the channel doubled in width to around 200 m over 30 years. This level of perturbation will have physical and biological influences many kilometres downstream.

Healthy sediments contain large numbers of organisms, living within the interstitial spaces if small, or burrowing if larger. Burrowing species, such as oligochaetes, are important in finer sediments because their burrowing and food storage activities irrigate the sediment and provide micro-habitats for micro-organisms (Maher *et al.*, 1999). If the sediment is disturbed, these organisms are removed and their re-establishment may be slow. The most economically desirable sediments are normally gravels, several centimetres in diameter, which may be naturally unstable and therefore

support a biota adapted to rapid re-establishment; continual extraction may, however, suppress even these organisms.

Gravel extraction can detrimentally affect breeding among fish species which spawn on gravel, including economically important salmonids (National Marine Fisheries Service, 1996). Obviously, removal of gravel of the size required will impact spawning directly. For example, the river blenny (*Salaria fluviatilis*) is endangered in the Iberian Peninsula partially as a consequence of instream gravel extraction. Breeding males choose relatively large stones as nest sites, and clutch size is correlated closely with the size of stones available. In areas where gravel extraction has occurred, reducing average particle size on the river bed, both nest density and clutch size are reduced (Cote *et al.*, 1999).

Other effects of gravel extraction on fish have been identified. Fine sediments reduce oxygen levels in the gravel, while high sediment loads may disrupt behavioural activities such as spawning and migration. Depressions left by extraction may divert the normal flow, creating areas of shallow water or low flow that impede fish migration. Gravel extraction also requires the initial removal of undesirable components, such as large woody debris, which provides habitat and stability to the river bed.

Effect on the sea bed

The impact of dredging on the sea bed is site-specific and depends on numerous factors, including extraction method, sediment type and mobility, bottom topography, and bottom current strength (de Groot, 1996b). The influences on the structure of the environment, particularly in the long term, are likely to be less profound than those on rivers, because only a tiny proportion of the entire habitat is being impacted at any one time. Despite this, localised effects do occur, and it is important to understand these effects and particularly the rate at which recovery can occur.

An area of sea bed off the English east coast was experimentally dredged by a commercial suction dredger (Kenny & Rees, 1994). About 50 000 t of aggregate was removed, representing around 70% of the sea bed area in the experimental plot cut down to an average depth of 30 cm. Benthic surveys of the experimental site and a nearby control site showed dredging to cause a significant reduction in the diversity, abundance and biomass of benthic organisms. Subsequent recolonisation of the dredged site by the dominant taxa proceeded relatively rapidly, although it had not fully recovered seven months later (Kenny & Rees, 1994). Side-scan sonar records and underwater cameras indicated a considerable amount of sediment transport during the first two winters following dredging, with the previously well-defined dredge tracks becoming in-filled with sand and gravel (Kenny & Rees, 1996). Even two years after the dredging event some of the rarer species had not recovered to pre-impact densities.

Figure 7.3 *Cotswold Water Park, England. (a) Major extraction in this area has resulted in a diverse range of shapes and sizes of flooded former gravel pits within a small area. (b) The large number of lakes has allowed a diverse range of uses. This lake is designated as a recreational site, whereas others are dedicated to conservation or ecotourism (photo by M. Dobson).*

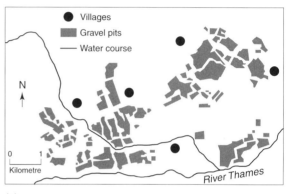

(a)

(b)

7.2e Gravel extraction and creative conservation

Dry-pit extraction and extraction from terrestrial deposits has an incidental benefit to aquatic systems. If gravel is extracted from below the water table, then the pit will flood, creating a new aquatic environment (Figure 7.3). In the UK, these have become major sites for waterbirds (Table 7.1), and a large proportion of sites identified as important for

Table 7.1 *Importance of gravel pits for waterbirds*

Species	Percent breeding (1980)	Percent wildfowl wintering (1979–82)
Canada goose	34.2	27.5
Great crested grebe	33.7	NA
Tufted duck	29.7	25.2
Little grebe	21.2	NA
Coot	20.8	NA
Teal	2.1	5.8
Mallard	18.7	12.1
Moorhen	18.0	NA
Mute swan	16.1	16.1
Pochard	10.0	29.2
Gadwall		26.9
Shoveler		11.8
Goosander		10.7
Goldeneye		4.2
Wigeon		1.7

Figures show the percentage of the total British population of each species on enclosed inland waters that was recorded from gravel pits.

conservation in the south and east of England are former gravel workings. By 1990 there were nearly 60 000 ha of exhausted workings, of which 15 000 had flooded, with a further 400–500 ha of flooded pits being added each year. This represents a large aquatic resource in an area naturally poor in lakes and which has lost most of its wetlands to drainage.

As a consequence of this, active gravel pits are often worked with the intention of optimising habitats for wildlife when extraction ceases (Andrews & Kinsman, 1990); indeed, this is generally a condition of new extraction licences being granted. A simple, vertical-sided pit with straight edges makes effective use of the resource for extraction, but leaves a depression of relatively low conservation value. However, by designing an extraction programme to produce a more complex depression, extraction can work towards an end product of high conservation value. Extracting to produce an uneven pit bed, for example, leaves a pit that contains a complex of different wetland habitats when flooded, while undulating banks

and gentle sloping at the edges of the pit create shelter and a variety bankside habitats (Figure 7.4).

7.3 Peat extraction

7.3a Occurrence and formation of peat

Peat is the accumulated remains of wetland plants, incompletely decomposed in waterlogged soils but highly compressed. Its high organic content makes it useful for fuel and as a fertiliser, while the concentration of plant remains gives it structural integrity and water-retaining properties, making it valuable as a horticultural compost.

Peat will form wherever permanently waterlogged conditions and rich plant growth coincide, the former suppressing decay of plant remains; it is constantly being created in appropriate locations. It is most abundant, however, in cool

Figure 7.4 *Gravel working to create appropriate topography for conservation. By selective extraction, features such as indented shorelines, islands and beds sloping at different angles can be created, at no extra cost (from Andrews & Kinsman, 1990).*

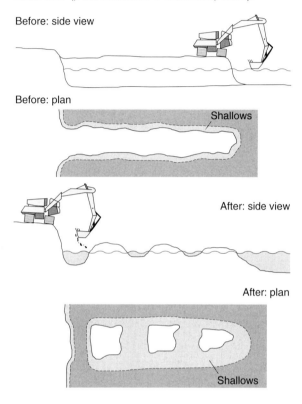

Before: side view

Before: plan

Shallows

After: side view

After: plan

Shallows

Table 7.2 *Distribution of the world's peat*

Continent	Maximum area (× 10⁶ ha)	Percent of total
Europe and former USSR	180.25	44.1
North America	171.13	41.9
Asia (excluding former USSR)	43.61	10.7
South America	6.17	1.5
Africa	4.86	1.1
Central America	2.61	0.6
Pacific	0.28	0.1

Figures give estimated maximum coverage prior to disturbance.
Source: Rieley *et al.* (1997).

temperate latitudes, where precipitation is high and evaporation rates are low. Eighty-six per cent of the world's peat occurs in the circumboreal zone of northern Europe, Russia and northern North America (Table 7.2), in which areas the dominant peat-forming plant is various species of *Sphagnum* moss. The growth form of *Sphagnum* facilitates rapid development of peat. *Sphagnum* grows as dense clumps which retain water very effectively – up to 98% by weight of a carpet of *Sphagnum* may be water (van Breeman, 1995). It has no roots, but continually grows from the top, while older parts of the stem die and become compressed by the weight of plant growth and water

above them. Its water-retaining properties allow *Sphagnum* to raise the water table above that of the surrounding land, creating raised bogs (see Dobson & Frid, 1998).

The bog consists of two layers – the surface layer, or acrotelm, containing living and recently dead plant parts, and the deeper layer, or catotelm, in which plant remains have been compressed by the weight of moss above them. The acrotelm is permeable and oxygenated, and maintains conditions required for growth of *Sphagnum* and other plant species that are able to grow in the bog environment. The catotelm, in contrast, is relatively impermeable and acts as a long term store of water. The water table remains close to the surface of the acrotelm in a healthy bog, but as different species have different growth forms and requirements, the surface of the bog is marked by a series of hummocks and hollows, allowing development of standing pools of water. Slight changes in topography and therefore in proximity to the water table favour different plant species, so a bog can support a diverse flora, along with a distinctive aquatic fauna adapted to the acidic conditions normally occurring in peat bogs.

7.3b Exploitation of peat

Peat is widely exploited both as a fuel and as compost, with around 90 million tonnes being excavated each year. *Sphagnum* bogs, furthermore, are a valuable source of high quality horticultural peat. Exploitation of peat is widespread and long established. In Ireland, for example, there is documentary evidence for its use dating back to the 8th century. This has resulted in large scale destruction of bogs, particularly the relatively small oceanic bogs in northwest Europe. The Netherlands has lost virtually all of its bogs, while in Britain only small remnants remain, virtually none of them undamaged. In Ireland, the area of relatively intact bog was reduced from 300 000 ha in the early 19th century to only 23 000 ha by the end of the 20th century (O'Connell, 1999). Not all destruction is caused by extractive activities; indeed, agricultural intensification and forestry are the major destructive agents worldwide, and peat cutting itself is more likely to degrade the bog than to destroy it completely. However, extraction is a major threat to remaining lowland raised bogs in northwest Europe.

Even if not completely destroyed, a small amount of peat cutting can cause irreparable harm to the structure and ecology of a peat bog. Waterlogging is essential for the ecosystems that peat bogs support, so any alteration in the rate of water loss destabilises them. Furthermore, a key regulator of water loss is the living layer of vegetation, whose damage can lead to alterations, reducing its suitability for the plant species of importance and therefore suppressing their recovery. Peat cutting inevitably damages the acrotelm.

Traditional peat extraction involves cutting blocks by hand from the active bog (Figure 7.5a). In any given year, the area impacted is likely to be low, as the process is time-consuming, and the mined area remains adjacent to unmined bog surface. Modern industrial scale extraction methods involve initial draining of the bog, removing any

chance of subsequent regeneration, followed by mechanised milling, which strips the upper layer from a bog (Figure 7.5b). In Ireland, peat is exploited as a commercial fuel, the state-owned peat producing and marketing company Bord na Móna supplying 14% of the country's electricity needs in 1991 by operating five peat-burning power stations. It extracts 5 million tonnes per year from its sites, systematically stripping the upper layer at a rate of 10 cm per year across the entire bog surface (McNally, 1997).

Once the hydrological balance of a bog has been altered by extraction, regeneration becomes difficult. Even where extraction has ceased, bog coverage will continue to decline and damaged sites gradually dry up. Raised bogs are particularly vulnerable to hydrological perturbation, because they maintain their own water table above that of the surrounding ground; breaking into the structure of the raised bog allows drainage to occur, threatening the entire system. Drainage not only threatens the continued survival of the wetland community, but also allows the rate of decomposition to increase as the peat becomes aerated. Only active management to maintain and restore water tables can save a raised bog that has been damaged by cutting.

Low intensity extraction by hand-cutting creates distinctive depressions in the surface of the bog, often enhancing wildlife diversity by creating different topographical layers relative to the water table (Figure 7.5a). If vegetation is left intact adjacent to the cutting, partial recovery can occur and the bog, although no longer in its pristine state, can maintain an ecologically healthy community. Milling, however, causes much more fundamental damage to the bog. Even if peat cutting ceases, restoration can be difficult because the undecomposed upper layer has been completely removed, exposing relatively decomposed peat, which is unsuitable for *Sphagnum* moss re-establishment (Van Seters & Price, 2001). Furthermore, the water balance will have been completely disrupted.

Figure 7.5 *Peat cutting. (a) Hand cutting can create a diverse wetland environment. (b) Milling destroys the entire surface layer of the bog. (Photos by M. Dobson)*

(a)

(b)

7.3c Protection and restoration of peatlands

The rarity of undamaged lowland raised bogs in northwest Europe has made them a major focus for conservation, with initiatives such as the 'Save the Bog Campaign' active in Ireland and the 'Peatlands Campaign' in the UK (Parkyn *et al.*, 1997). Restoration of sites is also being investigated, with some success in sites that have suffered minor degradation (O'Connell, 1999).

Once bogs have been worked out (**cutaway bogs**), they are no longer viable as peatland habitats, but provide an opportunity for creating other wetland habitats. Irish bogs exploited by Bord na Móna have an economic life of 45–50 years, after which it is not possible to extract any more peat. The company's intention is to use up to 50% of its cutaway bogs for creating wetland habitats (McNally, 1997). There is a clear precedent for this use of former extraction sites, both in gravel pits (**Section 7.2e**) and in peat cutting itself. The Norfolk Broads, a series of shallow lakes in eastern England, are amongst the most important wetland sites in the country but are entirely artificial in origin, created when medieval peat cuttings flooded.

7.4 Manganese nodules

Nodules, which are rocky lumps, rich in manganese, lie on the abyssal seafloor in a number of parts of the ocean. The existence of these nodules was first recorded by the *Challenger* expedition in 1874; on a number of occasions since schemes for their extraction have been elicited. The richest fields are found on the abyssal plains of the Pacific Ocean. The nodules vary in size from small boulders to the size of a potato; they are rich in metal ores, in particular manganese (up to 30%), copper (up to 1%), nickel (up to 1.25%) and cobalt (up to 0.25%). The nodules are accreting at rates of around 0.1 mm per 1000 years; while this may seem small, taken over the world ocean, it represents around 14.5 million tonnes per year. There are currently estimated to be 16×10^9 tonnes of nodules in the world ocean.

Since the 1960s over US$600 million has been spent by various consortia on locating the richest nodule fields and developing harvesting technology. To date extraction has been done by mechanical dredges (mesh bags towed on a frame) or suction heads. In spite of the proven reserves (about 20 times known land based reserves), the cost and the legal problems (most of the fields are outside Exclusive Economic Zones) have prevented any commercial extraction.

In the USA, consortia hold extraction licences for mineral deposits off Hawaii. These licences were granted after the completion of Environmental Impact Assessments. In general the dredging of the nodules will impact the seabed biota in a way analogous to the physical impacts of fishing dredges (see **Chapter 4**). However, deep sea organisms are not adapted to dealing with physical impacts in the way some shelf communities are (there are no storm waves in the deep sea) and they are generally very slow moving. This implies that impacts on the ecology of the system in the path of any dredge may be severe and long-lasting (Bluhm, 1994). Experimental studies associated with trial dredging have shown severe impacts in the dredge tracks but also substantial off-path effects associated with the sediment plume created by the dredging (Bluhm *et al.*, 1995; Morgan *et al.*, 1999). Economics are likely to favour intensive use of an area, although this may be the worst ecological strategy, as a good interspersion of impacted and pristine areas would help promote recovery. Given these considerations, there is a need for more detailed research into the scale of impacts, the recoverability of the benthos and the degree of inter-site variability in vulnerability before mining can be allowed (Nath & Sharma, 2000).

7.5 Oil exploration and production

Exploration for reserves of hydrocarbons in off-shore areas dates back to the 1930s. Advances in technology in the 1950s allowed massive development in this area from the 1960s, with fields such as the North Sea, the Arctic coast of Alaska and the Gulf of Mexico opening up. In the North Sea alone, a total of 1530 exploration, 997 appraisal and 2256 production wells were drilled between 1964 and 1993 (the three well types differ in their characteristics, including diameter and hence the amount of cuttings produced per metre drilled) (de Groot, 1996a).

7.5a Seismic survey effects

In order to locate reservoirs of oil, geologists carry out geological surveys, often using seismic techniques. Seismic surveys use the reflection of sound waves from subsurface features – changes in the density of rocks – to build up a picture of the underlying geology. The sound is generated by a variety of techniques ranging from simple airguns to explosive detonations. The reflections from these are recorded on arrays of microphones. In aquatic environments, the sound source may be towed behind a vessel or deployed on the sea/lake bed. The sudden blasts of sound can cause impacts on the biota of the area. Away from the sound source, the effects quickly dissipate; given that seismic surveys are generally one off events, the total impact is likely to be small in most cases. However, in certain circumstances the effects can be more significant. The groups most at risk from this type of disturbance are aquatic mammals (van Dessel, 1991), fish containing swim bladders (Engås *et al.*, 1993) and, in the case of bottom set devices, the benthos in the immediate vicinity. Mortality of mammals has been recorded but the principal effect is disorientation and the effects on natural behaviour. Fish kills are common close to the high output devices. The significance of these in the marine environment is small given the size and spatial extent of fish populations, but can be significant in enclosed areas and lake systems. In general, surveys are timed to occur in the seasons when mammals and fish are least abundant at the site, this being the only feasible way of gathering the data while minimising impacts on the biota.

7.5b Drilling effects

When the geological surveys have highlighted a potential oil/gas reservoir, a test well may be drilled. As the drill goes down, cuttings are washed back up to the platform and the cutting bit is cooled and lubricated (Figure 7.6a), normally by pumping 'mud' down the drill string. The 'mud' cools,

Figure 7.6 *(a) Cutting head detail. (b) Schematic of drilling and cuttings pile.*

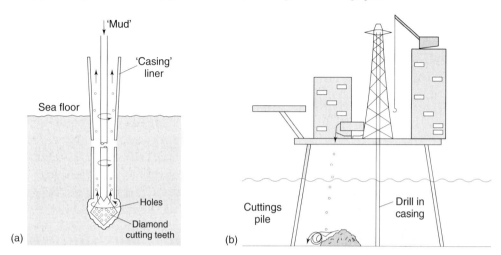

lubricates and flushes the cuttings back up to the platform where they are disposed of (Figure 7.6b). Pumping the 'mud' down under pressure serves to pressurise the drill head, so that when an oil/gas reservoir is breached the oil/gas does not 'blow-out' through the well in an uncontrolled way.

'Mud' is a complex chemical mixture including detergents and heavy metals in a lubricating solution, traditionally based on diesel oil as a carrier matrix. These oil based muds are, therefore, highly toxic. Given the expense of producing it there has always been an incentive to recover the 'mud' and reuse it. It is however impossible to remove all of the 'mud' from the rock cuttings before disposal.

The main environmental impact of offshore oil exploration results from the dumping of cuttings covered in oil based drilling muds to the sea bed at the platform site. This frequently results in an abiotic zone very close to the platform, a combination of the direct impact of dumping the cuttings and the toxic effects of the mud. This is surrounded by a zone containing opportunistic species, often dominated in temperate marine environments by species such as the polychaete worm *Capitella*, before there is a return to background communities. The return to a 'normal'

community typically occurs around 1 km down current from a well employing oil based muds (Figure 7.7). This basic pattern is repeated in the diversity within the community.

The toxic nature of oil based muds has led to the development of lower toxicity formulations. These are more expensive but the requirement for Environmental Impact Assessments for all new wells means that they are replacing oil based muds in all environmentally sensitive areas.

7.5c Platform waste

The sewage and waste water from a platform is also usually discharged 'over the side', but in offshore marine environments the scale of this discharge never exceeds the environment's capacity to assimilate the waste. In lakes, on board treatment is more usual prior to discharge.

Accidental releases from platforms also occur, for example 'blow-outs' and occasionally the accidental spillage of 'mud' from a platform. These cause impacts similar to those arising from oil spills caused by tanker accidents (see **Section 6.3c**).

Figure 7.7 *(a) Macrobenthic diversity and (b) three patterns of abundance shown by individual taxa on transects away from oil drilling platforms (from Kingston, 1987).*

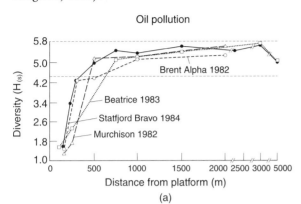

Oil pollution

(a)

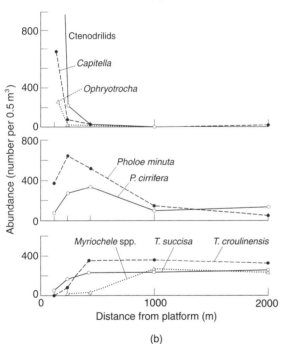

(b)

Figure 7.8 *Large scale commercial mining of sea salt. (a) Salt pans in Accra, Ghana being dried out by the sun. (b) The bagged product ready for export. (Photos by C. Frid)*

(a)

(b)

7.6 Salt extraction

In many tropical and subtropical areas salt is extracted directly from sea water. This has probably been going on since prehistoric times, when salt was a major trade item. In many parts of the world, the techniques used today remain essentially unchanged, although in some regions, such as the Gulf States, advanced technology is used to yield both salt and fresh water.

Most traditional salt extraction methods involve the use of shallow basins known as salt pans. These are filled with sea water and then sealed, allowing the water to be evaporated by the sun. When the water has been removed in this way, the remaining evaporites are mined. As the scale of human demand for salt has increased, so too has the scale of these operations. Where traditionally one or two ponds might be used, now there are salt pans several hectares in extent, and often several such pans built in sequence. These are surrounded by large banks, often carrying roadways, and linked by a series of channels and sluices.

The ecological effects of this process include mortality of all organisms impounded along with the water. Fish are often harvested as the water level drops, and these areas often form important feeding sites for wading birds. However, the largest impacts are associated with the gross habitat change from coastal wetlands to artificial impoundments (Figure 7.8).

7.7 Summary

- Mineral extraction from aquatic systems covers exploitation of two fundamental types of deposit: mineral deposits created by the aquatic environment itself and those which are simply located beneath aquatic environments.

- Gravels are extracted from river channels and coastal seas. Theoretically a renewable resource, over-extraction can lead to major disruption of both the immediate environment and of areas remote from the extraction point into which the sediment removed would normally be transported. There are more short-term effects associated with resuspension of fine sediments.

- Sediment extraction from glacial deposits on land provides the opportunity to create valuable wetland habitats in abandoned gravel pits.

- Peat bogs require an intact hydrology to persist. Peat extraction damages the living surface layer of peat bogs. Low intensity hand-cutting may create a high diversity environment. Industrial-scale cutting normally involves destroying all living components and severely limiting any chance of recovery.

- Peat and gravel are regenerating resources and therefore could be sustainable if carefully managed.

- Manganese nodules are not currently exploited commercially, due to the problems of working at great depths. Extraction may have profound effects upon deep sea benthos.

- Oil and gas exploration using seismic surveys can be damaging to living organisms. Lubricants used in the drilling process contain a range of toxic components, and seepage into the surrounding water is almost unavoidable.

- Salt is extracted from sea water by evaporation. Its main effect is loss of coastal habitats to create salt pans.

Chapter 8 Exploiting the General Aquatic Environment for Recreation

8.1 Introduction

8.1a The scale of recreation

Recreation refers to the use of aquatic systems or their biological components for pleasure, in which food gathering or economic gain, if it occurs, is secondary to the process of carrying out the activity. Whereas the use of aquatic systems for gain, particularly food, has occurred since the first appearance of humanity, their widespread exploitation for recreation is a relatively recent phenomenon. Angling for pleasure, as opposed to fishing for food, is one of the longest-established recreational activities, captured in European art for hundreds of years, while ancient Egyptian paintings demonstrate that hunting wetland mammals and birds is even older. For much of this time, however, they remained the preserve of a very small section of society; only during the 19th century did increasing wealth and leisure time for broad sections of European society facilitate an explosion of participants in harvesting for pleasure. Other activities, such as sailing or rowing, were enjoyed by very small numbers of people until the 20th century, while several currently popular water-based activities were non-existent until recent development of the appropriate tech-

nology. Canoeing as a leisure activity dates from the 1930s, while scuba diving required the development of the aqualung during the 1940s. Even the seaside holiday, with its associated exploitation of the beach and coastal seas, now well established throughout European culture, is a development of the 18th and 19th centuries.

Whatever their origins, recreational activities associated with water have burgeoned since 1950. As populations have increased and inhabitants of the developed world have gained access to greater disposable income and leisure time, so the impact of water-based activities on aquatic systems has increased. In the industrialised nations, this exploitation of the natural environment has developed gradually, but an important development towards the end of the 20th century involved long-distance travel. Participants in water-based activities are now more and more likely to pursue their hobbies in the developing world, carrying out activities alien to the indigenous population. The sudden translocation of intensive activity, such as scuba diving or water skiing, to areas where previously it was absent, is bound to place great strain on ecological processes in the water bodies affected.

Aquatic systems are coming under increasing pressure from recreational users. In the United Kingdom, around 100 million recreational visits were made to inland waterways during 1995, an average of nearly two per head of the population. Of these, 6.5 million visits involved direct use of the water (boating and angling), while the remainder were for terrestrially based activities in the vicinity of water bodies (Eaton, 1999). Similarly large numbers use coastal environments for recreation. Some aquatic habitats, such as rocky intertidal areas, often also receive high visitor numbers because of their important educational value (Frid & Evans, 1995). This 'visitor pressure' (a term which encompasses both visitor numbers and visitor behaviour, and implies disturbance to the natural environment) is becoming a topic of increasing concern; degradation of the natural environment through overuse is now recognised as a major international problem. The designation of protected areas may further exacerbate the problem, attracting large numbers of visitors, which can have deleterious impacts on the very biotic communities that the reserves are set up to protect (e.g. Brosnan, 1993). The problem of visitor pressure is compounded by the range of different activities pursued, often in the same area at the same time. The number of visitors a site receives is also variable, being influenced by season, time of day or week, weather, the presence of an ephemeral attraction and a host of other factors. These interacting processes make recreation management very difficult.

8.1b Types of recreational exploitation

The activities covered by the term 'recreation' are extremely diverse, their only common feature being the reason why they are pursued. Many recreational activities have effects similar to those of economic activities. Recreational boating, for example, is associated with pollution and physical damage in the same way as commercial use of boats, albeit on a generally very different scale. In other cases, the effect of the activity is indistin-

guishable from that of the economic activity – collecting lugworms on a beach for angling bait has exactly the same effect on their populations as their collection for sale. Some recreational activities, in contrast, are fundamentally different to commercial activities. Activities associated with conservation, or habitat creation, for example, where the recreational goal may simply be to enhance aesthetic appeal, have no equivalents in the economic or subsistence spheres.

Recreation brings people to water bodies, which leads to disturbance in the form of the activity itself, but also in the provision of visitor facilities and infrastructure. Recreational disturbance is of two types. Ephemeral disturbance is that caused by the recreational activity itself and lasts as long as the person is present, although its effects may last longer if biological or structural damage has been caused. Permanent disturbance is that caused by the development of infrastructure to support the recreational activity. Infrastructure development is concentrated on adjacent terrestrial environments, but can have important effects on aquatic systems, particularly through habitat loss and input of pollutants into the aquatic environment.

8.2 Impact of disturbance

8.2a Physical damage

Soft habitats

Merely visiting an aquatic site has an influence on the physical environment. The presence of people leads to trampling, which alters vegetation structure in wetlands. The passage of people across a saltmarsh, particularly the submergent marsh, leaves a trail discernible for weeks; it can take between 4 and 15 years for the vegetation to recover to its original level, and up to 50 years when the soil structure has been altered (Beeftink,

1979). In spite of such effects upon vegetation, however, saltmarsh fauna appear to be insensitive to human trampling (Chandrasekara and Frid, 1996a). The dominant macrobenthic fauna in salt-marsh are deposit-feeders, living at or very close to the sediment surface (Lopez and Levinton, 1987). The dense vegetation in these habitats prevents animals from being directly exposed and killed by the physical effects of the disturbance. The absence of effects on the vegetated marsh contrasts with the detrimental effects on the macrobenthic fauna in unvegetated tidal flat and tidal creek environments (Chandrasekara & Frid, 1997), implying that protection by vegetation is the key factor. The invertebrate community also shows 'resilience', meaning that it is able to recover rapidly from disturbance. In soft-bottom environments, this resilience is provided principally by the rapid movement of mobile adults from surrounding undisturbed areas (Sherman and Coull, 1980; Chandrasekara and Frid, 1996b).

Human trampling can create distinct footpaths (Vickery, 1995). People tend to follow paths strictly, and therefore the vegetation outside the paths remains relatively untrampled, confining direct influence to a narrow disturbed band. Unless a path is armoured with a solid surface, however, removal of vegetation and erosion will result in it becoming waterlogged, and people will walk around the edge of muddy areas. This can lead to an ever-widening band of disturbed habitat, resulting in degradation over a large area (Figure 8.1). It can be overcome in relatively dry environments, such as emergent salt marsh, by creating solid-surfaced paths, coupled with provision of information about the value and fragility of the surrounding habitat. In permanently flooded wetlands, if impacts are judged significant, it may be necessary to provide boardwalks to restrict visitor impacts to a limited area (Vickery, 1995). Unvegetated tidal flats and creeks may be damaged by trampling, but the dynamic nature of these environments ensures that effects are short term (see **Box 8.1**).

Figure 8.1 *Footpath across salt marsh, showing widening. (Photo by C. Frid)*

Rocky shores

Large numbers of visitors to temperate rocky shores lead to reduced algal diversity and abundance. Intertidal pathways, formed by repeated human use, have also been observed on rocky shores. Intertidal algae are, however, tolerant of some disturbance, and applying different intensities of trampling to rocky intertidal communities has shown that a 'threshold' trampling intensity can exist, below which there is no observable impact of trampling (Povey & Keough, 1991; Fletcher & Frid, 1996). Different species exhibit

Box 8.1   The impact of pilgrims on intertidal benthos

Intertidal sedimentary habitats are popular recreational areas among visitors and holidaymakers. Andersen (1995) reported that seashores in northern Europe rank among the most attractive venues for recreational visits. Many coastal areas of the world, particularly those with sandy beaches, are largely dependent on the tourist industry in terms of employment and income generation (Meijer, 1992). These habitats are also used as landing sites for fishing vessels or as sites for the deployment of fishing-gear, such as beach seines in many tropical countries.

Lindisfarne, or Holy Island, is an island off the coast of Northumberland in northeast England. At high tide it is isolated from the mainland, but at low tide the area between the island and the mainland drains to form an extensive area of intertidal mudflats dissected by creeks. A modern causeway carries the only road on to the island across these flats. Lindisfarne is the site of one of the earliest Christian settlements in Britain and following the beatification of St Cuthbert, who was buried there, the abbey became an important pilgrimage site. It remains so today and is also the end point of a long distance tourist trail – St Cuthberts Way, which links a number of historic and ecclesiastical sites in the England–Scotland border region. While most visitors drive on to the island across the modern causeway, many use the traditional Pilgrims Way and so come to the island in the same manner as the pilgrims of old. The Pilgrims Way is a path across the intertidal flats delimited by a series of wooden posts projecting from the mud. During a typical year about 2500 pilgrims walk to Lindisfarne along the Pilgrims Way, predominantly in the summer months between May and September.

The following account is of a survey which investigated the changes brought about in the benthic fauna of this path across the tidal flats, on the northeast coast of England (Chandrasekara & Frid, 1996a). The benthic community structure and the abundances of the benthic fauna on the path were compared with those in neighbouring untrampled areas in order to determine whether the continuous trampling disturbance had any effect on the soft-bottom benthic fauna.

The total number of infaunal species recorded during the summer and winter were 24 and 20 respect-

ively. *Capitella capitata* (Fabricius), *Clitellio arenarius* (Müller), *Tubificoides pseudogaster* (Dahl), *Enchytraeus buchholzi* (Vejdorsky), *Pygospio elegans* (Claperède), *Scoloplos armiger* (Müller), and the multispecies taxon Nematodes dominated at both times, accounting for 87.6% and 89.4% of the individuals during the summer and winter respectively.

MDS plots (see **Chapter 9** for explanation of these analyses) (Figure 9.6) clearly identified two groups in the summer data. All the samples from the footpath were tightly clustered together within the 87–94% similarity range; it was significantly different (ANOSIM, $r = 0.599$, $p = 0.001$) from the other cluster, indicating that repeated trampling has had a significant impact on the benthic community during the summer. The summer abundances of all the dominant taxa also varied significantly between the five locations along the transect. The abundances of *C. arenarius*, *E. buchholzi*, Nematodes, *T. pseudogaster* and *P. elegans* were significantly higher on the path (Chandrasekara & Frid, 1996a). In winter samples, while there were some differences in the fauna of the path, these were much less than in the summer samples.

All the dominant infaunal species recorded in this study are surface deposit-feeders. They live at or very close to the sediment surface (Brenchley, 1981; Lopez and Levinton, 1987; Unsal, 1988). Being species living close to the surface, it is possible that these soft-bodied infauna experience direct mortality from mechanical damage, such as crushing and fragmentation when the sediment is disturbed by trampling. The fauna may also experience indirect mortality as a result of burial in the mud (Rhoads and Young, 1970), through compaction of the sediment and reduction of its oxygen content, leading to asphyxiation, or displacement due to the collapsing of their burrows (Wynberg and Branch, 1994). The sediment disturbance could also bring the infauna onto the sediment surface, exposing them to avian predation (Jackson and James, 1979; Heiligenberg, 1987; Wynberg and Branch, 1991).

On the Pilgrims Way, the abundances of *Capitella capitata* and *Scoloplos armiger* were reduced towards the centre of the path during the summer. This suggests that exposure to high intensities of trampling was

Box 8.1 continued

detrimental to the survival of these species. Nevertheless, the results showed that the majority of species increased their abundances on the footpath during this time. The increased abundances of *Tubificoides pseudogaster*, *Clitellio arenarius*, *Enchytraeus buchholzi*, *Pygospio elegans* and Nematodes towards the centre of the path suggest that these species benefited from trampling. While there may be the loss of certain individuals due to the direct effects of trampling, there may be a process going on which compensates for the removal of these individuals. One possibility is that this compensatory process occurs through the rapid recruitment of adult stages to the disturbed areas (Thistle, 1980, 1981; Thrush, 1986; Thrush and Roper, 1988; Frid, 1989; Chandrasekara and Frid, 1996b).

A severe sediment disturbing agent like human trampling churns the upper sediment layer of the path, stimulating the bacterial growth on organic matter (Rhoads and Young, 1970). The subsequent bacterial degradation of organic matter may have made available a valuable commodity, food, for the opportunistic deposit feeding infauna (Lopez and Levinton, 1987). In addition, the bodies of animals dead or injured due to trampling may further enhance the food value of the sediment. This food resource may have attracted the adult stages of the above species to the trampled site, causing the observed increase in their abundances. Similar patterns of increasing abundance of fauna in disturbed patches in soft-bottoms have been observed previously (Simon and Dauer, 1977; Thistle, 1980; Wynberg and Branch, 1991).

The Pilgrims Way, which runs across the tidal flat, is inundated by the tide twice daily. The tidal flow provides a dispersal medium which the adult opportunists can use to move actively into disturbed sites from undisturbed areas (Thistle, 1980; McLusky *et al.*,

1983). Furthermore, the tide flowing over the tidal flat may erode the upper sediment layer, passively redistributing animals as bedload (Eckman, 1983) or moving them into the water column from where they can move actively with the inflowing water (Siegismund and Hylleberg, 1987). Olive (1993) has shown that reservoirs of adult stages suitable for colonisation are available in the intertidal soft-bottom environment.

It can be argued that the increases in abundances of some species on the Pilgrims Way should decline soon after the pilgrimage season stops as there is no permanent source of disturbance generating food for the infauna. In winter, when the trampling has virtually ceased, it appears that the community on the path had reverted to a structure similar to that in adjacent areas. The lack of significant differences in the winter abundances of the species which were impacted during the summer suggests that changes brought about in these populations were rapidly removed by natural processes as dispersal and sediment reworking (Probert, 1984).

Many studies (e.g. Ambrose, 1984; Wilson, 1986; Frid, 1989; Thrush *et al.*, 1992, 1995; Olive, 1993) have shown that disturbed patches in the soft-bottom marine intertidal environments are rapidly colonised. The rate of colonisation is dependent on the spatial variability of the patch, the nature of any resident fauna and the stage of the life cycle of the colonists. The disturbed area created by the Pilgrims Way is only 2 metres wide but it extends about 3.5 km. The sides of the patch are always in contact with the undisturbed areas, so it is possible that colonists can move through the sediment or creep over the sediment surface into the path. The flow of the tidal water containing potential recruits may also facilitate this process (McLusky *et al.*, 1983).

different susceptibilities to trampling. In general, foliose algal species (those possessing a point holdfast) are most susceptible to trampling, whereas algal turf tends to be more resistant, by virtue of its diffuse holdfast system and flatter growth profile. This differing susceptibility of algal

species means that the algal composition of rocky shores changes as public use intensifies. While heavily visited sites may be dominated by low profile turf algae, foliose species can survive and dominate sites subjected to less visitor pressure (Brosnan & Crumrine, 1994).

In contrast to the sessile component of the community, the effects of trampling on the mobile fauna of rocky intertidal areas have rarely been studied and existing knowledge of its impacts on these organisms is contradictory. Whereas Povey & Keough (1991) and Bally & Griffiths (1989) suggested that direct effects of trampling on mobile animals seemed unlikely, there is little doubt that some individuals can be crushed or dislodged underfoot.

In the western USA, where trampling damage to some shores is severe, management strategies with the aim of preserving biological communities have been suggested for heavily-visited sites in Oregon and California, often involving restricting public access in some way. This may be achieved spatially, for example, by the designation of pathways (Brosnan, 1993). Paths may already exist due to repeated trampling along the same route and have the advantage of confining the deleterious impacts of visitors to a small area of the site. In addition, visitors may be channelled using access points or topographical features (Ghazanshahi et al., 1983). Temporal restrictions of visitors can also be applied, limiting public access to the shore, both to allow recovery of damaged communities and also to minimise disturbance to shore communities at times when they are particularly sensitive. Brosnan (1993), for instance, suggested 'closed areas' to which access was to be denied for 3–4 months, in order to allow recovery of trampled algal communities. Such management strategies for visitor pressure have been largely reactive, responding to observed impacts on biotic communities, such as the formation of pathways through the algal canopy. However, more proactive measures, such as the provision of educational and interpretative facilities, can be highly effective at protecting natural areas from more damaging forms of public behaviour, at promoting interest in the coastal zone, and at encouraging public involvement in conservation issues (e.g. Carlson & Godfrey, 1989; Kaza, 1995).

Figure 8.2 *Recreational impacts by divers on coral reefs include (a) direct contact with coral colonies, causing mortality of the touched polyps and, in the case of heavy impacts, physical destruction of the colony; and (b) sediment kicked up by the fins can stress filter feeders including clams, sponges and corals and reduce turbidity and light levels, impacting photosynthesis in algae and corals. (Photos by Susan Clark)*

(a)

(b)

Coral reefs

Recreational scuba diving is a rapidly growing hobby and an increasingly significant component of international tourism, especially in regions such as the Red Sea, the Indo-Pacific islands, the Caribbean, the Florida Keys and Australia's Great Barrier Reef (Figure 8.2). Divers may impact the reef system as they gain access to the water by trampling across reef flats. This breaks off coral fragments, resulting in the accumulation of coral rubble (Hawkins & Roberts, 1993) and can precipitate significant alteration in the species composition of coral communities compared to untrampled sites (Kay & Liddle, 1989). Once in the water, they may continue to impact the system through physical damage to the corals or by increased coral mortality following contact. Observations of scuba divers show that impacts between divers and corals occur at average rates of 70–242 per hour with a maximum of 308 contacts per hour observed (Harriott *et al.*, 1997). While most of these contacts did not result in major damage, an average 0.6 to 1.9 corals were damaged per diver per 30-minute dive at four sites studied. Although such activities can lead to reduced ecological or amenity value of the areas dived, they are normally confined to small areas; furthermore, reefs show a high degree of resilience, an adaptation to major natural disturbances such as hurricanes, which also provides a capacity to recover from diver impacts.

Rivers, lakes and canals

The effect of trampling is a problem mostly confined to wetlands or intertidal habitats, while scuba diving and snorkelling are rarely pursued as recreational activities in fresh waters. However, shallow turbulent rivers attract boating activities – kayaking and rafting – which may cause damage to the river bed or banks. The recreational experience of such activities is enhanced by 'whitewater' conditions, in which occasional grounding or scraping of boat keels along the river bed is inevitable. The ecological effects of such events are almost unknown, but they tend to be concentrated in river stretches which experience a high level of turbulence and natural physical disturbance, and whose biota are adapted to tolerate or recover from occasional disruption of localised patches of the river bed. More damaging are

activities in which the river bed is deliberately modified. At one extreme, children building a dam across a stream will move sediment particles and alter flow patterns, while at the other, gold panning, a major recreational activity in mountainous regions of the western USA, involves deliberate collection and disruption of large volumes of riverine sediment. Although not as damaging as commercial scale extraction (**Section 7.2**), rivers may be impacted by large numbers of small scale panning activities during the summer months (Wright & Li, 1998).

Suitable stretches of river may be modified by placing baffles to enhance their value for kayaking (Figure 8.3). Such engineering alters the river channel, but is generally permanent, so detrimental effects, once construction work is completed, are probably minimal. Indeed, the river environment may be improved if it is organically polluted, as increased turbulence will enhance oxygenation of the water.

Boats are common on many inland water bodies. Their main physical influence is the creation of a wash, whose wave action can erode banks and disrupt emergent vegetation. Motor powered boats and jet skis are more damaging than those powered by wind or paddling, because their more rapid movement creates higher waves. The influence of boats is not entirely detrimental, however. Floral diversity in British canals is enhanced by moderate levels of boating traffic, maintaining an open channel. Little boat movement allows a small number of highly competitive species to dominate, while large amounts disturb the canal too much for macrophytes to establish. An average of 1–2 boats per day, however, keeps the channel open, allowing rare species to persist along with dominant ones (Murphy & Eaton, 1983).

Permanent disturbance

Many forms of water-based recreation, particularly those involving boats, require a permanent supporting infrastructure. Most types of boats

Figure 8.3 *Physical modification to create a kayaking course increases structural heterogeneity in the river channel. (Photo courtesy The Great Britain Junior Canoe Slalom Team)*

require moorings, and these are often concentrated into harbours or marinas, with breakwaters and piers. Once in place, such facilities further concentrate boating activity, and therefore side effects such as noise and pollution; they may also require dredging operations, increasing disruption to the local aquatic environment.

Infrastructure damage is most concentrated in the vicinity of the development. In some cases, however, its effects extend away from the source of the problem. Seagrass beds are highly productive environments in themselves, but also act as important nursery habitats and play a key role in stabilising soft marine sediments. Patches of exposed sand occur naturally in seagrass beds, and gradually migrate, erosion by wave action at one end being countered by rhizome growth at the other end. A permanent mooring creates a sand patch that expands at the eroding edge whilst being kept open by the mooring area at the other edge; therefore, sand patches increase in size and, where moorings are frequent, will coalesce. By this process, Rocky Bay on the western side of Rottnest Island, Western Australia, lost 18% of its seagrass bed cover between 1941 and 1992, two thirds of this since 1981. Concurrent with this loss was the increase in the number of boat moorings around the bay (Hastings *et al.*, 1995).

In some cases, infrastructure provided for eco-tourism with the intention of minimising or even reducing damage can have detrimental effects. In eastern Australia, several areas of mangrove forest have been made accessible to tourists through provision of boardwalks – raised walkways that allow people to walk in safety into the heart of the wetland. These are intended to ensure that mangroves are not damaged by trampling and, once installed, that further structural disturbance is minimised. However, other influences have been recorded. Mangals (the trees that dominate mangrove swamps) produce pneumatophores, vertical roots that grow into the air to ensure a constant supply of oxygen. The area around boardwalks in subtropical mangrove swamps in Queensland supports significantly reduced concentrations of pneumatophores compared to other areas in the wetland, with equivalent reductions in numbers of invertebrates that associate closely with these structures (Skilleter & Warren, 2000). Pneumatophore reduction is often a consequence of disturbance during boardwalk construction, so their abundance may recover in time, reducing these detrimental influences. In mangroves near Sydney, however, a change has been identified that is probably more difficult to reverse. Semaphore crabs (*Heloecius cordiformius*) are found at higher densities adjacent to boardwalks than elsewhere, apparently because the sediment is softer there, facilitating their burrowing activities (Kelaher *et al.*, 1998). High densities of crabs keep their food resources depleted and lead to further localised disturbance, as a consequence of their foraging activities. Unless crab densities are reduced, these detrimental influences will persist long after construction of the boardwalk is completed.

8.2b Behavioural disturbance

Mere presence of people can be a major disturbance to some groups of organisms, even if no structural damage ensues. The effect of intrusion is well studied among vertebrates, particularly

Figure 8.4 *Ecotourism. A birdwatcher visiting a seabird colony is not only observing the birds, but also interacting with them, probably causing them to modify their behaviour (Farne Islands, northern England). (Photo © S. Dobson)*

water birds (Carney & Sydeman, 1999). Two types of such disturbance can be distinguished. Where contact with wildlife is incidental to the leisure activity, the term 'recreators' has been applied (Carney & Sydeman, 1999). Recreators tend not to remain in close proximity to the disturbed population for extended periods, because either they or the animals move away. Ecotourists, in contrast, actively seek wildlife, may congregate in large numbers as close as possible to the animals of interest, and may remain for long periods (Figure 8.4). The activity is centred on wildlife and may, therefore, be very intrusive. Ecotourism is most effective as a leisure activity where animals are present in large numbers and unable to move, so breeding colonies of seabirds are particularly popular, attracting large numbers of visitors. In Florida, manatees (*Trichechus manatus*), which are susceptible to hypothermia during cold periods, congregate in warm water refugia during cold winter weather. One such place is Kings Bay, where a spring maintains water temperatures at 18–24°C, while air temperatures fluctuate between 0 and 29°C. This bay can support up to 300 man-

Figure 8.5 *Florida manatee, showing propeller wounds caused by collision with a boat. (Photo by Jeff Foott/bbcwild.com)*

Figure 8.6 *Diel activity of the Nile crocodile* (Crocodylus niloticus), *showing movement between land and water in response to air temperature e.g. at 12 noon, 70% of animals are in the water, 40% on land and the remainder are partially submerged ('Part') (from Cott, 1961).*

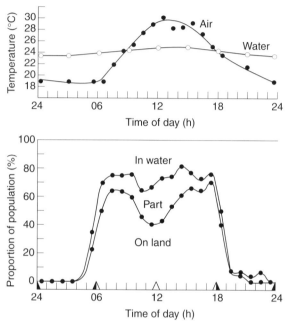

atees in a water body covering less than 10 km²; over 90 000 people visited the area in 1997 to see the manatees (Buckingham *et al.*, 1999). The disturbance caused to manatees by human presence, either as ecotourists or recreators, is frequently compounded by injuries they suffer when in collision with boats (Figure 8.5).

Disturbance can disrupt feeding or breeding activities of aquatic vertebrates, both of which will reduce fitness of the population. In the case of ecotourism, disturbance responses are usually an inevitable but unwanted effect of the observation – the ecotourist would prefer animals to stay rather than disperse. In some cases, however, eliciting a disturbance response adds to the experience. In Lake Baringo, Kenya, boats with guides can be hired by ecotourists to observe crocodiles (*Crocodylus niloticus*). During the day, crocodiles rest on the shore; the guides are able to spot them, then cause the boat to glide silently towards the animal. Eventually, the crocodile will suddenly move into the safety of the water, allowing the ecotourist to experience the speed at which crocodiles can move. Crocodiles, which are poikilothermic, move between water and land during the day as a means of maintaining their body temperature at its optimum level: land to warm up, water to cool down (Figure 8.6). Occasional disturbance

of this process is unlikely to have a long-term effect on an individual, but frequent disturbance may reduce its mobility, if it is unable to raise its body temperature to its optimum level, and therefore impair its fitness.

It is among birds that the effects of disturbance have been most intensively studied. Perhaps the most fundamental effect of disturbance is to reduce the amount of time the birds spend on feeding areas (Fitzpatrick & Bouchez, 1998). If disturbance is not of sufficient intensity to scare birds away from their feeding grounds, it may affect feeding efficiency as increased vigilance reduces foraging time. The presence of people on mudflats resulted in 34% reduction in the foraging time of the whimbrel, *Numenius phaeopus* on Inhaca Island, Mozambique (deBoer & Longamane, 1996). A study of the effects of human disturbance on foraging behaviour in the piping plover (*Charadrius melodus*) in New Jersey, USA, found that in areas with few people present, the birds spent 90% of their time foraging; with increasing numbers of people, this decreased to as little as 50% (Burger, 1994).

Impacts of human disturbance can be overcome if birds have access to alternative foraging habitats. Burger (1994) noted that the piping plover was capable of utilising several different habitats and thus was able to select the area with the fewest people, emphasising the importance of maintaining a diversity of habitats within the coastal zone. A similar effect was noted in the oyster-catcher (*Haematopus ostralegus*), which was able to forage on nearby grassland during the high tide period (Fitzpatrick & Bouchez, 1998). Oyster-catchers also spent time preening and resting upon arrival at a feeding area during the low tide period, suggesting that a reduction in the time spent on these activities may be used to compensate for lost foraging time. If the bird species is one that is limited to feeding in the intertidal zone only, then obviously the overall time available for compensation is controlled by tidal movements.

Behavioural disturbance is generally assumed to be of importance only to vertebrates. Invertebrates, being less aware of the environment outside their immediate vicinity, are unlikely to be influenced by distant activities of humans. There is, however, no evidence to corroborate or contradict this opinion. Aquatic insects are generally associated with the bed of a water body or with submerged vegetation, but the adult stage is terrestrial in most orders. Mayfly reproduction includes mating swarms; the mechanisms by which these are triggered and the habitat cues that are required are very poorly understood, so the influence of disturbance is unknown. Similarly, the effect of boating or other activities upon surface predators is completely unknown.

Scale of disturbance

In the case of vertebrates sensitive to human activity, even a small amount of activity can have major consequences; a single boat sailing on a small lake will be no less intrusive to sensitive bird species than several such boats, and a single power boat, emitting loud noise, can be more intrusive than several wind or canoe-powered vessels. Most recreational activities will, however, have an influence in proportion to the number of participants. A group of people or the presence of a dog is more disturbing than is a single person (Smit & Visser, 1993). However, large numbers of people involved in bait collecting activities, involving slow movement across a shore, will probably cause less disturbance than a single person walking a lively dog, or a family of holidaymakers whose party creates sudden movements or undue noise.

If disturbance is frequent, habituation may occur among vertebrates. The great crested grebe (*Podiceps cristatus*) is an aquatic water bird that nests in wetland fringes of open water bodies. A study of three lakes, all 20–30 ha in area, in Switzerland, demonstrated that its response to human disturbance is determined by the level of recreational activity allowed on a lake (Keller, 1989). Birds nesting on Gerzense, a lake where boating activity was banned, left their nests if an experimental boat approached to within 50 or even 100 m; on Moorsee, a lake where a maximum of 18 boats can operate at any one time, birds only flew when a boat approached within 0–20 m and on Burgäschisee, a lake where up to 100 boats may operate at one time, they flew when approached to within 0–10 m. On the latter two lakes, some individuals did not leave their nests at all, and defended them from the human intruder by pecking.

The manatees of King's Bay exhibit three responses to human activity. While some individuals avoid boats and scuba divers, others either ignore their presence or are attracted to people. A single individual may show all of these responses at different times (Buckingham *et al.*, 1999).

8.3 Recreational harvesting

For most recreational activities associated with water, such as boating, swimming or even eco-tourism, disturbance to aquatic organisms is an incidental or unwanted consequence of the activity being carried out. Recreational harvesting, in contrast, is deliberately aimed at disturbing specific aquatic organisms, and usually removing them. The most important targets for recreational harvesting are fish and wildfowl, although intertidal invertebrates are often heavily exploited to be used as bait for fishing.

8.3a Fishing

Recreational fishing is a major industry in the developed world, impacting both freshwater and coastal marine systems. As it is the process of capture (fishing) which is of equal or greater importance than the product of the catch (the fish), recreational fishing is almost exclusively by angling, in which the human participant catches (or attempts to catch) a single fish at a time, using a baited hook and line. There are an estimated 5000 angling clubs and 2.2 million freshwater anglers in England and Wales alone; expenditure on their hobby accounts for 0.5% of the UK's GDP (Hughes & Morley, 2000). Sea anglers total around 1 million in the UK, although most of these also practise freshwater angling.

A recurring theme of human exploitation of fresh waters has been the deliberate translocation of species of economic importance, predominantly fish. This is a long-established activity, to the extent that in some cases native and non-native distributions are unclear. Coarse fish are widely established throughout the British Isles, but geologically recent glacial activity and their inability to disperse through salt water mean that the only native populations are in the east of England, in catchments such as the Thames and the Humber. Historical records for Ireland suggest that most

introductions occurred around the 17th century, to enhance sport fishing (Wheeler, 1977). The rate of introductions increased, however, during the late 19th and early 20th centuries, with many sport fish, in particular, being translocated to different continents. Deliberate introduction as a management tool started in Greece in the 1920s, since when 23 species have been introduced, most of them deliberately, and a further 11 native species have been translocated to new sites within the country (Economidis *et al.*, 2000).

Translocation of fish was originally undoubtedly for food. Carp (*Cyprinus carpio*) is a native of central Asia and west of the Danube in central Europe. It had been transported to Italy by the 4th century and then throughout the rest of Europe over the next 1000 years, forming the basis of mass aquaculture (**Section 5.3a**) in many areas (Lever, 1977). This process continues today in the developing world. The primary reason for deliberate translocation in recent centuries has, however, become the development of sport fisheries. This is most widespread where economically viable – North America, Europe, Australasia – but has also occurred in East and South Africa, south Asia and South America. Fish are also introduced for ornamental reasons, into enclosed water bodies, although inevitably some may escape. Finally, the recreational activity of keeping fish in aquaria has resulted in some translocation. A stretch of the St Helens Canal in northwest England became famous for supporting breeding populations of two species of tropical fish, the guppy (*Poecilia reticulata*) from northern South America and a cichlid (*Tilapia zilli*) from East Africa. These species were able to persist in a temperate climate because the water at this point was being warmed by heated cooling water discharged from a glassworks. The fish became established following the closure of a tropical fish shop and the liberation of its stock into the canal in 1963. Several other tropical species persisted, although they apparently did not breed (Lever, 1977). This is an example of an introduction whose nature is unusual but of little conservation importance – the fish are confined to a highly modified environment in an artificial water body, and unable to expand their range. A more serious outcome of the trade in tropical fish has, however, been recorded from Lake Malawi, the source of many cichlid species. Most cichlids native to Lake Malawi are endemic and, more precisely, restricted in geographical distribution to a small section of the littoral zone of the lake. There are several cases, however, in which aquarium fish exporters have returned unsold catches to parts of the lake where they did not previously exist; the newcomers have become established and displaced closely related endemic species (Trendall, 1988).

In order for a food fishery based upon an introduction to be successful, the introduced species must establish itself rapidly and maintain a self-sustaining population, able to tolerate appropriate levels of harvesting. If intensive management is required, the fishery will not be viable. Freshwater recreational fisheries, in contrast, are rarely self-sustaining, but instead involve a range of management strategies including stocking and, in some cases, enhancing food resources for the fish.

Fishery management

Recreational fishing has, as its primary aim, catching fish rather than providing food. Indeed, much coarse fishing in the UK involves carp and their relatives (Cyprinidae), which are rarely eaten, so the fish is normally returned alive following capture. Marine sport fishing is based upon natural populations whose management is limited to timing and intensity of fishing effort. Freshwater fisheries are, however, usually managed intensively. Not all fish are considered suitable for sport fishing, so optimum management requires maximising number and size of target fish. Therefore, a freshwater fishery, in addition to structural infrastructure such as angling platforms and boat hire, normally has a 'biological infrastructure', manipulating numbers of organisms to optimise

the harvest. By far the most commonly employed method is artificial stocking, which has been standard fisheries management practice throughout the world for many years. Its role is to improve quantity and quality of fish, either for harvest or for sport fisheries. Stocking is carried out under four circumstances (Cowx, 1994), which may be described as mitigation, restoration, enhancement or new fishery.

Mitigation stocking is addition of fish to replace losses following activities such as dam construction or channelisation, its aim being to replace, as far as possible, what was lost as a consequence of the work. Restoration involves re-introduction of fish following removal of a problem that has led to their elimination, such as restocking upland lakes with salmonids after reversing acidification (see below).

Enhancement and new fishery stocking are more controversial in that they involve creating entirely new species assemblages or fish densities, rather than attempting to restore a former situation. Enhancement aims to improve stocks in response to a perceived poor quality. It can involve enhancing populations of indigenous species but, more commonly, incorporates introduction of non-native species. For example, the Flathead River-Lake system in Montana, USA, supports ten native species, to which 17 further species were introduced between 1898 and 1969 (Spencer *et al.*, 1991). New fishery stocking aims to develop a harvestable or sport fishery where previously one was absent; this may be through introducing fish to a formerly fishless water body, or attempting to increase diversity, improve yield or fill an apparently vacant niche. Examples of this include translocation of brown trout (*Salmo trutta*) and rainbow trout (*Oncorhynchus mykiss*) to fishless rivers in the highlands of Kenya and to rivers in New Zealand where native fish are unsuitable for sport fishing.

Stocking recreational fisheries ranges from single releases of new species to annual, or even more frequent, release of captive-bred fish to enhance densities of target species. Restocking to increase populations generally raises population densities above sustainable levels and can lead to pressure on food resources, although the aim of such programmes is to increase yields, which will therefore reduce densities once more. At its most extreme, it involves a 'put-and-take' system, whereby fully mature individuals are released at unsustainably high densities and then caught and removed by anglers within a few days. Occasionally stocking is carried out following eradication of non-game fish species (Ashley & Nordin, 1999), to reduce competition on target species.

More cost-effective than continual restocking is to enhance the survival of fish over a longer term. Enhancement techniques include increasing food supply, either by introducing appropriate fish prey or by fertilising to enhance phytoplankton production and therefore, indirectly, populations of invertebrates upon which the target fish will feed. Stocking and food enhancement are frequently combined, because the higher populations require more food. Enhanced numbers of fish attract predators, and predator control may have to be incorporated into the management programme.

Kootenay Lake in British Columbia illustrates two approaches that have been used to enhance food supplies for target fish (Ashley *et al.*, 1999). The lake supports three recreational fish species: kokanee salmon (*Oncorhynchus nerka*), Gerrard rainbow trout (*O. mykiss*) and bull trout (*Salvelinus confluentus*). Kokanee is a popular sport fish in an area heavily dependent upon recreational angling for its employment, and also comprises the primary food of Gerrard rainbow trout, an important trophy fish. However, kokanee populations declined rapidly when rivers feeding the lake were dammed, trapping nutrients and resulting in lower primary production. Therefore, fertilisation of the lake was attempted between 1992 and 1995, to enhance phytoplankton production and therefore numbers of zooplankton upon which kokanee feed. Nutrient addition led to enhanced algal and zooplankton densities, with evidence for some

recovery in fish populations. The manipulation was costly (around Can$500 000 per year over four years), but considered worthwhile to restore a fishery formerly worth around Can$2 million per year to the local economy.

Kootenay Lake was previously the site of another food manipulation – direct introduction of zooplankton to act as a prey species for target fish. The introduction of opossum shrimp to this and other lakes is considered in **Box 8.2**.

Consequences of introductions

Many introduced species do not establish, either because conditions are wrong or they are out-competed by natives. If they do establish, they may not spread beyond the region of introduction. Of 128 alien species now found in the USA, only four have become established in more than 40% of catchments, and 94 are found in fewer than 8%. Four of the ten most widespread species are Eurasian (carp, goldfish, brown trout and grass carp), while the rest are North American natives, translocated for sport or bait (Gido & Brown, 1999). In regions with diverse native fish assemblages, introductions may establish with little discernible effect upon the indigenous fauna. In the USA, introduction generally leads to a net increase in the number of fish species in a system; of 125 catchments studied by Gido & Brown (1999), 104 now support more species than before translocations occurred.

Normally, however, some effect of introduction is expected. If translocation is carried out to enhance stock of a species already present, then genetic differences between indigenous and introduced populations may become apparent, introduced individuals diluting fitness of natives (Cowx, 1994). If introductions are of species closely related to indigenous forms, hybridisation is possible (Wheeler, 2000).

Introduced fish may establish by eliminating closely related species. The brown trout (*Salmo trutta*) was extensively introduced into New

Box 8.2   Opossum shrimp and kokanee salmon

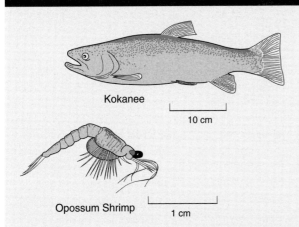

Kokanee

|— 10 cm —|

Opossum Shrimp |— 1 cm —|

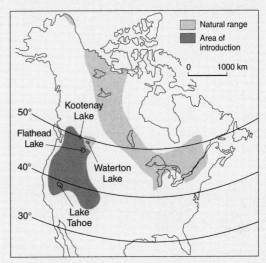

North American distribution of Opossum Shrimp

Natural range
Area of introduction

0   1000 km

50° — Kootenay Lake
Flathead Lake
40° — Waterton Lake
30° — Lake Tahoe

One of the most infamous introductions is that of the opossum shrimp (*Mysis relicta*), not as a target species for recreational harvesting but as a food source, with the aim of enhancing populations of target fish species. The opossum shrimp is a small crustacean (up to 3 cm long), native to a small number of deep oligotrophic lakes in previously glaciated areas of North America and Northern Europe. It is a predator of smaller crustaceans, particularly slow-moving cladocerans. It undergoes diel vertical migration (DVM), spending the day at depth and the night close to the surface. In its native habitat, *Mysis* is an important component of the diet of many fish and, for this reason, its introduction to other lakes, lacking a large planktonic crustacean, was recommended to enhance fisheries. The first introductions were to Kootenay Lake, British Columbia in 1949 and 1950, in an attempt to enhance yields of rainbow trout. However, another species, the kokanee salmon, benefited the most, individuals growing to unprecedented sizes. As a consequence of this, *Mysis* was introduced to over 100 lakes in the USA and Canada, along with several in Scandinavia (Lasenby *et al.*, 1986). The aim in North America was to repeat the success of Kootenay Lake by increasing sport fish yields. Unfortunately, however, the success achieved with the first introduction was rarely repeated elsewhere and, in some cases, exactly the opposite of what was expected occurred.

Lake Tahoe, on the border between California and Nevada, provides an interesting example. Kokanee

were first introduced in 1944, followed by intensive stocking in subsequent years, particularly during the 1960s. *Mysis* was introduced between 1963 and 1965, primarily in an attempt to stimulate yields of lake trout (*Salvelinus namaycush*), another introduced species. In this case, however, *Mysis* had a negative effect on kokanee. Establishment of *Mysis* and peak spawning runs of kokanee coincided with the disappearance of three cladoceran species (*Daphnia pulicaria*, *D. rosea* and *Bosmina longirostris*), which had previously been significant components of the zooplankton. Although the processes operating in this lake are complex, predation by *Mysis* and kokanee was almost certainly an important contributory factor in their disappearance (Morgan *et al.*, 1978). Before establishment of *Mysis*, *Daphnia* was the primary food source for kokanee, whose numbers declined following its disappearance. Some *Mysis* were eaten, but not enough to replace the lost food source.

Flathead Lake in Montana provides another somewhat different example. Kokanee were introduced into this lake in 1916, replacing the native cut-throat trout (*Oncorhynchus clarki lewisi*) as the dominant sport fish. *Mysis* was not introduced directly into Flathead Lake, but entered it in 1981 through passive dispersal following

Box 8.2   continued

its introductions into small lakes upstream. By 1986, it had reached a density of 126 m^{-2}, a high that coincided with the collapse of the kokanee fishery and the disappearance or precipitous decline in numbers of cladocerans and copepods (Spencer *et al.*, 1991). Reasons for these effects can be understood by comparing *Mysis* in Flathead Lake with an indigenous population in Waterton Lake on the Montana–Alberta border. In Waterton Lake, the dominant prey species, the cladoceran *Diaptomus sicilis*, coevolved with *Mysis* and appears to have developed an effective escape mechanism, keeping predation levels low. In Flathead Lake, in contrast, the dominant prey species is *Daphnia thorata*, which has no evolutionary experience of *Mysis*. Therefore, although cladocerans are less abundant in Flathead Lake, *Mysis* predation is more efficient; it consumes 3–4 times the energy consumed by individuals in Waterton Lake. As a consequence, whereas native populations of *Mysis* take a minimum of two years to complete their life cycle, the greater consumption in Flathead Lake allows the introduced population to complete its life cycle in one year, increasing fecundity. Kokanee were the dominant planktivores before the appearance of *Mysis*, with a population of around 1 million individuals. *Mysis* is, however, a better competitor for the large zooplankton on which kokanee feed and, through DVM, avoids itself becoming food for the salmon. *Mysis* is now an important component of the diet of two other introduced fish species, lake trout and lake whitefish (*Coregonus clupeaformis*), which feed at greater depths than kokanee and whose populations have increased dramatically. Attempts to restock kokanee have failed because

of the lack of zooplankton food and high levels of predation by lake trout.

Although the effects of *Mysis* on fish in Flathead Lake were predictable, in view of the experience of other lakes, it has had more far-reaching influences. Flathead kokanee spawn in McDonald Creek, a 5 km stream in Glacier National Park, about 100 km to the north of the lake. When their numbers were high, spawning attracted large numbers of bald eagles (*Haliaëtus leucocephalus*), whose diet is heavily dependent upon fish. Bald eagles began to aggregate in Glacier National Park in 1939, when kokanee were fully established, peaked during the early 1980s when kokanee populations were at their highest, and then declined rapidly as salmon stocks collapsed (Spencer *et al.*, 1991).

So why did introduction of *Mysis* succeed in stimulating kokanee in Kootenay Lake, but fail in Lake Tahoe and Flathead Lake? Kootenay Lake is a narrow lake with a discrete shallow west arm (mean depth 13 m) leading to its outlet. *Mysis* that enter this arm are unable to migrate to deeper areas to avoid predation and so are easily caught by kokanee. Furthermore, the greater primary productivity of Kootenay Lake resulted in densities of *Mysis* up to ten times those in Lake Tahoe. Elsewhere, Kootenay Lake is deep (mean depth 102 m) and kokanee have shown no increase in productivity. In recent years, nutrient levels have declined in Kootenay Lake, following construction of impoundments, and salmon stocks have dropped (Ashley *et al.*, 1999). Responses to this are considered further in **Section 8.3a**.

Zealand during the last half of the 19th century. The habitat was suitable for its ecological requirements and there are self-sustaining populations throughout New Zealand. In rivers where it is established, the native galaxiid fish (*Galaxias* spp.) almost inevitably disappear, partially through competition but also because trout eat large numbers of juvenile galaxiids. In rivers where trout are present, galaxiids are only found in headwaters,

above waterfalls large enough to inhibit the movement of fish from downstream (Townsend, 1996).

Very occasionally, a fish will fit into a vacant niche, with no apparent negative effects, as occurred when the cichlid *Oreochromis alcalicus grahami* was introduced into Lake Nakuru, Kenya, between 1959 and 1962 (Lévêque, 1995). In this case the introduction was not to create a sport fishery, but an attempt to combat mosquitoes by introducing an insectivorous fish species. It did, however, have an interesting recreational side effect in that it encouraged large numbers of fish-eating birds, such as pelicans and herons, to establish around the formerly fishless lake, increasing its ornithological interest to ecotourists.

Normally very little is known about ecological interactions beyond those with other fish. Effects of introductions on invertebrates, algae and nutrient dynamics are poorly understood. In one of the few cases studied, brown trout in New Zealand have been shown to exclude native crayfish (*Paranephrops* spp.) and large insect larvae. Crayfish use chemical cues to avoid predation by a native predator (eels) but have no such defences against trout, of which they have had no evolutionary experience. Trout increase standing crops of algae, partially by suppressing numbers of grazing insects, but also by altering their behaviour: in the presence of trout, some species spend less time foraging and more time hiding than in their absence (Townsend, 1996).

Habitat restoration

Much of the effort targeted at improving water quality in impacted lakes is prompted by a desire to restore fisheries. Fish are susceptible to acid conditions and are quickly eliminated when a water body becomes acidified. In Sweden, a major programme was initiated in 1977 to restore conditions in as many as possible of the estimated 90 000 km of rivers and 18 000 lakes that had become too acid for fish (Lessmark & Thörnelöf, 1986). This involved adding powdered limestone to affected water bodies, the calcium carbonate acting as an effective buffer and raising pH. The Swedish system adopted a method of adding limestone directly to the water, from the shore, boats, or in the case of very isolated water bodies, from helicopters. Although very effective in the short term, the lime lasts only as long as the flushing time of the lake and the process must be repeated typically every 2–3 years if conditions suitable for

fish are to be maintained. Liming has also been investigated as a management strategy in North America and elsewhere in northern Europe, although nowhere on the same scale as in Sweden.

Eutrophic lakes can suffer oxygen depletion in summer, a result of dense cyanobacterial blooms and thermal stratification; where extended periods of ice occur over water, these too can lead to oxygen depletion. Both will result in fish kills. If the fishery is economically valuable, lake aeration may be employed under such circumstances to maintain fish populations (Ashley & Nordin, 1999).

In some cases, the desire to restore fisheries has compromised research into the effects of liming. Loch Fleet in southern Scotland was the location of a major liming experiment. The lake became acid in the 1970s, following nearly two centuries of acidified rain eroding its buffering capacity. In 1986 a study was initiated assessing lime addition to the catchment rather than directly into the water. The operation was a success, in that application of lime to 20% of the catchment improved water quality to a level suitable for fish within two months, and a single application had an effect lasting several years (Figure 8.7a). Biologically, however, most macroinvertebrate groups showed an equivocal response, declining in numbers immediately following liming, following which some groups recovered while others did not. The results were, however, confounded by the reintroduction of trout as soon as possible after liming, so that chemical influences and the effects of predation were impossible to separate (Howells & Dalziel, 1992). The aim of the initiative was to raise pH to a level at which fish could be reintroduced, but this is not a true restoration because it created a condition in which calcium concentrations in the lake were much higher than had ever occurred naturally. Sediment core analysis was used to follow changes in diatom assemblages through much of the lake's history. It demonstrated that acid-sensitive species of diatoms had

Figure 8.7 *Restoration of Loch Fleet, Scotland. (a) pH and calcium concentrates immediately prior to and for several years following catchment liming. (b) Abundance of key diatom species at different times, inferred from the preserved remains in cores (from Howells & Dalziel, 1992).*

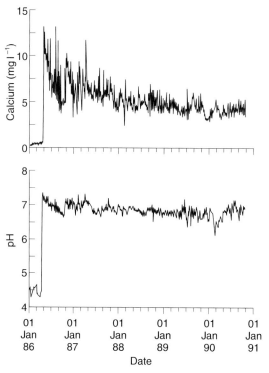

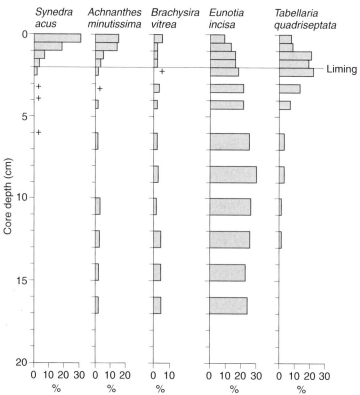

been lost when the lake acidified, and that many of these were able to re-establish following liming. However, several species never previously recorded also appeared. Among these was *Synedra acus*, a diatom known to be alkaliphilous – thriving in conditions of high alkalinity that had never previously occurred in Loch Fleet (Figure 8.7b). In this case, therefore, liming restored key conditions for fish, but only by creating an entirely new set of chemical and therefore biological conditions.

8.3b Bait collection

Intertidal populations may be harvested for human consumption, use as fishing bait, souvenirs, or for placing in aquaria (Underwood, 1993a & b). The collection of animals for human consumption may be carried out commercially, or as a subsistence activity, the effects of which are considered in **Section 4.4**. However, the collection of bait items by recreational sea anglers also represents a potential impact upon rocky shore communities. The large number of sea anglers represents a potentially large demand for bait items, both from rocky shores (peeler crabs and mussels) and soft sediments (lugworm and ragworm). Little research has been carried out into either the intensity of collection from rocky shores by the angling community, or the possible effects of this activity. The collection of bait animals from sediment shores has, in contrast, been extensively researched. Bait items are most commonly removed by digging, but a bait pump may be used to extract a core of sediment (Olive, 1993).

The intensity of bait digging activities in many areas has caused concern over the continued viability of targeted populations, and increased mortality of non-target species. Digging activities have a deleterious effect upon sediment habitats, due to physical disruption resulting in alterations to sediment surface topography, burial of fauna by resulting spoil mounds, deposition of anoxic subsurface sediments on the sediment surface, and

the resuspension of sediment bound heavy metals (McLusky *et al.*, 1983). Bait digging can also affect sediment chemistry by removing organisms which themselves influence factors such as sediment oxygenation, redox potential, and a whole suite of microbially mediated processes (Andersen & Kristensen, 1991).

The resilience of a given baitworm population to exploitation, or recovery time post-exploitation, varies between both locations and species (Olive, 1993). Several studies from various parts of England have demonstrated both the ability of one of the most exploited species – lugworm (*Arenicola marina*) – to recover and the conditions required for effective recovery. After having been exploited almost to absence in Lindisfarne National Nature Reserve, northeast England, significant increases in abundance were observed only six months after the introduction of a digging ban (Olive, 1993). The increase in numbers was attributed primarily to the migration of adult animals from adjacent unexploited areas. In contrast, a population of *A. marina* in the Swale Estuary, southeast England, had only recovered to 21% of control site abundance after six months (Havard & Tindall, 1994) (Figure 8.8), recovery in this site probably being hindered by continued digging activity in the area. A similar experimental study by Cryer *et al.* (1987), in which worms were removed from several plots, indicated that recolonisation was negligible over the six month study period. Cryer *et al.* (1987) suggest that the timing of digging may play an important role in recolonisation, with plots impacted in summer or autumn remaining at low densities until larval settlement and adult migration occurred the following spring. The biology of a given target species, particularly age at first reproduction, may vary between locations, and this in turn may be crucial in determining the resilience of that population to exploitation (Olive, 1993).

Bait digging may also have an effect upon other infaunal species. Digging for *A. marina* in the Swale Estuary resulted in an initial reduction in

Figure 8.8 *Impacts of digging on Arenicola marina cast numbers recorded from experimental and control plots in conservation areas, exploited areas, and outer area beyond the exploited area, seldom exploited due to distance from shore, over 180 days after digging had ceased (adapted from Havard & Tindal, 1994).*

Figure 8.9 *Changes in total invertebrate species number from experimental and control plots over 150 days after bait digging (adapted from Havard & Tindal, 1994; error bars omitted by authors).*

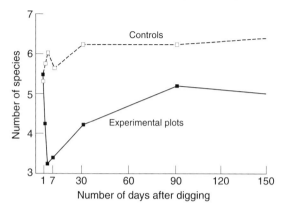

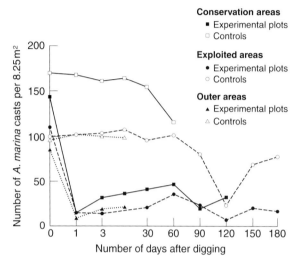

species richness in the exploited area, which recovered to only 78% of control values within 150 days of cessation of digging (Figure 8.9). Jackson & James (1979) reported that bait digging in north Norfolk, southeast England, resulted in heavy mortality of the cockle *Cerastoderma edule*. Both mechanical and manual harvesting of *A. marina* are employed in the Wadden Sea, Netherlands, with a significant deleterious impact upon infauna: for every gram of *A. marina* recovered, 1.9 g and 9–13.4 g of other infaunal species are removed by manual and mechanical harvesting, respectively (Heiligenberg, 1987). Mechanical digging in this area doubled the annual mortality rate for *A. marina*, and caused the near local extinction of *Mya arenaria*, which had initially represented 50% of the recorded biomass (Beukema, 1995). Recovery of total benthic biomass took five years after mechanical harvesting ceased.

8.3c Shooting

Shooting wildfowl (ducks, geese and swans) and wading birds is a major recreational pastime. During the 1999–2000 winter season, approximately 1.4 million hunters in the USA bagged an average of ten ducks and two geese each. There are clearly negative effects of this activity on wildfowl, both direct removal of individuals and disturbance to non-target individuals (Bell & Owen, 1990). Wildfowl are sensitive to disturbance, and will travel long distances to avoid hunting pressures. Pink footed geese (*Anser brachyrhynchus*) in Denmark roosted on a lake adjacent to feeding grounds when shooting was banned; when it was allowed, however, they used the same feeding grounds but flew up to 25 km to roost, increasing their daily energy expenditure by 20%. The same species in Scotland shifted its winter roosts in response to the hunting season. Initially, geese roosted on inland lakes, where the hunting season closes early, but following the complete cessation of hunting at the end of the season they moved their roost to an estuary (Fox & Madsen, 1997). There are many similar examples of wildfowl altering their behavioural patterns to avoid disturbance, all of which demonstrate the importance of permanent refuges from hunting.

Shooting can also be advantageous to target species if it stimulates preservation or creation of suitable habitat. During the 1930s, the number of wildfowl wintering in the USA, and therefore available to shoot, declined rapidly. Around 55% of North America's ducks breed in the prairie pothole region, an area of small discrete wetlands in south central Canada and adjacent areas of the USA. Realising that the decline in wildfowl numbers was a consequence of habitat loss, a group of American hunters set up an organisation to preserve wildfowl breeding areas, particularly in Canada. Ducks Unlimited, and its sister organisation Ducks Unlimited Canada (DUC), are now amongst the largest conservation organisations in North America. Their main activity has been to use funds raised in the USA to conserve or create wetlands in Canada suitable for ducks to breed. Between 1989 and 1997, DUC secured 39 000 ha of wetlands in the Canadian prairie pothole region. During this period, over 1700 ha of wetlands were restored by reversing drainage, 85% of these by DUC (Gray *et al.*, 1999). As a result of restoration programmes such as these, wildfowl numbers have risen, the 1998–99 total of 105 million migrating birds being the highest ever recorded.

Another contributory factor to wildfowl recovery in recent years has been the banning of lead shot. Ducks swallow small stones, their abrasive action in the birds' gizzards helping to break down food items and therefore assist in digestion. Unfortunately, lead shot, which is highly toxic, is the size that birds seek and is often easy to find on the surface of the ground or lake bed. Lead poisoning was first noticed in Texas in 1874, but the phenomenon was at least 20 years old by then. By the 1950s, 3000 tons of lead shot were being expended annually in the USA alone, killing as many as 2 million ducks per year by poisoning. After lead shot was banned in the USA in 1991, the number of deaths due to poisoning declined

by 64%. Much lead shot is, however, still in the environment and will remain there for many years. Canada banned it only in 1999, and in many parts of the world it is still legal.

8.4 Conservation, ecotourism and habitat creation

Whereas most recreational activities have, at best, neutral effects upon aquatic habitats, there is one activity that has a clearly beneficial effect. Ecotourism is a mechanism that allows a person to experience wildlife at first hand, but this derives from or generates a response that may be termed 'conservationism': the desire to preserve wildlife and their habitats. This recreational interest has ensured not only that aquatic habitats are protected by interested individuals or, more commonly, the societies that they support, but that new habitats are created, in some cases purely to enhance wildlife, in others to provide new opportunities to observe it. A good example of this is the Wildfowl and Wetlands Trust, an organisation founded in 1946 that has since set up 11 wetland sites throughout the UK. These sites have visitor centres and captive birds, some of them involved in breeding programmes. They also have large areas of wetland, some of which have been created from scratch and have since become major wintering areas for wildfowl.

Many conservationists will disapprove of the motives of DUC and similar organisations in creating wetlands to ensure plenty of ducks to shoot, but these wetlands also benefit other organisms, whose survival may have been jeopardised without the incentive to reverse drainage operations. Conservation projects are unlikely to succeed in the long term unless there is benefit to human society, whether this is economic or aesthetic.

The detrimental effects of ecotourism have been considered in **Section 8.2b**, but its positive features must also be emphasised. By creating a demand for aquatic habitats and wildlife it can, in the same way as hunting, facilitate their survival. Like hunting, it can also encourage restoration of aquatic habitats, or even their creation. The examples of Hula Lake, Israel, and Martin Mere, UK, illustrate this.

8.4a Hula Lake

Hula Lake and its associated wetlands formerly covered 400 km² in the headwaters of the River Jordan in northern Israel (Figure 8.10). They were drained during the 1950s to create arable land and to reduce the potential for malarial outbreaks, leaving only a small residual wetland area to the south of the original complex (Pollingher *et al.*,

Figure 8.10 *Location of the former Hula Lake and wetland, showing the recreated Lake Agmon (from Pollingher* et al., *1998).*

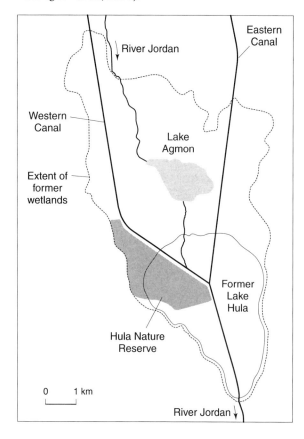

1998). Initially the drainage was successful from an agricultural perspective, but the area was plagued by a series of problems. Apart from the loss of endemic wetland species, subsidence, underground peat fires and dust storms contributed to gradual abandonment of the land. A further problem was increased nutrient loading into Lake Kinneret (Sea of Galilee) downstream, of particular importance because it is the main water supply reservoir for Israel. During the 1980s, a proposal was made to restore the water table to ameliorate these problems. Part of the restoration involved creating a small lake and wetland, its aim being to encourage ecotourism and thereby provide local communities with an alternative source of income. The new Lake Agmon was created in 1994 (Figure 8.10); it is very shallow (mean 0.8 m) and receives large amounts of agricultural runoff, so its water quality is not as good as had been hoped, but already it supports a proportion of the mixture of temperate and tropical fauna and flora that made its predecessor, Hula Lake, so attractive.

8.4b Martin Mere

Martin Mere was formerly a large, very shallow lake and wetland north of Liverpool in northwest England. It covered approximately 50 km², but was drained in 1692 by the simple process of digging a ditch to the sea, 2.5 km to the west. For much of the 20th century, the area had no standing water, but was an important feeding ground for wintering pink footed geese. In 1972, therefore, the Wildfowl and Wetlands Trust purchased a 150 ha farm at the southern end of the former wetland, and transfomed it into the current Martin Mere reserve. An 8 ha lake was created on the site and the surrounding land was managed to ensure suitable winter feed for waterfowl. The number of pink footed geese rose from a few hundred during the 1940s to 20 000 – one tenth of the world population, by the 1990s. Also wintering on the site are Bewick's swans (*Cygnus columbianus*) from Siberia and whooper swans (*C. cygnus*) from Iceland, internationally important populations of two species that were completely absent in 1972. The birds are provided with food in the form of waste potatoes, donated by local farmers, a strategy that serves both to keep birds close to observation hides during the day and to stop them from feeding on commercial crops in the area (Wisniewski, 1993). Creating a wetland reserve in an intensively farmed area has led to problems, including birds foraging on crops and poor water quality draining out of the site, a consequence of such a high density of wildfowl. It is, however, a major tourist destination and, as such, is economically very important to the area. By combining a wild habitat, involvement of local people and a strong educational role, Martin Mere demonstrates the role that ecotourism can play in conservation of wetlands and other aquatic systems.

8.5 Summary

- Recreation is a term encompassing many different activities, which together now have a major impact on inland and coastal water bodies.

- Recreational disturbance can cause physical damage, particularly in trampled coastal habitats. Scuba diving and boating activities can also damage benthic and shore environments.

- Many vertebrates suffer direct disturbance as a result of human activities. This is often accidental, but intrusion is compounded in some cases by ecotourism, as humans actively attempt to observe the organisms that they are disturbing. If disturbance is consistent and not directly threatening, animals can become adapted to it.

- Much recreation involves harvesting activities, including angling and shooting. The activity itself may have relatively little direct effect upon ecological communities, but associated activities such as stocking and bait collecting may severely impact aquatic habitats.

- Angling has stimulated translocations and introductions into fresh waters worldwide, often with profound effects upon community structure of water bodies into which fish have been introduced. Recreational angling is often accompanied by stocking target fish at artificially high densities, further reducing environmental quality.

- Bait collection for marine angling can severely deplete bait species in soft sediment coastal habitats.

- The economic value of recreational angling encourages restoration of water bodies that have become too degraded for fish.

- Ecotourism and the conservation ethic leads to habitat preservation and creation, a beneficial effect of recreation.

Part 4   Synthesis

Chapter 9 The Role of Ecological Science in Managing the Exploitation of Aquatic Systems

9.1 Introduction

Throughout this volume we have tried to present the interactions between humans and the aquatic environment in a structured way. In doing this it quickly becomes apparent that the same effect upon the ecology of a system can arise from a number of causes. For example, low oxygen levels may arise from disposal of an oxygen-demanding waste such as sewage. It can also arise from an algae bloom caused by eutrophication, through disposal of waste heat (cooling water), use of oxygen by the stock in an aquaculture facility, and the addition of nutrients and/or oxygen-demanding wastes from such a facility. In each case the effect of low oxygen levels on the fauna is the same, irrespective of the cause. Another example of multiple causes giving the same effect is that of introduced species, in which introductions may be deliberate for aquaculture or sport, accidental (along with aquaculture stock), or result from ballast water or vessel fouling vectors.

It is rare, however, for a human activity to impact on only one aspect of the ecology of the system. Sewage disposal reduces the levels of oxygen; however, it also adds fine particulate matter that may clog gills and smother sediments, and increases nutrient levels in the system. The scientific study of human interactions with their environment is complicated by these two facts, as

both make it more difficult to establish cause and effect and assign a magnitude to the impacts.

Impacts also rarely occur in isolation. Sewage causes pollution when it enters the environment in quantities that exceed the assimilative capacity. In these circumstances the sewage is likely to be derived from a conurbation and be accompanied by impacts such as other polluting inputs, canalisation, and loss of features of conservation value. Environmental management seeks to regulate these impacts in a coherent manner that protects the integrity of the ecosystem.

In many cases the uses to which man puts the aquatic ecosystem are conflicting. For example, coastal disposal of sewage may seriously affect the attractiveness of an area as a venue for recreation, while gravel extraction may destroy spawning beds for commercially important salmonid fish. In other cases the conflicts are minimal; deteriorating water quality, for example, as a consequence of waste disposal, may have little impact upon its role as a transport medium. However, it will have profound effects upon exploitable biota and even domestic water supply.

Considerable effort has been expended in recent years in developing approaches to the management of the environment that allow the consideration of these multiple uses. This is reflected in the development of a river basin (catchment) based approach to the management of fresh waters and the application of Integrated Coastal Zone Management for coastline and adjoining areas. It is not the role of the scientist to develop such management schemes (see Huxham & Sumner (2000) for a discussion of the role of science and its use and misuse in environmental management). In order to be effective, however, environmental management must be based on a sound understanding of the processes operating in the environment. This requires good science. Once a management scheme is developed it must be regularly reviewed to ensure that it is being effective in meeting its objectives. This is another key area in which scientists need to be involved if management strategies are going to work.

9.2 Monitoring the environment

One of the questions most frequently asked of environmental scientists, by industrialists, regulatory authorities, politicians, media and the public, is 'What impact has . . . had on the environment?'

And with the increasing number of schemes aimed at reducing human impacts, 'How long will it take to return to normal?' These questions appear quite straightforward but are in fact difficult to provide simple answers to. We can attempt to answer them at a variety of levels, the individual, the population and the community.

9.2a At the individual level

A number of parameters of an individual may be altered by an anthropogenic effect such as pollution, including, amongst others, its respiratory rate, feeding rate, longevity, and fecundity. How then do we know that these are different to normal? We need data on the 'normal' situation, either from laboratory trials or unaffected field sites. **Scope for growth** is a technique widely used to assess the integrated effects of environmental stress on an organism (see **Box 9.1**).

In using field sites we are faced with the problem of between site variability – sites will vary in any number of ways other than the presumed

---

Box 9.1   Scope for growth

Scope for growth techniques have been developed to provide an integrated assessment of the impact of a contaminant, or other stress on an organism's fitness. The procedure starts from the bio-energetic equation (see **Section 3.3**):

$$A = (P + G) - (R + U)$$

where A is the assimilated material, P is somatic production (tissue growth), G is gonadal production, R is respiration and U is excretion. The underlying assumption is that as an organism becomes stressed it directs a greater proportion of its resource into metabolism, for example detoxification pathways, and so has less available for growth.

The rate of growth is followed in test organisms exposed to different amounts or combinations of stress. The difference between the natural levels of

stress and the anthropogenically impacted systems gives a measure of impact and can be used to rank the various stress factors/concentrations used.

Scope for growth studies have been carried out on a range of organisms (e.g. Din & Ahamad, 1995; Alcaraz & Espina, 1997; Roast et al., 1999), but bivalves are commonly employed and in marine and estuarine studies the mussel, Mytilus edulis, is probably the most widely used (e.g. Anderlini, 1992; Gilek et al., 1992; Van Haren & Kooijman, 1993; Widdows & Page, 1993; Widdows et al., 1997).

The technique can be applied by measuring the growth rate of individuals in the field, but is most commonly applied in culture. In more advanced testing regimes, waters of differing properties may be supplied to clones or genetically pure strains of the test organism to provide comparable data sets.

impact. In order to establish a causal relationship to the effect, sites would need to be identical in all ways except the presence of the impact. In reality it is almost impossible to match sites in this way, so a practical alternative is to use a number of sites which are not impacted, but which provide some idea of the normal range of variation between sites in the parameter being considered. Laboratory studies can give us comparable data on the normal and stressed individuals that can be compared with the observed values from the field. There are, however, the usual problems associated with the artificiality of the laboratory situation and hence the applicability of laboratory data to the field situation.

If these problems can be overcome, we are left to test whether the value we obtained from our individual in the field differs from the norm, defined by values from the laboratory or 'clean' site(s). The formal comparison usually involves statistical testing procedures such as the Student's t-test or Mann–Whitney statistic. Such tests lack power as they essentially ask whether the value recorded (the impacted individual) comes from the distribution of values (the clean individuals). Much more powerful tests can be carried out when testing one population against another.

9.2b At the population level

While laboratory based toxicity studies are difficult to apply directly to the field situation, they do however allow us to identify species which are sensitive to or tolerant of particular contaminants. These are subsequently referred to as indicator species; changes in the biology of these in the field can be used as an assessment of pollution.

It is possible to identify indicator species from field survey data. However, it is not valid to identify indicator species from field data and then use the same data set to assess pollution – effectively a circular argument. Having identified *a priori* (in advance) the indicator species, changes in a number of parameters have been used in pollution assessment. The most commonly employed are abundance, biomass, for fish in particular the condition factor (weight divided by length cubed), and a range of reproductive parameters such as fecundity.

Whatever parameters are being used, the assessment involves a comparison of the parameter through space and/or time. For example, a transect of stations may be sampled across a dump site or sampling may be carried out both before and after a discharge is initiated. Data collected are then subject to appropriate statistical analysis, testing the null hypothesis that there is no statistically significant difference between sites and/or dates.

The advantage of this type of approach is the relative ease of application and the transparency of the result. The fact that the numbers of species x have declined in the vicinity of an outfall has immediate meaning. This is made even more powerful if the species concerned are ones of high public interest or economic value. Commonly used examples therefore include salmon, marine mammals, birds and mussels.

The main drawbacks with this approach are difficulty of selecting indicator species by means of independent and appropriate criteria, distinguishing change against background variability and establishing a cause–effect relationship. The latter can, to a certain extent, be addressed by choosing appropriate indicator species and parameters to follow. However, given the range and number of factors – physical, chemical and biological – that operate in the environment, establishing a strong cause–effect relationship with a single variable such as pollution will always remain difficult.

The high natural variability of marine systems operates both spatially and temporally (Figure 9.1) and at a range of scales. This needs a major consideration of the design of the sampling programme and is likely to involve considerable replicate sampling.

Figure 9.1 *A variety of marine time series showing the within year and internannual variability. (a) Phytoplankton density (monthly data) in the central western North Sea. (b) Total zooplankton (annual means) at a fixed station off northeast England. (c) North Sea cod biomass (annual data). (d) The density of breeding pairs of various seabird species on the coast of northeast England (derived from data in Clark & Frid, 2001).*

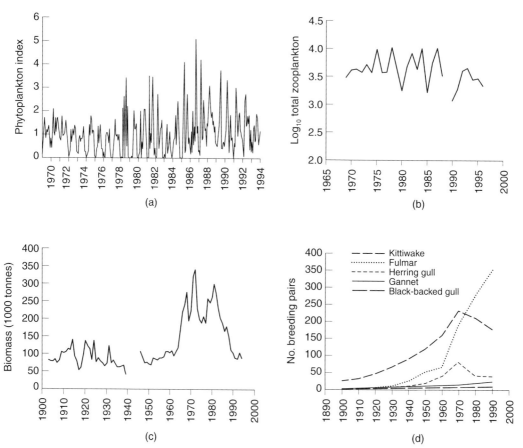

Two case studies already met in this volume seem to illustrate this approach. The first is changes in abundance of the indicator species *Capitella capitata* along the transect across the Tyne sewage sludge dump site (Figure 9.2). *Capitella* is frequently used as an indicator of organic enrichment, although it also responds to any disturbance in soft sediment systems. As such one should be cautious in attributing cause and effect – although in the specific case of sewage sludge disposal, magnitude of the response and its gener-

ality increase one's confidence in the appropriateness of the indicator.

The second case is imposex in dog whelks (see **Section 6.3c**). At least initially there was good evidence for imposex having been induced by exposure to TBT. Thus incidence of imposex has been used both spatially to show centres of and the extent of TBT contamination, and more recently to show how levels have changed since the introduction of legislation covering TBT use (Figure 9.3).

Figure 9.2 *Abundance of the polychaete* Capitella *on a transect from the Tyne sewage site in March 1999, three months after sludge dumping ended. Station 1 was the centre of the site, Stations 1 and 2 lie within the dump site and Station 7 is the furthest down current (Bustos-Baez & Frid, unpublished data).*

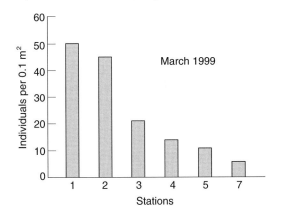

9.2c At the community level

We use the term community in its widest sense, to include any data about the assemblage of species present, and do not infer any biological interaction between them. As such communities represent the real world, their analysis includes all the species present, rather than just 'selected indicators'.

The usual question asked is 'Is there any evidence that the community present at any site/time is different from "normal" as a result of human activities?' This usually involves examination of the community at either a number of sites varying in their presumed exposure or sites at varying times in the region in which the impacts have changed. The basic data are usually species abundance or presence/absence data, and occasionally species biomass data. These can be analysed by means of univariate or multivariate statistical procedures or presented graphically.

Univariate approaches

Univariate analysis involves the use of the species abundance (or biomass) data to calculate some single numeral descriptions of the community. The most commonly used is 'diversity'. A number of indices are available for this and for other parameters such as species richness, evenness or dominance are also used. *A priori* selection of a 'keystone' species in the community means that analysis of its abundance or biomass, etc. could also be taken as a univariate descriptor of the community. A simplification of this technique is to use presence–absence data, rather than more complex measures of abundance. This is a technique that is successfully employed in several biotic index scoring systems used to assess river quality (**Box 9.2**). Between site/time differences in these measures can be subjected to statistical testing in the same way as population parameters for indicator species. However care should be taken in the selection of appropriate statistical methods as such indices are unlikely to fulfil the assumptions of many parametric statistical tests.

There exists a considerable literature on the diversity of aquatic communities and the effects of pollution on that diversity. Ecological theory would suggest that under highly polluted conditions diversity would be low, and in the extreme case zero as abiotic conditions prevail. However, at intermediate levels of pollution diversity may increase from background; as sensitive species decline in the face of increasing stress, opportunistic species invade, thus potentially raising diversity. Both the response to such effects and the absolute value of each diversity measure vary between habitats. It is therefore impossible to set criteria against which to judge any measured diversity except by reference to 'control' areas or historic data.

Graphical analysis

Graphical methods attempt to yield a visual representation of the community. Two methods are commonly employed, k-dominance curves and species-abundance plots. In the former, species abundance is ranked, converted to percentage

Figure 9.3 *The incidence of imposex in dogwhelks around the southwest of England. The RPSI (Relative Penis Size Index) is the ratio of the size of the female penis to the male; a high RPSI indicates heavy impact. The spatial patterns of impact in 1985 before regulations controlling TBT were introduced can be seen in the first column of data. Note the concentration near estuaries containing small craft moorings. The reduction in imposex following regulations is shown for 1994 and 2001 data. In 1994, some sites which had been heavily impacted in 1985 showed local extirpations, but other sites were showing recovery; by 2001 recovery was underway at most sites (Birchenough & Evans, unpublished data).*

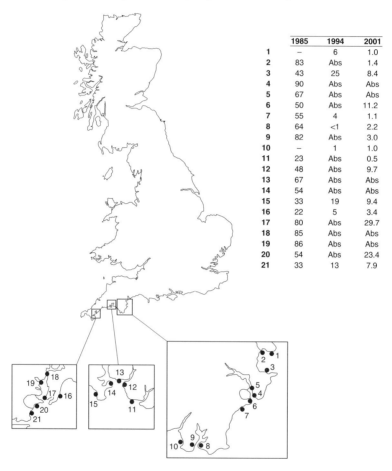

	1985	1994	2001
1	–	6	1.0
2	83	Abs	1.4
3	43	25	8.4
4	90	Abs	Abs
5	67	Abs	Abs
6	50	Abs	11.2
7	55	4	1.1
8	64	<1	2.2
9	82	Abs	3.0
10	–	1	1.0
11	23	Abs	0.5
12	48	Abs	9.7
13	67	Abs	Abs
14	54	Abs	Abs
15	33	19	9.4
16	22	5	3.4
17	80	Abs	29.7
18	85	Abs	Abs
19	86	Abs	Abs
20	54	Abs	23.4
21	33	13	7.9

abundance (i.e. proportion of individuals in sample belonging to that species), and the cumulative percentage plotted against rank. This plot often has the analogous plot for cumulative biomass superimposed to give the ABC (abundance–biomass comparison) curve. Polluted sites are indicated by the k-dominance curve lying above the biomass curve across the entire range (Figure 9.4a), this being produced by the increase

in small-bodied opportunistic species at polluted sites and the loss of large bodied 'climax' species.

Another approach using species abundance plots, or plots of the distribution of individuals amongst species, are histograms of the number of species falling into each abundance class, with these scaled geometrically. It is argued that at polluted sites, many species represented by a few

Box 9.2 Biotic indices – the BMWP and ASPT

A biotic index is a mechanism that uses a scoring system for organisms, different taxa being weighted according to their sensitivity to pollution. The standard biological assessment of river water quality used in the United Kingdom is the BMWP (Biological Monitoring Working Party) index, based upon benthic invertebrate assemblages (Armitage *et al.*, 1983). It is designed to allow rapid collection of data in the field and therefore equally rapid assessment of quality and the development of any problems. Standardised kick samples are taken from the bed of the river. Each taxon captured in the sample is identified, normally to family level, and is allocated a score (from 1 to 10) based upon its known tolerance to oxygen stress, less tolerant taxa receiving higher scores. Even if several taxa are distinguishable within a subdivision used for the classification (for example, two or more species within a single stonefly family), only one score is registered for that family. The individual scores for each taxon are then summed to give a total score for the entire sample, its BMWP score. The lower limit for this score is 0 (no organisms present), with a higher score signifying increasing water quality.

Unlike diversity indices, the BMWP is a qualitative system, based upon presence–absence and with no consideration of abundance. There are disadvantages to this approach, but these are outweighed by the speed of assessment and the ease of use. Furthermore, by using the entire macroinvertebrate assemblage in a sample, it is highly unlikely that large numbers of high scoring taxa will be present in low numbers in a polluted site, so the potential error is reduced.

Very clean waters may support few species for reasons unrelated to chemical water quality; for example, temperature is a limiting factor at high altitude. Therefore, a polluted site with many low scoring species may have a BMWP score higher than that of a clean site with a few species, all of which are oxygen sensitive. To overcome this problem, the ASPT (average score per taxon) can be calculated. The BMWP score is simply divided by the number of taxa used to calculate it, giving a value between 0 (no organisms present) and 10 (all organisms score 10). An example is given in the table.

Taxon	BMWP score	Site A	Site B
Heptageniidae	10	√	
Leuctridae	10	√	
Perlodidae	10	√	
Hydropsychidae	5		√
Elminthidae	5	√	
Tipulidae	5		√
Simuliidae	5		√
Baetidae	4		√
Sialidae	4		√
Sphaeridae	3		√
Glossiphoniidae	3		√
Asellidae	3		√
Chironomidae	2	√	√
Oligochaeta	1		√
Total (BMWP score)		37	35
ASPT		7.4	3.5

In this example, there is little difference between the two sites in terms of BMWP score, but the ASPT is very different, demonstrating that site A, despite its lower species richness, is probably the cleaner of the two sites.

Such systems have their limitations. Their reliance upon oxygen stress limits their value to assessment of water bodies impacted in this way, although admittedly this is one of the most widespread symptoms of human misuse of rivers. Stoneflies (Plecoptera) are very sensitive to oxygen levels, most families scoring 10 in the BMWP system, but many families are highly tolerant of low pH, so a BMWP will not register acidification problems.

The biotic index score, in common with diversity measurements, is meaningless by itself. Only in comparison with a score from another site, or from the same site on a different date, does it have any value.

Further details of this and other techniques may be found in Metcalfe (1989) and Mason (1996).

Figure 9.4 *Graphical analysis of impacted macrobenthic communities at stations across the Clyde sewage sludge dump site. Stations 1 and 12 are the clean ends of the transect and 6 and 7 are in the centre of the dump site. (a) K-dominance curve (abundance) and the distribution of biomass give an Abundance-Biomass comparison (ABC) plot (redrawn from Clarke & Warwick, 1994).*

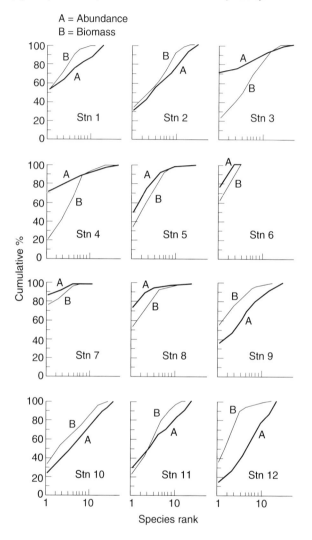

individuals in the 'normal' community will be replaced by a few species in high abundance (Figure 9.4b).

Testing for statistical differences between the plots requires (a) replication and (b) either their reduction to some single parameter descriptor of the plot (i.e. effectively thus becoming a univari-ate method), or the use of the replicate data and some appropriate computational simulation to generate a series of plots in order to define prob-ability of getting the observed result by chance (e.g. bootstrapping). However, such methods are good at presenting the magnitude of the difference between sites.

Figure 9.4 *(cont'd) (b) Geometric (×2) species abundance plots. Note the presence of many rare species at the unpolluted sites; their loss leads to a flatter curve at the impacted sites (redrawn from Clarke & Warwick, 1994).*

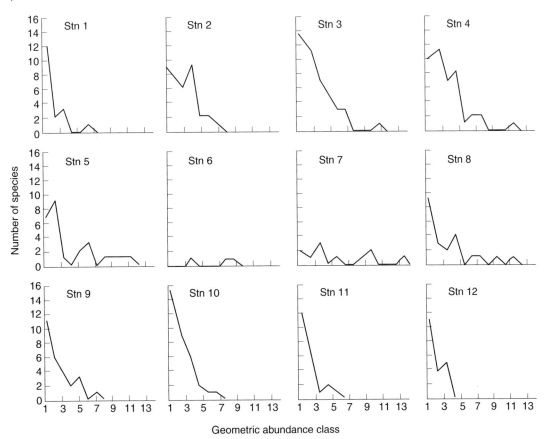

Geometric abundance class

Multivariate analysis

Multivariate techniques are based on an assessment of the extent to which samples show species at similar levels of abundance. They are usually formulated on some measure of similarity and involve such repetitive levels of calculation that computers are essential. Three methods commonly appear in the scientific literature.

The first, cluster analysis, uses similarity between samples, and groups of samples, to build a dendrogram. Samples with similar structures (distribution of individuals between species) are linked closely together in the dendrogram (Figure 9.5a). Depending on the similarity coefficient

used, it may be possible to attach a level of statistical significance to the delimitation of groups of stations/samples.

Non-metric multi-dimensional scaling – usually referred to as MDS – also starts from the calculation of similarities, but then 'maps' the samples, usually in two or three dimensions, in a way that rank order of distances between samples on the 'map' agrees with the rank order of the similarities. The resulting plot gives a very clear visual impression of the similarity of the samples (Figure 9.5b).

The third commonly used technique is principal components analysis. This too usually ends up with a plot of samples in a two or three dimensional space. The axes are defined in such a way

Figure 9.5 *Marine macrofaunal benthos from five sites in the Frierfjord. Four replicates were taken at each site. (a) A dendrogram showing the similarity in species abundance between the samples. (b) A non-metric multi-dimensional scaling ordination of these data. The more similar two samples are to each other in their composition, the closer they appear in the plot. (c) Principal components analysis (PCA) of 11 environmental measures (i.e. % C, % N, depth and the concentration of eight metals) from stations across the Clyde sewage sludge dump site. The first two principal components (plotted) account for 88% of the variation in these variables. Note how the most impacted stations (6 and 7) and distinguished along axis 1 (redrawn from Clarke & Warwick, 1994).*

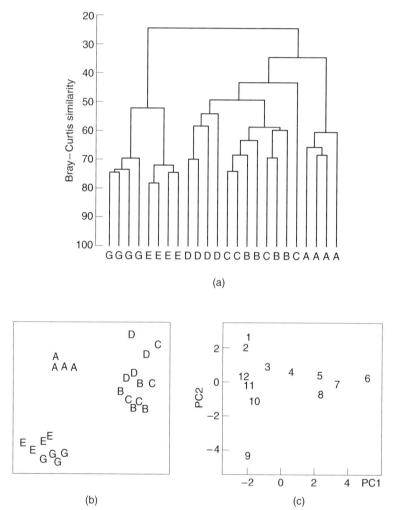

(a)

(b)

(c)

that Principal Component 1 explains the greatest amount of variation in the total data set, with axis 2 computed to explain as much as possible of the remaining variance. As such, the axes have units and can be correlated both with original variables (species abundance) and with other data, such as environmental measures. As the first ordination method to be devised, principal components analysis is frequently encountered in the literature and it is well suited for the analysis of environmental data sets (Figure 9.5c). However, the technique is problematic when applied to marine biological data, with its high predominance of rare species.

While the graphical output from all three methods easily allows the discrimination of samples into groups, assigning statistical significance to these groups is only possible through simulation methods. These essentially repeatedly sample from the data to build up a probability distribution around the pattern of the observed data.

Establishing causality

For all the community analyses described, establishing that pollution is responsible for any observed pattern is only possible through experimental manipulations either in microcosms or the field. However, if analysis of environmental data, such as levels of contaminants, gives a similar pattern to that produced by the biological data, a strong inference of a causal mechanism can be proposed.

For univariate measures, including univariate descriptions of graphical representations, regression and correlation techniques are appropriate. With the multivariate techniques two approaches are possible: simple superimposition of environmental values on the plot of the sample's distribution, or separate ordination of the environmental data and comparison of the pattern produced with that from the biological ordination (Figure 9.6).

9.3 Assessing recovery

With increasing regulation of pollution, many sites formerly used for waste disposal have ceased to be used. This raises the question: when will they have recovered? The first problem is to define 'recovered'. The initial response is to define this as the set of conditions that occurred before it was impacted. However, the aquatic environment is highly variable, both spatially and temporally; it is therefore very unlikely that a site would return to the pre-impact state (Figure 9.7).

Figure 9.6 *Clyde sewage sludge dumpsite macrofauna. (a) MDS ordination of species abundance. On this can be superimposed environmental data, with the size of the circle scaling with increasing concentration of (b) carbon, (c) manganese and (d) lead. The isolation of station 6 seems to be a function of the high levels of carbon and lead (redrawn from Clarke & Warwick, 1994).*

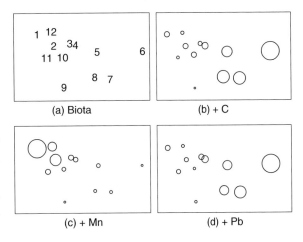

More realistic is to define recovery as having occurred when the impacted site is no longer distinguishable from nearby comparable areas that have never been impacted. However, this requires a matched control area, which is not always easy to find. Once such sites have been decided upon, any of the measures described above could be used. However, as one may be dealing with increasingly subtle changes as the community approaches the 'normal' state, multivariate techniques are probably the most powerful.

The rate of recovery in aquatic environments is frequently assumed to be rapid. In particular the massive numbers of larvae produced by many species, coupled with the dispersal achieved through plankton or drift often means colonists arrive rapidly into a region. This has often been the case following the dramatic consequences of oil spills on rocky shores (**Chapter 6**). However, in some areas suitable recruits of certain species seem to be in short supply or they are excluded from the

Figure 9.7 *A hypothetical pattern of change in some environmental value (i.e. species diversity, abundance of a key species, oxygen saturation, etc.) with and without the impact of a development. The impact is the difference between the observed with development value and what the value would have been without the development. In this case note that if only data were available from the beginning of the series (Time A) and the end (Time C), one would conclude that the development had caused an increase in the environmental parameter. If there was a one off survey just prior to the development (Time B) and then again later (Time C), one would conclude no change. In fact the development has caused a decrease in the environmental measure.*

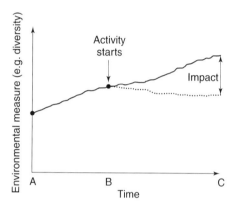

recovering community by earlier colonists such that recovery is not complete a decade or more after the impact ceases. Inter-annual variability in breeding, fluctuations in currents, chance encounters with suitable habitats and a host of other processes all contribute to the high natural variability in colonisation and will hence influence the dynamics of recovery. While the wide geographic range and dispersive powers of most marine and lacustrine species will ensure that they eventually recolonise suitable environments, the time scales are likely to be highly variable. The more limited intercatchment dispersal abilities of many river organisms make recovery rates difficult to predict, even where potential sources of colonists occur in river sites in close proximity.

9.3a Restoring damaged ecosystems

Recognition of the need to restore ecosystems that have been degraded to the point whereby the natural processes that support the ecosystem have failed has lead to the development of the field of restoration ecology. This is a new discipline, especially in the aquatic sciences (Frid & Clark, 1999), but it is rapidly evolving. The last decade has seen increased ecological understanding of complex biological interactions and increasing interest in the application of new technologies to the rehabilitation and restoration of aquatic habitats and their associated organisms (Cairns, 1986; Thayer, 1992). There exists a growing body of information, theoretical and case studies, ranging from small experimental projects to large-scale field applications (e.g. ecological restoration of wetlands and salt marshes) (see Cairns, 1986, 1988; Berger, 1990 for reviews).

Restoration ecology can be defined as 'the full or partial replacement of structural or functional characteristics that have been extinguished or diminished and the substitution of alternative qualities or characteristics than the ones originally present with the proviso that they have more social, economic or ecological value than existed in the disturbed state' (Cairns, 1988). Restoration and rehabilitation are not the same thing. Restoration is any activity that aims to return a system to the condition it was in prior to the disturbance (regardless of whether this was pristine). Rehabilitation is defined as any activity which aims to convert a degraded system to a stable alternative use which is designed to meet a particular management objective. Under these definitions, restoration and rehabilitation activities form the components which are the basis of restoration ecology. Restoration is generally regarded as having a well defined end point (i.e. the state before the system was degraded); however, this implies a stable state which is not always applicable to natural systems, and in practice 'restoration' rarely means returning an ecosystem to its original condition.

Perhaps the main challenge for ecological restoration is the need to integrate scientific knowledge on ecosystem dynamics with economic and social pressures that have a direct impact on the system. For example, eutrophication of coastal lagoons most often stems from an increase in nutrients entering the lagoon but which originate long distances from it (generally terrestrial sources within the drainage basin, such as urban settlements and agricultural practices in the surrounding area). Thus analysis of the problems may require extensive studies on industrial and agricultural practices at a geographically large distance, as well as tourism and recreational use in the vicinity of the water body itself.

Nevertheless, identifying the source of the problem is in most cases achievable. Restoring the degraded system into a functioning ecosystem represents a greater challenge. Ecological restoration usually involves both preventive measures and rehabilitative actions aimed at restoring damaged ecosystems. This entails changes in human behaviour and attitudes as well as practical, cost-effective, technical solutions.

All ecosystems have an innate ability to recover provided that: (i) the factors causing damage or disturbance are eliminated; (ii) the physical environment matches the needs of the organisms; and (iii) there exists a suitable supply of colonists. In some systems natural processes can act to restore the physical structure of the habitat. The potential for such natural recovery is greater in physically dynamic systems such as open coastal seas (Hawkins *et al.*, 1999) and low-order streams (Dobson and Cariss, 1999), both of which support species assemblages adapted to natural disturbance processes. Recovery processes, however, can take decades and management objectives may seek earlier recovery through active restoration. In less physically dynamic systems, natural processes are either insufficient to re-create the natural features or will only do so over geological time scales. In such systems, active restoration may be the only option (e.g. Clark and Edwards, 1999; Kemp *et al.*, 1999).

The open nature of many aquatic communities means that restoration programmes that recreate the physical structure required of the habitat may then allow natural colonisation to proceed. Such schemes, often referred to as passive restoration, will fail if suitable sources of colonists no longer exist. Therefore all schemes should make an assessment of the available sources of colonists and the existence of the necessary colonisation corridors.

Biological restoration is more complex than physical restoration. It is impossible to transplant a complete functioning community. Transplantation of 'key species' – either important biological structuring species such as vegetation or ecologically important species – may result in only transient success, if the transplanted individuals are not replaced as they die.

The global ecosystem has been estimated to provide around US\$$33 \times 10^{12}$ to the global economy, of which some US\$$21\,659 \times 10^{9}$ is contributed by aquatic systems (Costanza *et al.*, 1997). There is clearly, therefore, an economic case for protecting these systems to ensure their long-term sustainable use. If they become degraded the case for restoration has to be made. Restoration projects are often controversial given their long-term and experimental nature, the risk of failure and the high costs (Shonman, 1990). Considerable debate arises whenever funding becomes available for mitigation of environmental damage, often focusing on whether resources are best targeted into additional protection measures or into the application of ecological restoration measures.

9.4 The place of science in environmental management

Environmental management can be seen as providing a response to uncertainty in the face of risk and responses driven by values (rather than technical data) (Huxham and Sumner, 2000). The risk derives from the potential negative consequences of our actions (species extinction, global warming), the uncertainty from our inability to understand fully the interactions in the environment and predict the consequences of our actions. Society will always express value judgements as to the desirability of various outcomes or management measures.

Science advises managers and needs to inform society as a whole – in order to inform the valuation process. This requires a range of communication skills that are not normally part of the professional scientist's trade. We can consider the control of marine pollution as an example. Under the usual definitions, an unnatural substance in the sea is a contaminant; it only becomes a pollutant when it has some undesirable effect. This raises the question about how to define 'undesirable' and who should decide what level of change is undesirable. These are sociological/political questions rather than scientific ones.

The primary scientific questions in pollution relate to measuring environmental change and attributing cause and effect. Such studies, as we have seen, are often difficult to carry out and require considerable time, and expense, in gathering the data. Political decisions often have to be taken rapidly and scientists need to be ready to offer advice based on experience and the current state of knowledge, even if this is clearly incomplete. This flies in the face of normal 'scientific training', where one always emphasises the uncertainty of the results – politicians and the public usually cannot deal with uncertainty, even in the scientific sense of the word; they are seeking advice and expect to get a clear answer.

We must also recognise that pollution control measures are not always undertaken on a rational basis. Control or clean-up measures may be instigated as a result of public pressure without regard for the actual environmental damage of the pollutant or the clean-up – oil spills are a classic example of needing to be seen to be doing something, even if improving the visual aspect of the environment only happens at the cost of

causing more severe ecological damage (see **Section 6.7b**). Alternatively, when the economic costs to industry of cleaning up a discharge are high, the uncertainty in the scientific data may be exploited to justify the continuation of the cheapest option – even when the prudent response may be to apply the precautionary principle (see **Section 6.7c**).

9.4a Balancing socio-economic and political needs with science

Decisions about management objectives must consider a range of needs of society as a whole (Huxham & Sumner, 2000). Difficult questions must be asked involving financial cost and human livelihood. For example, all societies must now ask how much they are willing to pay directly for waste disposal, and how much they are willing to see the environment degraded as a result. As over-harvesting becomes common, societies that support fishing industries need to determine what price, financial and ecological, a catch of fish is worth, and how important it is to maintain a fishing community with a history extending over centuries but whose principal stocks have collapsed.

There is considerable potential for confusion about the respective roles for technical experts, policy makers, and advocates (i.e. non-governmental organisations) of many sectors including users and non-users. It is important that the role of science is understood in the larger process of selecting and implementing environmental management regimes. The term technical expert is used here, to emphasise that this includes not just biological, physical and chemical scientists but also social scientists.

The selection of properties of ecosystems that are essential to their conservation is the responsibility of technical experts, as is the selection of measures/indicators of those properties. It would also be appropriate for technical experts to have a

key, but not exclusive, role if the relative importance of ecosystem properties or their indicators is
to be set. Technical experts are the appropriate
group to assign priorities based on the degree to
which conservation of the ecosystem depends on
each of various properties of the system, as well
as to assign priorities among indicators based on
their reliability and sensitivity. It is not for the
environmental scientist to determine the final
ranking of properties and indicators based on
human values. Social scientists may work with
policy makers and the public to clarify public
opinion on such rankings.

9.4b Integrated environmental management

The mass of influences on aquatic environments, caused by the large diversity of end users,
means that management will only be effective if all
impacts are taken into account, rather than trying
to manage each separately. A relatively new way
of approaching management is, therefore, from
the perspective of an ecologically, rather than
politically, meaningful unit. Conflicting uses of the
water body can then be addressed together, with
a compromise management strategy developed.

Integrated river basin management (catchment management)

Integrated river basin management (IRBM), also
known as integrated catchment management
(ICM), may be defined as the planning and management of natural resources within a river basin
in order to achieve sustainable use. It takes into
account the interdependency of natural and
human factors within a catchment. All processes
operating within a given catchment are linked by
the river system, so that effective sustainable use
of resources needs to take into consideration the
influence of its exploitation upon other processes
within the catchment. Thus, changing land use in

the headwaters of a river system may influence
fisheries further downstream, while extraction or
any input of pollutants will influence water quantity and quality, respectively, throughout the rest
of the catchment.

IRBM is being adopted in many parts of the
world. In England and Wales, for example, the
Environment Agency produces catchment management plans, detailing uses of aquatic systems
and any processes or feature that may compromise the current state of the water body and
therefore influence its use. It is most successful,
however, in areas where political boundaries do
not divide the catchment, as its success requires
integration of all human use. The most complex
challenge in IRBM is devising workable management strategies for river systems that cross
international frontiers: each country has its own
priorities and will work for the good of its own
people before those of other countries. A state that
controls the catchment of a river may, therefore,
damage ecological function and economic processes further downstream if its own use of water
is excessive.

Integrated coastal zone management

Many different user groups use the world's coasts.
This has lead to a proliferation of organisations
with responsibilities or interests in the management of the coast. Coastlines are generally large
and extensive so there has been the tendency to
manage them in smaller units. These units develop
their management plans and in theory these can
then be arranged hierarchically to form the management plan for the entire coast.

Integrated coastal zone management (ICZM),
often simply referred to as coastal zone management (CZM), seeks to provide an integrating
framework to resolve conflicts between sectors or
user groups and ensure smooth integration of area
plans into a national coastal policy framework
(French, 1997). Human activities, ranging from
planning considerations for new buildings, through

Table 9.1 *The co-objectives of integrated coastal zone management as outlined in Agenda 21 of the UN Rio 'Earth Summit', 1992*

1 To provide for an integrated coastal policy and decision-making process in order to promote compatibility and balance of coastal uses. This also requires the inclusion and co-operation of government departments and agencies that have control over specific aspects of the coast.

2 To apply preventive and precautionary approaches in development, including prior assessment and systematic observation of the impacts of major projects.

3 To promote the development and application of techniques that reflect changes in value resulting from uses in coastal areas. These changes include pollution; loss of value due to erosion; loss of natural resources and habitat destruction, on which assigning a cost is difficult; and the increase in value due to development of the hinterland.

4 To liaise with all interested groups, to provide access to relevant information, and to provide opportunities for consultation and participation in planning and decision-making processes associated with the development of management plans.

waste water disposal to recreational zoning, should take place within this framework and so give due regard to environmental considerations including ecosystem health, the needs of erosion and accretion and coastal stability. The use of ICZM is specifically mentioned in the text of the Agenda 21 document produced during the UN Environment Conference in Rio de Janeiro in 1992. Agenda 21 commits the signatory coastal states to the implementation of ICZM initiatives as a means of providing effective and sustainable management and protection of coastal areas and marine ecosystems (Table 9.1).

The 'coastal zone', as used in ICZM, is not defined. It is often taken to include the zone between the landward extent of saline influence on vegetation to the extent of the territorial waters (12 nautical miles from the coast). The limits should recognise the reality of the sediment dynamics and geomorphology of the region and its ecological habitats. Socio-economic aspects of planning should then be fitted around the environmental constraints. There is therefore a key role in defining these for the aquatic environmental sciences.

To date, many of the CZMs proposed have been developed by panels of experts, and in developing countries these experts have often been overseas experts brought in to 'advise' by international aid agencies. While such structures benefit from a high level of technical expertise, they are often perceived as imposition from 'on high'. Effective ICZM requires all those with interests and aspirations to be committed to the plan; this is best achieved by involvement of these 'stakeholders' in the development of the plan. They will require guidance and technical advice but the final plan must be something that the community feels it can literally 'sign-up' to.

9.4c Ecosystem health and ecosystem quality objectives

Ecosystem health is a widely used term in assessing human impacts on ecosystems, albeit one with various, often imprecise, definitions. A healthy ecosystem is one with the following characteristics: an absence of stress, as defined by accepted indicators; an inherent ability to respond to stress and to recover to its previous form (resilience); an absence of known risk factors, such as industrial discharges (Norris & Thomas, 1999). It can also be defined in terms of vigour (as measured by an activity such as productivity), organisation (with increasing species diversity and connectivity signifying increasing health) and resilience. However, these criteria alone are insufficient, because high productivity and resilience are often features of aquatic systems with heavy organic or nutrient pollution loads (Karr, 1999). The component relating to absence of obvious detrimental human impacts is therefore an important part of the definition. This, in turn, requires undisturbed systems with which a given site can be compared, which leads to practical problems of finding such sites or, as is normally the case, predicting in their absence what they would have been like before human intervention. In practice, aquatic ecosystem health is often adequately defined in terms of sustainability; if multiple uses, such as extraction, pollution input, fisheries, and recreation can be managed sustainably, the system is probably healthy. This is particularly the case if management input can be minimised, because a healthy ecosystem should require little restorative intervention.

However it is defined, ecosystem health is a valuable phrase because it uses terminology that can be understood by administrators and the wider public. Defining the quality of a water body in terms of an analogy with human health emphasises the large number of interconnected processes that together create the whole. It also requires water users to view the aquatic system as a whole rather than in terms of the feature that each one requires (large volumes for irrigation, high productivity for coarse fisheries, etc.). Integrated river basin management (see **Section 9.4b**) is an application of this approach to management of freshwater systems.

In the late 1990s a number of initiatives were begun which sought to develop integrated environmental management through the application of ecological quality objectives (EcoQOs). These can be seen as a development of the EQO (environmental quality objectives) approach developed in the EU under the various water directives. The requirement for the development of EcoQOs arises from the need to bring forward an 'ecosystem approach' to environmental management. This is a key part of the adoption of the Convention on Biological Diversity (CBD) signed at the UN Rio

Conference and adopted as a basis for management by the EU and the Intermediate Ministerial Meeting of North Sea ministers.

Unfortunately the term 'ecosystem approach' has been used in a wide variety of contexts and been imparted with a range of definitions. The OSPAR (see **Section 6.7c**) has begun the development of this approach in the North Sea. The starting point for the development of an ecosystem approach to environmental management is to define the 'overall structure and function' desired for the ecosystem being considered. This is the desired ecological quality (EcoQ). The development of the concept of this 'desired ecosystem' is a societal decision, although science has some key roles (see **Section 9.4**). This desired overall state of the ecosystem must be expressed as a series of clear statements that will constitute the list of EcoQs. Next it is necessary to identify at least one measure for each EcoQ. There remains the question of the necessary and sufficient number of indicators needed to ensure protection of the ecosystem. Next the desired value for the various measures of the system, which correspond back to the 'desired ecosystem' initially specified by society, must be determined. These values comprise the suite of EcoQOs.

There is no inherent need for EcoQOs to be set always to the condition where anthropogenic influences are minimal. In fact this would imply no use of environmental services such as waste treatment or food production. Rather, the 'appropriate' values for the EcoQOs are determined by the overall desired ecosystem.

The appropriate measures and quantitative values for the EcoQs and EcoQOs will vary among systems and depend on the priority given to various issues, such as recreation versus waste disposal, or food production versus mineral extraction. The desire for an 'ecosystem approach' requires that EcoQOs should be determined in an integrated manner, so that they are mutually achievable and collectively sufficient to ensure conservation of the ecosystem.

9.4d Environmental economics

The measured gross national product (the total value of all economic transactions) in the global economy is around US$18 $\times$ 10^{12}, but aquatic ecosystems are estimated as providing US$21 659 $\times$ 10^9 (all the values quoted are standardised to 1994 US$ values) worth of goods and services. While a small proportion of this economic contribution, fisheries for example, is actually traded, the majority is actually invisible in traditional economic measures. For example, lakes and rivers contributed an estimated US$133 $\times$ 10^6 in the form of waste treatment, while marine ecosystems provided over US$26 $\times$ 10^9 worth of nutrient cycling.

Concern about the inability of traditional economics to incorporate the environment fully has lead to the development of the field of environmental economics (Pearce and Barbier, 2000). This aims to develop methods whereby the true value of goods and services provided by the environment is incorporated, the intention being that cost–benefit and similar analyses will then reflect the true cost of the activities.

A development of this approach has been the consideration of environmental taxes, the principle being that the more polluting a substance is the more it is taxed on disposal. This then provides a monetary incentive for organisations to minimise pollution but allows them freedom to decide what is the best method – building treatment plants, changes to the production process or waste minimisation (Pearce *et al.*, 1989).

9.5 Looking to the future – sustainability and biodiversity issues

Human populations are growing rapidly – quadrupling in the last 100 years and doubling from 3 billion in 1960 to 6 billion in 1999 (12 October 1999 being officially declared 6 billion day by the UN). While growth is predicted to slow, some predictions suggest that global populations will fail to stabilise during this century. Even relatively conservative estimates put the world's population at 8.04 billion by 2025 and 9.4 billion in 2050. At the same time as the population has been growing so have economic aspirations of people throughout the world. This will result in a greater requirement for resources, and hence impact, per person than at any time in the past. It is simply not possible to support, even for a short time, these levels of resource use in the current way and so these changes must be accompanied by novel ways of using and managing the environment. The UN Rio Convention set out a framework for this with the adoption of 'sustainable development' as a guiding principle and the commitment to preserve natural levels of biological diversity and ecological functionality. Ecological science has a major role to play in this evolving agenda.

Following on from the Convention on Biological Diversity, the 1995 Jakarta Mandate on Marine and Coastal Biological Diversity highlighted five (now six) thematic areas, second amongst them being sustainable use. The text specifically requires that 'the present mono-species approach to modelling and assessment should be augmented by an ecosystem process-oriented approach, based on research of ecosystem processes and functions, with an emphasis on identifying ecologically critical processes that consider the spatial dimension of these processes'. The ecosystem approach is further defined as 'the application of appropriate scientific methodologies focused on levels of biological organisation, which encompass the essential structure, processes, functions and interactions among organisms and their environment. It recognises that humans, with their cultural diversity, are an integral component of many ecosystems'. This again emphasises the need to consider not just protection of the full inventory of taxa present, but also protection of ecological processes and explicitly the spatial elements of these processes.

9.5a Sustainability

There are few who would argue against the principles of sustainability, and yet our society's track record is not good; after over 100 years of science-based fisheries management, more fish stocks are at risk of collapse than ever before. Hopefully this book has shown that this problem is in part scientific – the difficulties of developing a predictive capability in a multi-species, highly variable system – and partly societal, including the actual development and implementation of an appropriate management regime.

It should also be clear that up to now we have generally attempted to manage one activity at a time. This ignores the ecological and environmental linkages in aquatic ecosystems. The development of ecosystem approaches, integrated river basin management and integrated coastal zone management, seek to address this. Science has a role to play in delivery of these.

9.5b Biological diversity

Natural biodiversity is akin to our global capital; we may spend this capital at our ultimate peril. Organisms provide us with a host of services – food, waste removal and natural remedies (many commercial drugs are based on substances first isolated from biotic systems). We recognise that we have not yet fully recorded all the organisms on the planet, especially in aquatic environments with their greater sampling difficulties. Loss of biological diversity therefore threatens to remove these services from us and prevent us finding new useful compounds from bio-prospecting.

Just maintaining the full list of species on the planet may not provide us with a global biosphere that functions as we would desire. Ecological functionality is notoriously difficult to measure directly, but includes considerations such as ecosystem resilience, productivity, trophic structure and throughput (i.e. internal consumption to available yield of carbon). Ecological functions are essential for the environment to provide many of the services we gain from it – nutrient cycling for example. Therefore, we need to understand not only the diversity of organisms on the planet but the nature and scale of their interactions.

9.6 Concluding thoughts

Sound environmental management requires excellent science to underpin it. It also requires integration of social considerations. Society needs to be informed in advance of the full consequences of its choices, so that an informed society might express its desires. Therefore scientific advances must focus on developing a true predictive capability in the environmental sciences, a major intellectual challenge given the uncertainties and complexities (i.e. non-linearities) of these multidimensional systems. Furthermore, scientists must ensure that this information is communicated effectively to the wider community of stakeholders, i.e. those with an interest in a particular system.

In this volume we have seen a number of examples of failures of environmental systems through misuse by humans: fisheries, introduced species and pollution are key examples. We have also seen some examples of the application of new approaches where sound science and wide social involvement is leading to better environmental protection. Examples include the estuarine clean-up programmes in areas such as the Thames and Mersey, and the development of integrated ecosystem management regimes for areas such as the North Sea.

9.7 Summary

- Science has a key role to play in managing human impacts on the environment. Principal amongst these are providing assessments of the extent of human induced change and monitoring of the exploited environment. To provide a context to these data, information must also be available on the scale of natural fluctuations.

- Ecological data on the state of an ecosystem can be presented as univariate statistics, such as diversity indices, graphically or subject to multivariate analysis. In some regions a number of biotic indices have been produced. These provide useful measures of the state of the environment, but need to be developed and tested for each biogeographic area.

- Human activities have degraded aquatic environments. The new discipline of restoration ecology seeks to use ecological principles to restore ecosystems. Restoration of aquatic systems is often viable only in cases where biotic systems are important structuring agents. In most cases the natural dispersal of aquatic organisms leads to rapid recolonisation. Rarely however does a system revert to its pre-impacted condition.

- Science has a major role to play in the development of integrated ecosystem management systems. However, while science can advise on the natural state of an ecosystem, the degree of degradation, the sustainable levels of exploitation and the management regime needed to achieve this, it is for society to decide what configuration of the environment it desires. There will always be trade-offs between uses.

- Sustainability and biodiversity are now central tenets of international agreements. For some systems we have a good understanding of the science which underpins the setting of sustainable levels of exploitation. In multispecies systems we are far less certain of our ability to make such predictions. For most aquatic systems, focused studies on mechanisms responsible for maintaining biological richness and ecological functioning remain a priority.

References

Addessi, L. (1994) Human disturbance and long term changes on a rocky intertidal community. *Ecological Applications*, **4**, 786–797.

Agostinho, A.A., Miranda, L.E., Bini, L.M., Gomes, L.C., Thomaz, S.M. and Suzuki, H.I. (1999) Patterns of colonization in neotropical reservoirs, and prognoses on aging. In Tundisi, J.G. and Straskraba, M. (eds) *Theoretical Reservoir Ecology and its Applications*, pp. 227–265. Leiden: Backhuys.

Alcaraz, G. and Espina, S. (1997) Scope for growth of juvenile grass carp *Ctenopharyngodon idella* exposed to nitrite. *Comparative Biochemistry and Physiology C: Pharmacology, Toxicology and Endocrinology*, **116**, 85–88.

Ali, M.M., Murphy, K.J. and Langendorff, J. (1999) Interrelations of river ship traffic with aquatic plants in the River Nile, Upper Egypt. *Hydrobiologia*, **415**, 93–100.

Allen, K.R. (1980) *Conservation and Management of Whales*. Seattle: University of Washington Press.

Alverson, D.L., Freeburg, M.H., Murawski, S.A. and Pope, J.G. (1995) A global assessment of fisheries by-catch and discards. *FAO Fisheries Technical Paper*, **339**, Rome: FAO.

Ambrose, W.G. (1984) Role of predatory polychaetes and epibenthic predators on the structure of a soft-bottom community in Maine Estuary. *Journal of Experimental Marine Biology and Ecology*, **81**, 115–145.

Anderlini, V.C. (1992) The effect of sewage on trace-metal concentrations and scope for growth in *Mytilus edulis aoteanus* and *Perna canaliculus* from Wellington Harbor, New Zealand. *Science of the Total Environment*, **125**, 263–288.

Andersen, F.O. and Kristensen, E. (1991) Effects of burrowing macrofauna on organic matter decomposition in coastal marine sediments. *Symposium of the Zoological Society of London*, **63**, 69–88.

Andersen, U.V. (1995) Resistance of Danish coastal vegetation types to human trampling. *Biological Conservation*, **71**, 223–230.

Anderson, I. (1992) End of the line for deadly stowaways. *New Scientist*, 24 October, 12–13.

Andrew, N.L. and Pepperell, J.G. (1994) The by-catch of shrimp trawl fisheries. *Oceanography and Marine Biology Annual Review*, **30**, 527–565.

Andrews, J. and Kinsman, D. (1990) *Gravel Pit Restoration for Wildlife. A Practical Manual*. Sandy: Royal Society for the Protection of Birds.

Anon. (1994) Report of the Study Group on the North Sea plaice box. ICES, C.M. 1994/Assess: 14.

Anon. (1996) Manual of Methods of Measuring the Selectivity of Towed Fishing Gears. *ICES Co-operative Research Report*, 215, ICES, Copenhagen.

Anon. (1997) Incidental impacts of gill nets (Fantared). EC (DGXIV) contract 94/095.

Anon. (1998) Report of the Working Group on the assessment of demersal stocks in the North Sea and Skagerrak. ICES CM 1998/Assess. **7**, ICES, Copenhagen.

Anon. (1999) ICES CM 1999/D6, ICES, Copenhagen.

Armitage, P.D. (1977) A quantitative study of the invertebrate fauna of the River Tees below Cow Green Reservoir. *Freshwater Biology*, **6**, 229–240.

Armitage, P.D., Moss, D., Wright, J.F. and Furse, M.T. (1983) The performance of a new biological water quality score system based on macroinvertebrates over a wide range of unpolluted running water sites. *Water Research*, **17**, 333–347.

Ashley, K. and Nordin, R. (1999) Lake aeration in British Columbia: applications and experiences. In Murphy, T. and Mumawar, M. (eds) *Aquatic Restoration in Canada*, pp. 87–108. Leiden: Backhuys.

Ashley, K., Thompson, L.C., Sebastian, D., Lasenby, D.C., Smokorowski, K.E. and Andrusak, H. (1999) Restoration of kokanee salmon in Kootenay Lake, a large intermontane lake, by controlled seasonal application of limiting nutrients. In Murphy, T. and Mumawar, M. (eds) *Aquatic Restoration in Canada*, pp. 127–269. Leiden: Backhuys.

Attrill, M.J., Ramsay, P.M., Thomas, R.M. and Trett, M.W. (1996) An estuarine biodiversity hot-spot. *Journal of the Marine Biological Association of the United Kingdom*, **76**, 161–175.

Aure, J. and Stigebrandt, A. (1990) Quantitative estimates of the eutrophication effects of fish farming on fjords. *Aquaculture*, **90**, 135–156.

Auster, P.J., Malatesta, R.J. and Larosa, S. (1995) Patterns of microhabitat utilization by mobile megafauna on the southern New England (USA) continental shelf and slope. *Marine Ecology Progress Series*, **127**, 77–85.

Baird, D., McGlade, J.M. and Ulanowicz, R.E. (1991) The comparative ecology of six marine ecosystems. *Philosophical Transactions of the Royal Society of London*, **333**, 15–29.

Bally, R. and Griffiths, C.L. (1989) Effects of human trampling on an exposed rocky shore. *International Journal of Environmental Studies*, **34**, 115–125.

Bamber, R.N. (1980) The properties of fly ash as a marine sediment. *Marine Pollution Bulletin*, **11**, 323–326.

Bamber, R.N. (1984) The benthos of a marine fly-ash dumping ground. *Journal of the Marine Biological Association of the United Kingdom*, **64**, 211–226.

Barnes, N. and Frid, C.L.J. (1999) Restoring shores impacted by colliery spoil dumping. *Aquatic Conservation Marine and Freshwater Ecosystems*, **9**, 75–82.

Bate, G.C. and Adams, J.B. (2000) The effect of a single freshwater release into the Kromme Estuary. V: Overview and interpretation for the future. *Water SA*, **26**, 329–332.

Beck, C.A. and Barros, N.B. (1991) The impact of debris on the Florida manatee. *Marine Pollution Bulletin*, **22**, 508–510.

Beddington, J.R. (1995) The primary requirements. *Nature*, **374**, 213–214.

Beeftink, W.G. (1979) The structure of saltmarsh communities in relation to environmental disturbances. In Jefferies, R.L. and Davy, A.J. (eds) *Ecological Processes in Coastal Environments*, pp. 77–93. Oxford: Blackwell.

Bell, D.V. and Owen, M. (1990) Shooting disturbance – a review. In Matthews, G.V.T. (ed.) *Managing Waterfowl Populations*, pp. 159–171, Special Publication 12, Slimbridge: International Wetlands and Waterfowl Research Bureau.

Bentley, M.G. and Pacey, A.A. (1992) Physiological and environmental control of reproduction in polychaetes. *Oceanography and Marine Biology Annual Review*, **30**, 443–481.

Berger, J.J. (ed.) (1990) *Environmental Restoration. Science and Strategies for Restoring the Earth*, Selected papers from the Earth Conference, University of California, Berkeley, 1988. Washington, DC: Island Press.

Beukema, J.J. (1995) Long-term effects of mechanical harvesting of lugworms *Arenicola marina* on the zoobenthic community of a tidal flat in the Wadden Sea. *Netherlands Journal of Sea Research*, **33**, 219–227.

Beverton, R.J.H. and Holt, S.J. (1957) On the dynamics of exploited fish populations. *Fisheries Investigations*, Series II, **19**, 1–533.

Bilby, R.E., Fransen, B.R. and Bisson, P.A. (1996) Incorporation of nitrogen and carbon from spawning coho salmon into the trophic system of a small stream: evidence from stable isotopes. *Canadian Journal of Fisheries and Aquatic Sciences*, **53**, 164–173.

Bluhm, H. (1994) Monitoring megabenthic communities in abyssal manganese nodule sites of the East Pacific Ocean in association with commercial deep sea mining. *Aquatic Conservation Marine and Freshwater Ecosystems*, **4**, 187–201.

Bluhm, H., Schriever, G. and Thiel, H. (1995) Megabenthic recolonization in an experimentally disturbed abyssal manganese nodule area. *Marine Georesources and Geotechnology*, **13**, 393–416.

Bolch, C.J. (1993) Chemical and physical treatment options to kill toxic dinoflagellate cysts in ships' ballast water. *Journal of Marine Environmental Engineering*, **1**, 23–29.

Boon, P.J. (1993) Distribution, abundance and development of Trichoptera larvae in the River North Tyne following the commencement of hydroelectric power generation. *Regulated Rivers: Research and Management*, **8**, 211–224.

Boylston, J.W. (1996) Ballast water management for the control of non-indigenous species. *SNAME Transactions*, **104**, 391–417.

Bradley, R.S., Diaz, H.F., Eischeid, J.K., Jones, P.D., Kelly, P.M. and Goodess, C.M. (1987) Precipitation fluctuations over northern hemisphere land areas since the mid-19th century. *Science*, **237**, 171–175.

Brenchley, G.A. (1981) Disturbance and community structure: an experimental study of bioturbation in marine soft-bottom environments. *Journal of Marine Research*, **39**, 767–790.

Briand, F. and Cohen, J.E. (1984) Community food webs have scale-invariant structure. *Nature*, **307**, 264–266.

Briand, F. and Cohen, J.E. (1987) Environmental correlates of food-chain length. *Science*, **238**, 956–960.

Briggs, M.K. and Cornelius, S. (1998) Opportunities for ecological improvement along the lower Colorado River and delta. *Wetlands*, **18**, 513–529.

Brockmann, U., Billen, G. and Gieskes, W.W.C. (1989) North Sea nutrients and eutrophication. In Salomons, W., Bayne, B.L., Duursma, E.K. and Förstner, U. (eds) *Pollution of the North Sea: an Assessment*, pp. 348–389. Berlin: Springer-Verlag.

Brosnan, D.M. (1993) The effect of human trampling on biodiversity of rocky shores: monitoring and management strategies. *Recent Advances in Marine Science and Technology*, **92**, 333–341.

Brosnan, D.M. and Crumrine, L.L. (1994) Effects of human trampling on marine rocky shore communities. *Journal of Experimental Marine Biology and Ecology*, **177**, 79–97.

Brown, E.E., La Plante, M.G. and Covey, L.H. (1969) *A Synopsis of Catfish Farming*. Athens, GA: University of Georgia.

Brown, J.R., Gowen, R.J. and McLusky, D.S. (1987) The effect of salmon farming on the benthos of a Scottish sea loch. *Journal of Experimental Marine Biology and Ecology*, **109**, 39–51.

Bryan, G.W. and Langston, W.J. (1992) Bioavailability, accumulation and effects of heavy metals in sediments with special reference to United Kingdom estuaries – a review. *Environmental Pollution*, **76**, 89–131.

Buckingham, C.A., Lefebvre, L.W., Schaefer, J.M. and Kochman, H.I. (1999) Manatee response to boating activity in a thermal refuge. *Wildlife Society Bulletin*, **27**, 514–522.

Burger, J. (1994) The effect of human disturbance on foraging behaviour and habitat use in piping plover (*Charadrius melodus*). *Estuaries*, **17**, 695–701.

Cairns, J. (1980) *The Recovery Process in Damaged Ecosystems*. Ann Arbor, MI: Science Publishers.

Cairns, J. (1986) Restoration, reclamation and regeneration of degraded and destroyed ecosystems. In Soulé, M.E. (ed.) *Conservation Biology*. Sunderland, MA: Sinaeur.

Cairns, J. (1988) Restoration ecology: the new frontier. In *Restoration of Damaged Ecosystems*, Vol. 1, pp. 1–12. Boca Raton, FL: CRC Press.

Camphuysen, C.J., Calvo, B., Durinck, J., *et al.* (1995) Consumption of discards in the North Sea. Final report EC DG XIV research contract BIOECO/ 93/10. Netherlands Institute for Sea Research, Texel, NIOZ Rapport 1995–5.

Carlson, L.H. and Godfrey, P.J. (1989) Human impact management in a coastal recreation and natural area. *Biological Conservation*, **49**, 141–156.

Carney, K.M. and Sydeman, W.J. (1999) A review of human disturbance effects on nesting colonial waterbirds. *Waterbirds*, **22**, 68–79.

Carpenter, S.R. (1988) *Complex Interactions in Lake Communities*. New York: Springer-Verlag.

Carpenter, S.R. and Kitchell, J. (1985) Consumer control of lake productivity. *Bioscience*, **38**, 764–769.

Carpenter, S.R., Kitchell, J.F. and Hodgson, J.R. (1985) Cascading trophic interactions and lake productivity. *Bioscience*, **35**, 634–639.

Caspers, H. (1984) Spawning periodicity and habitat of the Palolo worm *Eunice viridis* (Polychaeta, Eunicidae) in the Samoan Islands. *Marine Biology*, **79**, 229–236.

Castilla, J.C. and Duran, L.R. (1985) Human exclusion from the rocky intertidal zone of central Chile: the effects on *Concholepas concholepas* (Gastropoda). *Oikos*, **45**, 391–399.

Chandra Prakesh, P.S., Sinha, R.K. and Reddy, A.K. (1990) Economic viability of aquaculture in sewage. *Journal of Environmental Biology*, **1**, 1, 7–14.

Chandrasekara, W.U. and Frid, C.L.J. (1996a) Effects of human trampling on tidalflat infauna. *Aquatic Conservation: Marine and Freshwater Ecosystems*, **6**, 299–311.

Chandrasekara, W.U. and Frid, C.L.J. (1996b) The effects of relic fauna on initial patch colonisation in a British saltmarsh. *Netherlands Journal of Sea Research*, **30**, 49–60.

Churchill, J.H. (1989) The effect of commercial trawling on sediment resuspension and transport over the Middle Atlantic Bight continental shelf. *Journal of the Marine Biological Association of the United Kingdom*, **9**, 841–864.

Clark, R.A. and Frid, C.L.J. (2001) Long term changes in the North Sea ecosystem. *Environmental Reviews*. In press.

Clark, R.B., Frid, C.L.J. and Attrill, M. (1997) *Marine Pollution*. Oxford: Clarendon Press.

Clark, S. and Edwards, A.J. (1999) An evaluation of artificial reef structures as tools for marine habitat rehabilitation in the Maldives. *Aquatic Conservation: Marine and Freshwater Ecosystems*, **9**, 5–22.

Clarke, K.R. and Warwick, R.M. (1994) *Change in Marine Communities: An Approach to Statistical Analysis and Interpretation*. PRIMER-E Ltd. and Plymouth Marine Laboratory.

Colborn, T., von Saal, F.S. and Soto, A.M. (1993) Developmental effects of endocrine disrupting chemicals in wildlife and humans. *Environmental Health Perspectives*, **101**, 378–384.

Collie, J.S., Escanero, G.A. and Valentine, P.C. (1997) Effects of bottom fishing on the benthic megafauna of Georges Bank. *Marine Ecology Progress Series*, **155**, 159–172.

Collier, M.P., Webb, R.H. and Andrews, E.D. (1997) Experimental flooding in Grand Canyon. *Scientific American*, January 1997, 66–73.

Connell, J.H. (1978) Diversity in tropical rain forests and coral reefs. *Science*, **199**, 1302–1310.

Costanza, R., d'Arge, R., de Groot, R., *et al.* (1997) The value of the world's ecosystem services and natural capital. *Nature*, **387**, 253–260.

Cote, I.M., Vinyoles, D., Reynolds, J.D., Doadrio, I. and Perdices, A. (1999) Potential impacts of gravel extraction on Spanish populations of river blennies *Salaria fluviatilis* (Pisces, Blennidae). *Biological Conservation*, **87**, 359–367.

Cott, H.B. (1961) Scientific results of an inquiry into the ecology and economic status of the Nile crocodile. *Transactions of the Zoological Society of London*, **29**, 211–356.

Courtenay, W.R. Jr and Meffe, G.K. (1989) Small fishes in strange places: a review of introduced poeciliids. In Meffe, G.K. and Snelson, F.F. Jr (eds) *Ecology and Evolution of Livebearing Fishes (Poeciliidae)*, pp. 319–331. Englewood Cliffs, NJ: Prentice Hall.

Cowx, I.G. (1994) Stocking strategies. *Fisheries Management and Ecology*, **1**, 15–30.

Cryer, M., Whittle, G.N. and Williams, R. (1987) The impact of bait collection by anglers on marine intertidal invertebrates. *Biological Conservation*, **42**, 83–93.

Cushing, D.H. (1977) The problems of stock and recruitment. In Gulland, J.A. (ed.) *Fish Population Dynamics*, pp. 116–133. London: John Wiley.

Cyrus, D.P., Martin, T.J. and Reavell, P.E. (1997) Saltwater intrusion from the Mzingazi River and its effects on adjacent swamp forest at Richards Bay, Zululand, South Africa. *Water SA*, **23**, 101–108.

D'Ancona, U. (1954) Fishing and fish culture in brackish-water lagoons. *FAO Fisheries Bulletin*, **7**, 147–172.

Daan, N., Bromley, P.J., Hislop, J.R.G. and Nielson, N.A. (1990) Ecology of North Sea fish. *Netherlands Journal of Sea Research*, **26**, 343–386.

Dall, W., Hill, B.J., Rothlisberg, P.C. and Sharples, D.J. (1990) The biology of the Penaeidae. *Advances in Marine Biology*, **27**, 1–484.

Dayton, P.K. (1985) Ecology of kelp communities. *Annual Review of Ecology and Systematics*, **16**, 215–245.

Dayton, P.K. and Tegner, M.J. (1984) Catastrophic storms, El-Nino, and patch stability in a Southern California kelp community. *Science*, **224**, 283–285.

Dayton, P.K., Tegner, M.J., Parnell, P.E. and Edwards, P.B. (1992) Temporal and spatial patterns of disturbance and recovery in a kelp forest community. *Ecological Monographs*, **62**, 421–445.

Dayton, P.K., Thrush, S.F., Agardy, M.T. and Hofman, R.J. (1995) Environmental effects of marine fishing. *Aquatic Conservation: Marine and Freshwater Ecosystems*, **5**, 205–232.

de Boer, W.F. and Longamane, F.A. (1996) The exploitation of intertidal food resources in Inhaca Bay, Mozambique, by shorebirds and humans. *Biological Conservation*, **78**, 295–303.

de Groot, S.J. (1984) The impact of bottom trawling on benthic fauna of the North Sea. *Ocean Management*, **9**, 177–190.

de Groot, S.J. (1996a) Quantitative assessment of the development of the offshore oil and gas industry in the North Sea. *ICES Journal of Marine Science*, **53**, 1045–1050.

de Groot, S.J. (1996b) The physical impact of marine aggregate extraction in the North Sea. *ICES Journal of Marine Science*, **53**, 1051–1053.

de Veen, J.F. (1976) On changes in some biological parameters in the North Sea sole. *Journal du Conseil Permanent International pour l'Exploration de la Mer*, **37**, 60–90.

Depledge, M.H., Aagaard, A. and Gyorkos, P. (1995) Assessment of trace-metal toxicity using molecular, physiological and behavioral biomarkers. *Marine Pollution Bulletin*, **31**, 19–27.

Desse, J. and Desse-Berset, N. (1993) Pèche et surpèche en Méditerranée: le temoinage des os. In Desse, J. and Audoin-Rouzeau, F. (eds) *Exploration des Animaux Sauvages à Travers le Temps*, pp. 327–339. Juan-les-Pins. Editions APDCA.

Diaz, R.J. and Rosenberg, R. (1995) Marine benthic hypoxia – a review of its ecological effects and the behavioral responses of benthic macrofauna. *Oceanography and Marine Biology*, **33**, 245–303.

Din, Z.B. and Ahamad, A. (1995) Changes in the scope for growth of blood cockles (*Anadara granosa*) exposed to industrial discharge. *Marine Pollution Bulletin*, **31**, 406–410.

Dobson, M. and Cariss, H. (1999) Restoration of afforested upland streams – what are we trying to achieve? *Aquatic Conservation: Marine and Freshwater Ecosystems*, **9**, 133–140.

Dobson, M. and Frid, C. (1998) *Ecology of Aquatic Systems*. Harlow: Addison Wesley Longman.

Dumont, H.J. (1999) The species richness of reservoir plankton and the effect of reservoirs on plankton dispersal (with particular emphasis on rotifers and cladocerans). In Tundisi, J.G. and Straskraba, M. (eds) *Theoretical Reservoir Ecology and its Applications*, pp. 477–491. Leiden: Backhuys.

Dye, A.H. (1992) Experimental studies of succession and stability in rocky intertidal communities subject to artisanal shellfish gathering. *Netherlands Journal of Sea Research*, **30**, 209–217.

Easa, M.E.S., Shereif, M.M., Shaaban, A.I. and Mancy, K.H. (1995) Public health implications of waste water reuse for fish production. *Water Science and Technology*, **32**, 145–152.

Eaton, J.W. (1999) Water-resource allocation for recreation, leisure and amenity. *J Chart Inst Water E* 13: 235–240.

Eckman, J.E. (1983) Hydrodynamic processes affecting benthic recruitment. *Limnology and Oceanography*, **28**, 241–257.

Economidis, P.S., Dimitriou, E., Pagoni, R., Michaloudi, E. and Natsis, L. (2000) Introduced and translocated fish species in the inland waters of Greece. *Fisheries Management and Ecology*, **7**, 239–250.

Edwards, R. (1997) New fish farm pesticides to flood Scottish lochs. *New Scientist*, **153**, 10.

Elliott, J.M., Lyle, A.A. and Campbell, R.N.B. (1997) A preliminary evaluation of migratory salmonids as vectors of organic carbon between marine and freshwater environments. *Science of the Total Environment*, **194–195**, 219–223.

Ellis, D.V. (1987) A decade of environmental impact assessment at marine and coastal mines. *Marine Mining*, **6**, 385–417.

Ellis, D.V. and Hoover, P.M. (1990) Benthos recolonizing mine tailings in British Columbia fjords. *Marine Mining*, **9**, 441–457.

Elner, R.W. and Vadas, R.L. (1990) Inference in ecology: the sea urchin phenomenon in the northwestern Atlantic. *American Naturalist*, **136**, 108–125.

Eng, C.H., Paw, J.N. and Guarin, F.Y. (1989) The environmental impact of aquaculture and the effects of pollution on coastal aquaculture development in southeast Asia. *Marine Pollution Bulletin*, **20**, 335–343.

Engås, A., Løkkeborg, S., Ona, E. and Soldal, A.V. (1993) Effects of seismic shooting on catch and catch-availability of cod and haddock. *Fisken og Havet*, **9**, 1–117.

Eno, N.C., Clark, R.A. and Sanderson, W.G. (eds) (1997) *Non-native Marine Species in British Waters: a Review and Directory*, Peterborough: Joint Nature Conservancy Council.

Erbe, C. and Farmer, D.M. (2000) Zones of impact around icebreakers affecting beluga whales in the Beaufort Sea. *Journal of the Acoustical Society of America*, **108**, 1332–1340.

Erman, A. (1971) *Life in Ancient Egypt*. New York: Dover Publications.

Ervik, A., Johannessen, P. and Aure, J. (1985) Environmental effects of marine Norwegian fish farms. ICES CM1985/F:37.

European Commission (DG XIV) (1998) *Statistical Bulletin of Fisheries*. European Commission (DG XIV), Brussels.

European Commission (2000) Directive 2000/60/EC of the European Parliament and of the Council of 23 October 2000 establishing a framework for community action in the field of water policy. *Official Journal of European Communication*, **L327**, 1–72.

Evans, S.M. (1999) TBT pollution: the catastrophe that never happened. *Marine Pollution Bulletin*, **38**, 629–636.

Evans, S.M., Evans, P.M. and Leksono, T. (1996) Widespread recovery of dogwhelks, *Nucella lapillus* (L.) from tributyltin contamination in the North Sea and the Clyde Sea. *Marine Pollution Bulletin*, **32**, 263–269.

Evans, S.M., Hunter, J.E., Elizal and Wahju, R.I. (1994) Composition and fate of the catch and bycatch in the Farne Deep (North Sea) *Nephrops* fishery. *ICES Journal of Marine Science*, **51**, 155–168.

FAO (1995) Global fishery production in 1993, FAO Report. FAO, Rome.

FAO (1998a) Fishery fleet statistics. *Bulletin of Fishery Statistics*, **35**. Rome: FAO.

FAO (1998b) *World Review of Fisheries and Aquaculture*. Rome: FAO.

Fitzpatrick, S. and Bouchez, B. (1998) Effects of recreational disturbance on the foraging behaviour of waders on a rocky beach. *Bird Study*, **45**, 157–171.

Fletcher, H. and Frid, C.L.J. (1996) The impact and management of visitor pressure in rocky intertidal algal communities of northeast England. *Aquatic Conservation: Marine and Freshwater Ecosystems*, **6**, 287–297.

Folke, C. and Kautsky, N. (1989) The role of ecosystems for a sustainable development of aquaculture. *Ambio*, **18**, 234–243.

Folke, C. and Kautsky, N. (1992) Aquaculture with its environment: prospects for sustainability. *Ocean and Coastal Management*, **17**, 5–24.

Fox, A.D. and Madsen, J. (1997) Behavioural and distributional effects of hunting disturbance on waterbirds in Europe: implications for refuge design. *Journal of Applied Ecology*, **34**, 1–13.

Fraser, W.R., Trivelpiece, W.Z., Ainley, D.G. and Trivelpiece, S.G. (1992) Increases in Antarctic penguin populations: reduced competition with whales or a loss of sea ice due to environmental warming? *Polar Biology*, **11**, 525–531.

French, P.W. (1997) *Coastal and Estuarine Management*. New York: Routledge.

Frid, C.L.J. (1989) The role of recolonisation processes in benthic communities, with special reference to the interpretation of predator-induced effects. *Journal of Experimental Marine Biology and Ecology*, **126**, 163–171.

Frid, C.L.J. and Clark, R.A. (2000) Long term changes in North Sea benthos: discerning the role of fisheries. In Kaiser, M.J. and de Groot, S.J. (eds) *Effects of Fishing on Non-target Species and Habitats*, pp. 198–216. Oxford: Blackwell.

Frid, C.L.J. and Clark, S. (1999) Restoring aquatic ecosystems. *Aquatic Conservation: Marine and Freshwater Ecosystems*, **9**, 1–4.

Frid, C.L.J. and Evans, P.R. (1995) Coastal habitats. In Sutherland, W.J. and Hill, D.A. (eds) *Managing Habitats for Conservation*, pp. 59–83. Cambridge: Cambridge University Press.

Frid, C.L.J. and Mercer, T.S. (1989) Environmental monitoring of caged fish farming in macrotidal environments. *Marine Pollution Bulletin*, **20**, 379–383.

Frid, C.L.J., Buchanan, J.B. and Garwood, P.R. (1996) Variability and stability in benthos: twenty-two years of monitoring off Northumberland. *ICES Journal of Marine Science*, **53**, 978–980.

Frid, C.L.J., Hansson, S., Ragnarsson, S.A., Rijnsdorp, A. and Steingrimmsson, A.A. (1999) Changing levels of predation on benthos as a result of exploitation of fish populations. *Ambio*, **28**, 578–582.

Frid, C.L.J., Harwood, K.G., Hall, S.J. and Hall, J.A. (2000) Long-term changes in the benthic communities on North Sea fishing grounds. *ICES Journal of Marine Science*, **57**, 1303–1309.

Furevik, D.M. and Fosseidengen, J.E. (2000) Investigation of naturally and deliberately lost gill nets in Norwegian waters. *ICES Annual Science Conference 2000*, Brugge. ICES, Copenhagen.

Gabrielides, G.P., Golik, A., Loizides, L., Marino, M.G., Bingel, F. and Torregrossa, M.V. (1991) Man made garbage pollution on the Mediterranean coastline. *Marine Pollution Bulletin*, **23**, 437–441.

Gaines, S., Brown, S. and Roughgarden, J. (1985) Spatial variation in larval concentrations as a cause of spatial variation in settlement for the barnacle, *Balanus glandula*. *Oecologia*, **67**, 267–272.

Garthe, S. and Damm, U. (1997) Discards from beam trawl fisheries in the German Bight (North Sea). *Archives of Fisheries and Marine Research*, **45**, 232–242.

Garthe, S., Campyhuysen, C.J. and Furness, R.W. (1996) Amounts of discards by commercial fisheries and their significance for food for seabirds in the North Sea. *Marine Ecology Progress Series*, **136**, 1–11.

Gerlach, S.A. (1988) Nutrients – an overview. In Newman, P.J. and Agg, A.R. (eds) *Environmental Protection of the North Sea*, pp. 147–175. Oxford: Heinemann.

GESAMP (1990) *The State of the Marine Environment*. Oxford: Blackwell.

Gessner, M.O., Chauvet, E. and Dobson, M. (1999) A perspective on leaf litter breakdown in streams. *Oikos*, **85**, 377–384.

Ghazanshahi, J., Huchel, T.D. and Devinny, J.S. (1983) Alteration of Southern California rocky shore ecosystems by public recreational use. *Journal of Environmental Management*, **16**, 379–394.

Gibbins, C.N. and Heslop, J. (1998) An evaluation of inter-basin water transfers as a mechanism for augmenting salmonid and grayling habitat in the River Wear, North-East England. *Regulated Rivers Research and Management*, **14**, 357–382.

Gibbs, P.E., Bryan, G.W. and Pascoe, P.L. (1991) TBT-induced imposex in the dogwhelk, *Nucella lapillus* – geographical uniformity of the response and effects. *Marine Environmental Research*, **32**, 79–87.

Gido, K.B. and Brown, J.H. (1999) Invasion of North American drainages by alien fish species. *Freshwater Biology*, **42**, 387–399.

Gilek, M., Tedengren, M. and Kautsky, N. (1992) Physiological performance and general histology of the blue mussel, *Mytilus edulis* L, from the Baltic and North Seas. *Netherlands Journal of Sea Research*, **30**, 11–21.

Gislason, H. and Rice, J. (1998) Modelling the response of size and diversity spectra of fish assemblages to changes in exploitation. *ICES Journal of Marine Science*, **55**, 362–370.

Godoy, C. and Moreno, C. (1989) Indirect effects of human exclusion from the rocky intertidal in southern Chile: a case of cross-linkage between herbivores. *Oikos*, **54**, 101–106.

Golik, A. and Krom, M.D. (1989) Fly ash disposal in the Mediterranean Sea. *IOLR Technical Report* H1/90.

Goodsell, J.A. and Kats, L.B. (1999) Effect of introduced mosquitofish on pacific treefrogs and the role of alternative prey. *Conservation Biology*, **13**, 921–924.

Gowen, R.J. and Bradbury, N.B. (1987) The ecological impact of salmonid farming in coastal waters: a review. *Oceanography and Marine Biology Annual Review*, **25**, 563–575.

Gowen, R.J., Brown, J.R., Bradbury, N.B. and McLusky, D.S. (1988) Investigations into benthic enrichment, hypernutrification and eutrophication associated with mariculture in Scottish coastal waters (1984–1988). Report to Highlands & Islands Development Board, Crown Estate Commissioners, Nature Conservancy Council, Countryside Commission for Scotland and the

Scottish Salmon Growers Association from Dept Biological Sciences, University of Stirling, Stirling.

Graham, N. (1997) Reduction of by-catch in the brown shrimp, *Crangon crangon* fisheries of the Wash and Humber estuaries. PhD thesis. University of Lincolnshire and Humberside.

Grant, A., Hateley, J.G. and Jones, N.V. (1989) Mapping the ecological impact of heavy metals on the estuarine polychaete *Nereis diversicolor* using inherited metal tolerance. *Marine Pollution Bulletin*, **20**, 235–238.

Gray, B.T., Coley, R.W., MacFarlane, R.J., Puchniak, A.J., Sexton, D.A. and Stewart, G.R. (1999) Restoration of prairie wetlands to enhance bird habitat: a ducks unlimited Canada perspective. In Murphy, T. and Mumawar, M. (eds) *Aquatic Restoration in Canada*, pp. 171–194. Leiden: Backhuys.

Gray, J.S. and Ventilla, R.J. (1973) Growth rates of sediment living marine protozoans as a toxicity indicator for heavy metals. *Ambio*, **2**, 118–121.

Greenstreet, S.P.R. (1996) Estimating the daily consumption of production of fish in the North Sea in each quarter of the year. *Scottish Fisheries Research Report*, **55**.

Greenstreet, S.P.R., Bryant, A.D., Broekhuizen, N., Hall, S.J. and Heath, M.R. (1997) Seasonal variation in the consumption of food by fish in the North Sea and implications for food web dynamics. *ICES Journal of Marine Science*, **54**, 243–266.

Guillemain, M., Fritz, H. and Guillon, N. (2000) The use of an artificial wetland by shoveller *Anas clypeata* in western France: the role of food resources. *Revue d'Ecologie de la Terre et la Vie*, **55**, 263–274.

Gundlach, E.R. and Hayes, M.O. (1978) Vulnerability of coastal environments to oil spill impacts. *Marine Technology Society Journal*, **12**, 18–27.

Güttinger, H. and Stumm, W. (1992) An analysis of the Rhine pollution caused by the Sandoz chemical accident, 1986. *Interdisciplinary Science Reviews*, **17**, 127–135.

Hakulinen, T., Anderson, A.A., Malker, B., Rikkala, E., Shou, G. and Tulinius, H. (1996) Trends in cancer incidence in the Nordic countries. A collaborative study of the five Nordic cancer requisites. *Acta Pathologica et Microbiologia Immunologica Scandinava*, **94A**, Suppl. 288, 10151.

Hall, S.J. (1999) *The Effects of Fishing on Marine Ecosystems and Communities*. Oxford: Blackwell.

Hall, S.J. and Raffaelli, D. (1991) Food web patterns – lessons from a species rich web. *Journal of Animal Ecology*, **60**, 823–842.

Hall, S.J. and Raffaelli, D.G. (1993) Food webs: theory and reality. *Advances in Ecological Research*, **24**, 187–239.

Hanson, J.A. and Goodwin, H.L. (eds) (1977) *Shrimp and Prawn Farming in the Western Hemisphere*. Strondsburg: Dowden, Hutchinson & Ross.

Harries, J.E., Sheahan, D.A., Jobling, S., Matthiessen, P., Neall, P., Sumpter, J.P., Tylor, T. and Zaman, N. (1997) Estrogenic activity in five United Kingdom rivers detected by measurements of vitellogenesis in caged male trout. *Environmental Toxicology and Chemistry*, **16**, 534–542.

Harriott, V., Davis, J.D. and Banks, S.A. (1997) Recreational diving and its impact on marine protected areas in eastern Australia. *Ambio*, **26**, 173–179.

Hastings, K., Hesp, P. and Kendrick, G.A. (1995) Seagrass loss associated with boat moorings at Rottnest Island, Western Australia. *Ocean and Coastal Management*, **26**, 225–246.

Havard, M.S.C. and Tindall, E.C. (1994) The impacts of bait digging on the polychaete fauna of the Swale Estuary, Kent, U.K. *Polychaete Research*, **16**, 32–36.

Havinga, H. and Smits, A.J.M. (2000) River management along the Rhine: a retrospective view. In Smits, A.J.M., Nienhuis, P.H. and Leuven, R.S.E.W. (eds) *New Approaches to River Management*, pp. 15–32. Leiden: Backhuys.

Hawke, C.J. and José, P.V. (1996) Reed bed management for commercial and wildlife interests. Sandy: Royal Society for the Protection of Birds.

Hawkins, J.P. and Roberts, C.M. (1993) Effects of recreational scuba diving on coral reefs: trampling on reef-flat communities. *Journal of Applied Ecology*, **30**, 25–30.

Hawkins, S.J. and Southward, A.J. (1992) Lessons from the Torrey Canyon oil spill, recovery and stability of rocky shore communities. In Thayer, G.D. (ed.) *Restoring the Nation's Marine Environment*, pp. 584–631. Symposium on Marine Habitat Restoration, NOAA, USA. Maryland Sea Grant Publications.

Hawkins, S.J., Allen, J.R. and Bray, S. (1999) Restoration of temperate marine and coastal ecosystems: nudging nature. *Aquatic Conservation: Marine and Freshwater Ecosystems*, **9**, 23–46.

Haynes, D. (1997) Marine debris on continental islands and sand cays in the far northern section of the Great Barrier Reef Marine Park, Australia. *Marine Pollution Bulletin*, **34**, 276–279.

He, P. and Foster, D. (2000) Reducing seabed contact of shrimp trawls. ICES Fishing Technology and Fish Behaviour Working Group Meeting (Working paper). Haarlem, The Netherlands.

Heiligenberg, T.V.D. (1987) Effects of mechanical and manual harvesting of lug worms *Arenicola marina* L. on the benthic fauna of tidal flats in the Dutch Wadden Sea. *Biological Conservation*, **39**, 165–177.

Herrando-Pérez, S. and Frid, C.L.J. (1998) The cessation of long term fly ash dumping: effects on macrobenthos and sediments. *Marine Pollution Bulletin*, **36**, 780–790.

Hickel, W., Mangelsdorf, P. and Berg, J. (1993) The human impact in the German Bight: eutrophication during three decades (1962–1991). *Helgoländer Meeresuntersuchungen*, **47**, 243–263.

Hilderbrand, G.V., Hanley, T.A., Robbins, C.T. and Schwartz, C.C. (1999) Role of brown bears (*Ursus arctos*) in the flow of marine nitrogen into a terrestrial ecosystem. *Oecologia*, **121**, 546–550.

Hockey, P.A.R. and Bosman, A.L. (1986) Man as an intertidal predator in Transkei: disturbance, community convergence, and management of a natural food resource. *Oikos*, **46**, 3–14.

Hockey, P.A.R., Bosman, A.L. and Siegfried, W.R. (1988) Patterns and correlates of shellfish exploitation by coastal people in Transkei – an enigma of protein production. *Journal of Applied Ecology*, **25**, 353–363.

Hoff, J.T., Thompson, J.A.J. and Wong, C.S. (1982) Heavy metal release from mine tailings into sea water – a laboratory study. *Marine Pollution Bulletin*, **13**, 283–286.

Horsman, P.V. (1982) The amount of garbage pollution from merchant ships. *Marine Pollution Bulletin*, **13**, 167–169.

Howells, G. and Dalziel, T.R.K. (eds) (1992) *Restoring Acid Waters: Loch Fleet 1984–1990*. London: Elsevier.

Huang, C. and Sih, A. (1991) Experimental studies on direct and indirect interactions in a three trophic-level stream system. *Oecologia*, **85**, 530–536.

Hughes, S. and Morley, S. (2000) Aspects of fisheries and water resources management in England and Wales. *Fisheries Management and Ecology*, **7**, 75–84.

Huxham, M. and Sumner, D. (eds) (2000) *Science and Environmental Decision Making*. Harlow, UK: Pearson Education.

Hynes, H.B.N. (1960) *The Biology of Polluted Waters*. Liverpool: Liverpool University Press.

ICES (1998) Report of the Working Group on the Ecosystem Effects of Fishing Activities. ICES, Copenhagen, ICES CM 1998/ACFM/ACME:1 Ref.:E.

ICES (2000) Fishing Technology and Fish Behaviour Working Group meeting (Working paper). Haarlem, The Netherlands.

Inoue, H. (1972) On water exchange in a net cage stocked with the fish, hamachi. *Bulletin of the Japanese Society of Scientific Fisheries*, **38**, 167–176.

Isaksen, B., Valdemarsen, J.W., Larsen, R.B. and Karlsen, L. (1992) Reduction of fish by-catch in shrimp trawl using a rigid separator grid in the aft belly. *Fisheries Research*, **13**, 335–352.

Jackson, M.J. and James, R. (1979) The influence of bait digging on cockle *Cerastoderma edule* population in North Norfolk. *Journal of Applied Ecology*, **18**, 559–569.

Jennings, S. and Reynolds, J.D. (2000) Impacts of fishing on diversity: from pattern to process. In Kaiser, M.J. and de Groot, S.J. (eds) *The Effects of Fishing on Non-target Species and Habitats*, pp. 235–250. Oxford: Blackwell.

Jones, J.A.A. (1997) *Global Hydrology. Processes, Resources and Environmental Management*. Harlow: Addison Wesley Longman.

Jones, P. (1998) Submarine detection tests and whale strandings. *Marine Pollution Bulletin*, **36**, 506–507.

Jonsson, M., Malmqvist, B. and Hoffsten, P-O. (2001) Litter breakdown rates in boreal streams:

does shredder species richness matter? *Freshwater Biology*, **46**, 161–171.

Kaiser, M.J. and de Groot, S.J. (eds) (2000) *Effects of Fishing on Non-target Species and Habitats*. Oxford: Blackwell Scientific.

Kaiser, M.J. and Spencer, B.E. (1994) Fish scavenging behaviour in recently trawled areas. *Marine Ecology Progress Series*, **112**, 41–49.

Kaiser, M.J. and Spencer, B.E. (1996) Behavioural responses of scavengers to beam trawl disturbance. In Greenstreet, S.P.R. and Tasker, M.L. (eds) *Aquatic Predators and their Prey*, pp. 116–123. Oxford: Blackwell.

Kalk, M. (ed.) (1995) *A Natural History of Inhaca Island, Mozambique*, University of Witwatersrand Press.

Karr, J.R. (1999) Defining and measuring river health. *Freshwater Biology*, **41**, 221–234.

Kaster, J.L. and Jacobi, G.Z. (1978) Benthic macroinvertebrates of a fluctuating reservoir. *Freshwater Biology*, **8**, 283–290.

Kay, A.M. and Liddle, M.J. (1989) Impact of human trampling in different zones on a coral reef flat. *Environmental Management*, **13**, 509–520.

Kaza, S. (1995) Community involvement in marine protected areas. *Oceanus*, **31**, 75–81.

Kelaher, B.P., Underwood, A.J. and Chapman, M.G. (1998) Effect of boardwalks on the semaphore crab *Heloecius cordiformis* in temperate urban mangrove forests. *Journal of Experimental Marine Biology and Ecology*, **227**, 281–300.

Keller, V. (1989) Variations in response of great crested grebes *Podiceps cristatus* to human disturbance – a sign of adaptation? *Biological Conservation*, **49**, 31–45.

Kemp, J., Harper, D.M. and Crosa, G.A. (1999) Use of 'functional habitats' to link ecology with morphology and hydrology in river rehabilitation. *Aquatic Conservation: Marine and Freshwater Ecosystems*, **9**, 159–178.

Kenny, A.J. and Rees, H.L. (1994) The effects of marine gravel extraction on the macrobenthos early post-dredging recolonization. *Marine Pollution Bulletin*, **28**, 442–447.

Kenny, A.J. and Rees, H.L. (1996) The effects of marine gravel extraction on the macrobenthos – results 2 years post-dredging. *Marine Pollution Bulletin*, **32**, 615–622.

Kenny, D. and Loehle, C. (1991) Are food webs randomly connected? *Ecology*, **72**, 1794–1799.

Keough, M.J., Quinn, G.P. and King, A. (1993) Correlations between human collecting and intertidal mollusk populations on rocky shores. *Conservation Biology*, **7**, 378–390.

Kersten, M., Dicke, M., Kriews, M., Naumann, K., Schmidt, D., Schulz, M., Schwikowski, M. and Steiger, M. (1989) Distribution and fate of heavy metals in the North Sea. In Salomons, W., Bayne, B.L., Duursma, E.K. and Förstner, U. (eds) *Pollution of the North Sea: an Assessment*, pp. 300–347. Berlin: Springer-Verlag.

Khalil, M.T. and Hussein, H.A. (1997) Use of waste water for aquaculture: an experimental field study at a sewage-treatment plant, Egypt. *Aquaculture Research*, **28**, 859–865.

Kim, J.P. and Burggraaf, S. (1999) Mercury bioaccumulation in the rainbow trout (*Oncorhynchus mykiss*) and the trout food web in lakes Okareka, Okaro, Tarawera, Rotomakana and Rotorua, New Zealand. *Water, Air and Soil Pollution*, **115**, 535–546.

Kingsford, M.J., Pitt, K.A. and Gillanders, B.M. (2000) Management of jellyfish fisheries, with special reference to the order Rhizostomeae. *Oceanography & Marine Biology: Annual Review*, **38**, 85–156.

Kingston, P.F. (1987) Field effects of platform discharge on benthic macrofauna. *Philosophical Transactions of the Royal Society, London, Series B*, **316**, 545–565.

Kitchener, A.C. and Conroy, J.W.H. (1997) The history of the Eurasian beaver *Castor fiber* in Scotland. *Mammal Review*, **27**, 95–108.

Knight, J., McCarron, S.G., McCabe, A.M. and Sutton, B. (1999) Sand and gravel aggregate resource management and conservation in Northern Ireland. *Journal of Environmental Management*, **56**, 195–207.

Kress, N., Golik, A., Galil, B. and Krom, M.D. (1993) Monitoring the disposal of coal fly ash at a deep water site in eastern Mediterranean sea. *Marine Pollution Bulletin*, **26**, 447–456.

Kyle, R., Pearson, B., Fielding, P.J., Robertson, W.D. and Birnie, S.L. (1997) Subsistence shellfish harvesting in the Maputuland marine reserve in northern Kwazulu-Natal, South Africa: rocky shore organisms. *Biological Conservation*, **82**, 183–192.

Laist, D.W. (1987) Overview of the biological effects of lost and discarded plastic debris in the marine environment. *Marine Pollution Bulletin*, **18**, 319–326.

Lancaster, J. and Robertson, A.L. (1995) Micro-crustacean prey and macroinvertebrate predators in a stream food web. *Freshwater Biology*, **34**, 123–134.

Langford, T.E. (1983) *Electricity Generation and the Ecology of Natural Waters*. Liverpool: Liverpool University Press.

Lasenby, D.C., Northcote, T.G. and Fürst, M. (1986) Theory, practice and effects of *Mysis relicta* introductions to North American and Scandinavian lakes. *Canadian Journal of Fisheries and Aquatic Sciences*, **43**, 1277–1284.

Lasiak, T. (1991) The susceptibility and or resilience of rocky littoral mollusks to stock depletion by the indigenous Coastal People of Transkei, Southern Africa. *Biological Conservation*, **56**, 245–264.

Lasiak, T.A. and Field, J.G. (1995) Community-level attributes of exploited and non-exploited rocky infratidal macrofaunal assemblages in Transkei. *Journal of Experimental Marine Biology and Ecology*, **185**, 33–53.

Law, R. and Grey, D.R. (1989) Life history, evolution and sustainable yields from populations with age-specific cropping. *Evolutionary Ecology*, **3**, 343–359.

Laws, R.M. (1985) The ecology of the Southern Ocean. *American Scientist*, **73**, 26–40.

Le Cren, E.D. (1973) Some examples of the mechanisms that control population dynamics of salmonid fish. In Bartlett, M.S. and Hiorns, R.W. (eds) *The Mathematical Theory of the Dynamics of Biological Populations*, pp. 125–135. London: Academic.

Ledger, M.A. and Hildrew, A.G. (2001) Recolonization by the benthos of an acid stream following a drought. *Archiv für Hydrobiologie*. In press.

Lesage, V., Barrette, C., Kingsley, M.C.S. and Sjare, B. (1999) The effect of vessel noise on the vocal behavior of Belugas in the St. Lawrence River estuary, Canada. *Marine Mammal Science*, **15**, 65–84.

Lessmark, O. and Thörnelöf, E. (1986) Liming in Sweden. *Water, Air and Soil Pollution*, **31**, 809–815.

Lévêque, C. (1995) Role and consequences of fish diversity in the functioning of African freshwater ecosystems: a review. *Aquatic Living Resources*, **8**, 59–78.

Lévêque, C. (1997) *Biodiversity Dynamics and Conservation. The Freshwater Fish of Tropical Africa*. Cambridge: Cambridge University Press.

Lever, C. (1977) *The Naturalized Animals of the British Isles*. London: Hutchinson.

Lewin, R. (1986) Supply side ecology. *Science*, **234**, 25–27.

Lewis, J.R. (1964) *The Ecology of Rocky Shores*. London: English Universities Press.

Lindeboom, H.J. and de Groot, S.J. (1998) The effects of different types of fisheries on the North Sea and Irish Sea Benthic Ecosystems. NIOZ Report 1998-1, RIVO-DLO Report C003/98.

Ling, S.W. (1977) *Aquaculture in Southeast Asia – A Historical Review*. Seattle, USA: University of Washington Press.

Littlepage, J.L., Ellis, D.V. and McInterney, J. (1984) Marine disposal of mine tailings. *Marine Pollution Bulletin*, **15**, 242–244.

Lopez, G.R. and Levinton, J.S. (1987) Ecology of deposit-feeding animals in marine sediments. *Quarterly Review of Biology*, **62**, 235–260.

Lowry, N. (1995) The effect of twine diameter on trawl cod-end selectivity. ICES Fishing Technology and Fish Behaviour Working Group Meeting (Working paper). Aberdeen, UK. ICES, Copenhagen.

Lucas, Z. (1992) Monitoring persistent litter in the marine environment on Sable Island, Nova Scotia. *Marine Pollution Bulletin*, **24**, 192–199.

Lumb, C.M. (1989) Self-pollution by Scottish salmon farms? *Marine Pollution Bulletin*, **20**, 375–379.

Lutz, R.A. (1985) Mussel aquaculture in the United States. In Huner, J.V. and Brown, E.E. (eds) *Crustacean and Mollusc Aquaculture in the United States*, pp. 311–363. Westport, USA: A.V. Publishers.

Lye, C.M., Frid, C.L.J., Gill, M.E. and McCormick, D. (1997) Abnormalities in the reproductive health of flounder *Platichthys flesus* exposed to effluent from a sewage treatment works. *Marine Pollution Bulletin*, **34**, 34–41.

Lye, C.M., Frid, C.L.J. and Gill, M.E. (1998) Seasonal reproductive health of flounder Platichthys flesus exposed to sewage effluent. *Marine Ecology Progress Series*, **170**, 249–260.

MAFF (1997) *United Kingdom Sea Fisheries Statistics*. London: HMSO.

Magurran, A.E. (1988) *Ecological Diversity and its Measurement*. London: Croom Helm.

Maher, W., Batley, G.E. and Lawrence, I. (1999) Assessing the health of sediment ecosystems: use of chemical measurements. *Freshwater Biology*, **41**, 361–372.

Mann, K.H. and Lazier, J.R.N. (1991) *Dynamics of Marine Ecosystems: Biological–Physical Interactions in the Ocean*. Oxford: Blackwell.

Mason, C.F. (1990) Biological aspects of freshwater pollution. In Harrison, R.M. (ed.) *Pollution: Causes, Effects and Control*, pp. 99–125. London: Royal Society of Chemistry.

Mason, C.F. (1996) *Biology of Freshwater Pollution*, 3rd edition. Harlow: Pearson Higher Education.

Mato, Y., Isobe, T., Takada, H., Kanehiro, H., Ohtake, C. and Kaminuma, T. (2001) Plastic resin pellets as a transport medium for toxic chemicals in the marine environment. *Environmental Science and Technology*, **35**, 318–324.

Matthiessen, P., Allen, Y.T., Allchin, C.R., *et al.* (1998) Oestrogenic endocrine disruption in flounder (*Platichthys flesus* L.) from United Kingdom estuarine and marine waters. CEFAS Lowestoft, *Science Series Technical Report*, **107**.

Mayer, L.M., Schick, R.H., Findlay, R.H. and Rice, D.L. (1991) Effects of commercial dragging on sedimentary organic matter. *Marine Environmental Research*, **31**, 249–261.

McKay, D.W. and Fowler, S.L. (1997) Review of winkle, *Littorina littorea*, harvesting in Scotland. *Scottish Natural Heritage Review*, **69**, 32.

McLusky, D.S., Anderson, F.E. and Wolfe-Murphy, S. (1983) Distribution and population recovery of *Arenicola marina* and other benthic fauna after bait digging. *Marine Ecology Progress Series*, **11**, 173–179.

McNally, G. (1997) Peatlands, power and post-industrial use. In Parkyn, L., Stoneman, R.E. and Ingram, H.A.P. (eds) *Conserving Peatlands*, pp. 245–251. Wallingford: CAB International.

McNeil, W.J. (1979) Review of transportation and recruitment of anadromous species. In Tully, T.V.R. and Dill, W.A. (eds) *Advances in Aquaculture*, pp. 547–554. Oxford: Fishing News Books.

McVey, J.P. (ed.) (1983) *Handbook of Mariculture, Volume 1. Crustacean Aquaculture*. Boca Raton, FL: CRC Press.

Meijer, L.W. (1992) Recreation in the Dutch Wadden dune areas, a curse or a blessing. In Hilgerloh, G. (ed.) *Dune Management in the Wadden Sea Areas*, pp. 55–61. Proceedings of the Third Trilateral Working Conference, Nordney, Germany.

Messieh, S. (1991) Fluctuations in Atlantic herring succession in relation to organic enrichment and pollution of the marine environment. *Oceanography and Marine Biology Annual Review*, **16**, 229–311.

Metcalfe, J. (1989) Biological water quality assessment of running waters based on macroinvertebrate communities: history and present status in Europe. *Environmental Pollution*, **60**, 101–139.

Michael, R.G. (1986) *Ecosystems of the World; Managed Aquatic Ecosystems*. Amsterdam: Elsevier.

Mills, E.L., Leach, J.H., Carlton, J.T. and Secor, C.L. (1993) Exotic species in the Great Lakes: a history of biotic crises and anthropogenic introductions. *Journal of Great Lakes Research*, **19**, 1–54.

Mills, S. (1989) Salmon farming's unsavoury side. *New Scientist*, **122**, 58–60.

Milne, P.H. (1970) Fish farming: a guide to the design and construction of net enclosures. Dept Agriculture & Fisheries Scotland, *Marine Research*, **1**.

Miserendino, M.L. and Pizzolón, L.A. (2000) Macroinvertebrates of a fluvial system in Patagonia: altitudinal zonation and functional structure. *Archiv für Hydrobiologie*, **150**, 55–83.

Moreau, J. and Costa-Pierce, B. (1997) Introduction and present status of carp in Africa. *Aquaculture Research*, **28**, 717–732.

Moreno, C.A., Lunecke, K.M. and Lepez, M.I. (1984) Man as a predator in the intertidal zone of southern Chile. *Oikos*, **42**, 155–160.

Moreno, C.A., Lunecke, K.M. and Lopez, M.I. (1986) The response of an intertidal *Concholepas concholepas* (Gastropoda) population to protection from Man in southern Chile and the effects on benthic sessile assemblages. *Oikos*, **43**, 359–364.

Morgan, C.L., Odunton, N.A. and Jones, A.T. (1999) Synthesis of environmental impacts of deep seabed mining. *Marine Georesources and Geotechnology*, **17**, 307–356.

Morgan, M.D., Threlkeld, S.T. and Goldman, C.R. (1978) Impact of the introduction of kokanee (*Oncorhynchus nerka*) and opossum shrimp (*Mysis relicta*) on a subalpine lake. *Journal of the Fisheries Research Board of Canada*, **35**, 1572–1579.

Morrison, B.R.S. (1990) Recolonisation of four small streams in central Scotland following drought conditions in 1994. *Hydrobiologia*, **208**, 261–267.

Moss, B. (1983) The Norfolk Broadland: experiments in the restoration of a complex wetland. *Biological Reviews*, **58**, 521–561.

Mudge, G.P. (1983) The incidence and significance of ingested lead pellet poisoning in British wildfowl. *Biological Conservation*, **27**, 333–372.

Murphy, K.J. and Eaton, J.W. (1983) Effects of pleasure-boat traffic on macrophyte growth in canals. *Journal of Applied Ecology*, **20**, 713–729.

Nath, B.N. and Sharma, R. (2000) Environment and deep-sea mining: a perspective. *Marine Georesources and Geotechnology*, **18**, 285–294.

National Marine Fisheries Service (1996) *NMFS National Gravel Extraction Policy*. United States Department of Commerce.

Nehring, S. (1998) 'Natural' processes in the Wadden Sea – challenging the concept of an intact ecosystem. *Ocean Challenge*, **8**, 27–29.

Newell, R.C., Maughan, D.W., Trett, M.W., Newell, P.F. and Seiderer, L.J. (1991) Modification of benthic community structure in response to acid-iron wastes discharge. *Marine Pollution Bulletin*, **22**, 112–118.

Newman, P.C. (1987) *Caesars of the Wilderness – Company of Adventurers*. Viking.

Nicholas, A.P., Woodward, J.C., Christopoulos, G. and Macjlin, M.G. (1999) Modelling and monitoring river response to environmental change: the impact of dam construction and alluvial gravel extraction on bank erosion rates in the Lower Alfios Basin, Greece. In Brown, A.G. and Quine, T.A. (eds) *Fluvial Processes and Environmental Change*, pp. 117–137. London: John Wiley.

Norris, R.H. and Thomas, M.C. (1999) What is river health? *Freshwater Biology*, **41**, 197–209.

Norse, E.A., Rosenbaum, K.L., Wilcove, D.S., Wilcox, B.A., Romme, W.H., Johnston, D.W. and Stout, M.L. (1986) *Conserving Biological Diversity in our Natural Forests*. Washington, DC: The Wilderness Society.

North Sea Task Force (1993) *North Sea Quality Status Report 1993*. Fredensborg: Olsen & Olsen.

O'Connell, C. (1999) Ireland's raised bogs. *British Wildlife*, **11**, 100–107.

Olive, P.J.W. (1993) Management of the exploitation of lugworm *Arenicola marina* and the ragworm *Nereis virens* (polychaeta) in conservation areas. *Aquatic Conservation: Marine and Freshwater Ecosystems*, **3**, 1–24.

Paine, R.T. (1974) Intertidal community structure. Experimental studies on the relationship between a dominant competitor and its principal predator. *Oecologica*, **15**, 93–120.

Paine, R.T. (1980) Food webs: linkage, interaction strength and community infrastructure. *Journal of Animal Ecology*, **49**, 667–685.

Paine, R.T. (1988) Food webs: road maps of interactions or grist for theoretical development? *Ecology*, **69**, 1648–1654.

Paine, R.T. (1992) Food-web analysis through field measurements of per capita interaction strength. *Nature*, **355**, 73–75.

Pandit, A.K. (1999) *Freshwater Ecosystems of the Himalaya*. Carnforth: Parthenon.

Parkyn, L., Stoneman, R.E. and Ingram, H.A.P. (eds) (1997) *Conserving Peatlands*. Wallingford: CAB International.

Pascoe, S. (1997) By-catch management and the economics of discarding. *FAO Fisheries Technical Paper*, **370**. Rome: FAO.

Pauly, D. and Christensen, V. (1995) Primary production required to sustain global fisheries. *Nature*, **374**, 255–257.

Pauly, D., Christensen, V., Dalsgaard, J., Forese, R. and Torres, F. (1998) Fishing down marine food webs. *Science*, **279**, 860–863.

Pearce, D.W. and Barbier, E. (2000) *Blueprint for a Sustainable Economy*. London: Earthscan.

Pearce, D.W., Markandya, A. and Barbier, E. (1989) *Blueprint for a Green Economy: a Report*. London: Earthscan.

Pearce, F. (1995) How the Soviet Seas were lost. *New Scientist*, 11 November, 39–42.

Pearson, H.W. (1996) Expanding the horizons of pond technology and application in an environmentally conscious world. *Water Science and Technology*, **33**, 1–9.

Pearson, H.W., Avery, S.T., Mills, S.W., Njaggah, P. and Odiambo, P. (1996) Performance of the Phase II Dandora Waste Stabilization Ponds the largest in Africa: the case for anaerobic ponds. *Water Science and Technology*, **33**, 91–98.

Pearson, T.H. and Rosenberg, R. (1978) Macrobenthic succession in relation to organic pollution of the marine environment. *Oceanography and Marine Biology Annual Review*, **16**, 229–311.

Pedersen, T.F. (1983) Dissolved heavy metals in a lacustine mine tailings deposit – Buttle Lake, British Columbia. *Marine Pollution Bulletin*, **14**, 249–254.

Petitjean, M.O.G. and Davies, B.R. (1988) Ecological impacts of inter-basin water transfers: some case studies, research requirements and assessment procedures in southern Africa. *South African Journal of Science*, **84**, 819–828.

Pillay, T.V.R. (1993) *Aquaculture: Principles and Practices*. Oxford: Fishing News Books.

Pimm, S.L. and Hyman, J.B. (1987) Ecological stability in the context of multispecies fisheries. *Canadian Journal of Fisheries and Aquatic Sciences*, **44**, 84–94.

Pinnegar, J.K. and Polunin, N.V.C. (2000) Contributions of stable-isotope data to elucidating food webs of Mediterranean rocky littoral fishes. *Oecologia*, **122**, 399–409.

Pitcher, T.J. and Hart, P.J.B. (1983) *Fisheries Ecology*. Dordrecht: Kluwer Academic Publishers.

Polis, G.A. and Strong, D.R. (1996) Food web complexity and community dynamics. *American Naturalist*, **147**, 813–846.

Pollingher, U.K., Zohary, T. and Fishbein, T. (1998) Algal flora in the Hula Valley – past and present. *Israel Journal of Plant Sciences*, **46**, 155–168.

Pope, J.G. and Macer, C.T. (1996) An evaluation of the stock structure of North sea cod, haddock and whiting since 1920, together with a consideration of the impacts of fisheries and predation effects on their biomass and recruitment. *ICES Journal of Marine Science*, **53**, 1157–1169.

Pope, J.G., Stokes, T.K., Murawski, S.A. and Idoine, J.S. (1988) A comparison of fish size composition in the North Sea and on Georges Bank. In Wolff, W., Soeder, J.C. and Drepper, F.R. (eds) *Eco-dynamics – Contributions to Theoretical Ecology*, pp. 146–152. Berlin: Springer-Verlag.

Por, F.D. (1978) *Lessepsian migration. The influx of Red Sea biota into the Mediterranean by way of the Suez Canal*. Ecological Studies, **23**. Berlin: Springer-Verlag.

Povey, A. and Keough, M.J. (1991) Effects of trampling on plant and animal populations on rocky shores. *Oikos*, **61**, 355–368.

Prins, T.C. and Smaal, A.C. (1990) Benthic–pelagic coupling: the release of inorganic nutrients by an intertidal bed of *Mytilus edulis*. In Barnes, M. and Gibson, R. *Trophic Interactions in the Marine Environment*, pp. 89–103. Aberdeen: Aberdeen University Press.

Probert, P.K. (1984) Disturbance, sediment stability and trophic structure of soft-bottom communities. *Journal of Marine Research*, **42**, 893–921.

Pruter, A.T. (1987) Sources, quantities and distribution of persistent plastics in the marine environment. *Marine Pollution Bulletin*, **18**, 305–310.

Raffaelli, D. and Hall, S.J. (1992) Compartments and predation in an estuarine food web. *Journal of Animal Ecology*, **61**, 551–560.

Ramsay, K., Kaiser, M.J. and Hughes, R.N. (1996) Changes in the hermit crab feeding patterns in response to trawling disturbance. *Marine Ecology Progress Series*, **144**, 63–72.

Reeders, H.H., Devaate, A.B. and Slim, F.J. (1989) The filtration rate of *Dreisena polymorpha* (Bivalvia) in 3 Dutch lakes with reference to biological water-quality management. *Freshwater Biology*, **22**, 133–141.

Reeve, M.R. and Cosper, E. (1972) Acute effects of heated effluents on the copepod *Acartia tonsa* from a sub-tropical bay and some problems of assessment. In Ruivo, M. (ed.) *Marine Pollution and Sea Life*, pp. 250–252. London: FAO and Fishing News Books Ltd.

Reise, K. (1982) Long-term changes in the macro-benthic invertebrate fauna of the Wadden Sea – are polychaetes about to take over? *Netherlands Journal of Sea Research*, **16**, 29–36.

Renberg, I. and Battarbee, R.W. (1990) The SWAP palaeolimnological programme: a synthesis. In Mason, B.J. (ed.) *The Surface Waters Acidification Programme*, pp. 281–300. Cambridge: Cambridge University Press.

Resources Agency of California (1997) *California's Ocean Resources: an Agenda for the Future.* Sacramento: Resources Agency of California.

Revill, A. (1996) The UK (East Coast) brown shrimp fishing fleet. Inventory of vessels and gear and analysis of effort. Supplement to RIVO Report CO54/97.

Revill, A. (1999) The biological and economic consequences of discarding in the North Sea *C. crangon* fisheries. PhD thesis, University of Lincolnshire and Humberside, UK.

Rhoads, D.C. and Young, D.K. (1970) The influence of deposit-feeding organisms on sediment stability and sediment structure. *Journal of Marine Research*, **28**, 150–178.

Rice, J. and Gislason, H. (1996) Patterns of change in the size spectra of numbers and diversity of the North Sea fish assemblage, as reflected in surveys and models. *ICES Journal of Marine Science*, **53**, 1214–1225.

Richardson, P.L. (1983) Gulf Stream rings. In Robinson, A.R. (ed.) *Eddies in Marine Science*, pp. 19–45. New York: Springer-Verlag.

Richardson, W.J. and Wursig, B. (1997) Influences of man-made noise and other human actions on cetacean behaviour. *Marine and Freshwater Behaviour and Physiology*, **29**, 183–209.

Rieley, J.O., Page, S.E. and Shepherd, P.A. (1997) Tropical bogforests of South East Asia. In Parkyn, L., Stoneman, R.E. and Ingram, H.A.P. (eds) *Conserving Peatlands*, pp. 35–41. Wallingford: CAB International.

Riemann, B. and Hoffmann, E. (1991) Ecological consequences of dredging and bottom trawling in the Limford, Denmark. *Marine Ecology Progress Series*, **69**, 171–178.

Rijnsdorp, A.D. and Miller, R.S. (1996) Trends in population dynamics and exploitation of North Sea plaice (*Pleuronectes platessa* L.) since the late 1980s. *ICES Journal of Marine Science*, **53**, 1170–1184.

Rijnsdorp, A.D. and van Leeuwen, P.I. (1996) Changes in growth of North Sea plaice since 1950 in relation to density, eutrophication, beam-trawl effort, and temperature. *ICES Journal of Marine Science*, **53**, 1199–1213.

Roast, S.D., Widdows, J. and Jones, M.B. (1999) Scope for growth of the estuarine mysid *Neomysis integer* (Peracarida: Mysidacea): effects of the organophosphate pesticide chlorpyrifos. *Marine Ecology Progress Series*, **191**, 233–241.

Rogers, S.I. (1997) A review of closed areas in the United Kingdom Exclusive Economic Zone. *CEFAS Technical Report*, **106**.

Rosenthal, H., Weston, D., Gowen, R. and Black, E. (1987) Report of the ad hoc study group on environmental impact of mariculture. ICES CM 1987/F:2.

Ross, A. (1988) *Controlling Nature's Predators on Fish Farms.* Ross-on-Wye: Marine Conservation Society.

Ross, A. (1989) Nuvan use in salmon farming: the antithesis of the precautionary principle. *Marine Pollution Bulletin*, **20**, 372–374.

Ross, A. and Horsman, P.V. (1988) *The Use of NUVAN 500EC in the Salmon Farming Industry.* Ross-on-Wye: Marine Conservation Society.

Rosseland, B.O. and Staurnes, M. (1994) Physiological mechanisms for toxic effects and resistance to acidic water: an ecophysiological and eco-toxicological approach. In Steinberg, C.E.W. and Wright, R.F. (eds) *Acidification of Freshwater Ecosystems; Implications for the Future*, pp. 227–246. Chichester: John Wiley.

Rowe, G.T., Clifford, C.H., Smith, K.L. Jr and Hamilton, P.L. (1975) Benthic nutrient regeneration and its coupling to primary productivity in coastal waters. *Nature*, **255**, 215–217.

Rumohr, H. and Krost, P. (1991) Experimental evidence of damage to benthos by bottom trawling with special reference to *Arctica islandica*. *Meersforschung*, **33**, 340–345.

Rupp, H.R. (1996) Adverse assessments of *Gambusia affinis*: an alternate view for mosquito control practitioners. *Journal of American Mosquito Control*, **12**, 155–159.

Sainsbury, J.C. (1986) *Commercial Fishing Methods: an Introduction to Vessels and Gear*. Oxford: Fishing News Books.

Salomons, W., Bayne, B.L., Duursma, E.K. and Förstner, U. (eds) (1989) *Pollution of the North Sea: an Assessment*. Berlin: Springer-Verlag.

Sattaur, O. (1989) The threat of the well-bred salmon. *New Scientist*, **122**, 54–58.

Savchenko, V.K. (1995) *The Ecology of the Chernobyl Catastrophe*. London: Parthenon.

Schaefer, M.B. (1954) Some aspects of the dynamics of populations important to the management of commercial fisheries. *Bulletin of the Inter-American Tropical Tuna Commission*, **1**, 27–56.

Schindler, D.W., Mills, K.H., Malley, D.F., *et al.* (1985) Long-term ecosystem stress: the effects of years of experimental acidification of a small lake. *Science*, **228**, 1395–1401.

Schloesser, D.W. and Nalepa, T.F. (1994) Dramatic decline of native unionid bivalves in offshore waters of western Lake Erie after infestation by the zebra mussel, *Dreissena polymorpha*. *Canadian Journal of Fisheries and Aquatic Sciences*, **52**, 1449–1461.

Schwinghamer, P., Guigne, J.Y. and Siu, W.C. (1996) Quantifying the impact of trawling on benthic habitat structure using high resolution acoustics and chaos theory. *Canadian Journal of Fisheries and Aquatic Sciences*, **53**, 288–296.

Serchuk, F.M., Kirkegaard, E. and Daan, N. (1996) Status and trends of the major roundfish, flatfish and pelagic fish stocks in the North Sea: thirty year overview. *ICES Journal of Marine Science*, **53**, 1130–1145.

Sherman, K.M. and Coull, B.C. (1980) The response of meiofauna to sediment disturbance. *Journal of Experimental Marine Biology and Ecology*, **46**, 59–71.

Shiganova, T.A. and Bulgakova, Y.V. (2000) Effects of gelatinous plankton on Black Sea and Sea of Azov fish and their food resources. *ICES Journal of Marine Science* **57**, 641–648.

Shonman, D. (1990) Resolving coastal conflicts: environmental Protection Guarantees. In *Environmental Restoration. Science and Strategies for Restoring the Earth*, **99**, pp. 321–327. Selected papers from the Earth Conference, University of California, Berkeley, 1988. Washington, DC: Island Press.

Shushkina, E.A., Musaeva, E.I., Anokhina, L.L. and Lukasheva, T.A. (2000) The role of gelatinous macroplankton, jellyfish aurelia, and ctenophores mnemiopsis and beroe in the planktonic communities of the Black Sea. *Oceanology*, **40**, 809–816.

Siegismund, H.R. and Hylleberg, J. (1987) Dispersal-mediated coexistence of mud snails (Hydrobiidae) in an estuary. *Marine Biology*, **94**, 395–402.

Simon, J.L. and Dauer, D.M. (1977) Re-establishment of a benthic community following natural defaunation. In Coull, B.C. (ed.) *Ecology of Marine Benthos*. Columbia: University of South Carolina Press.

Skilleter, G.A. and Warren, S. (2000) Effects of habitat modification in mangroves on the structure of mollusc and crab assemblages. *J Exp Mar Biol Ecol*, **244**, 107–129.

Smit, C.J. and Visser, G.J.M. (1993) Exploitation of disturbance on shorebirds: a summary of existing knowledge from the Dutch Wadden Sea and Delta area. *Wader Study Group Bulletin*, **68**, 6–19.

Snaddon, C.D. and Davies, B.R. (1998) A preliminary assessment of the effects of a small South African inter-basin water transfer on discharge and invertebrate community structure. *Regulated Rivers: Research and Management*, **14**, 421–441.

Solbe, J.F. de L.G. (1982) Fish farm effluents: a United Kingdom survey. In Alabaster, J.A. (ed.) *Report of the EIFAC Workshop on Fish Farm Effluents, EIFAC Technical Paper* **41**, 29–55.

Sørensen, J. (1978) Capacity for denitrification and reduction of nitrate to ammonia in a coastal marine environment. *Applied Environmental Microbiology*, **35**, 301–305.

Southwood, T.R.E. (1978) *Ecological Methods*. London: Chapman and Hall.

Sparholt, H. (1990) An estimate of the total biomass of fish in the North Sea. *Journal du Conseil International pour l'Exploration de la Mer*, **46**, 200–201.

Spencer, C.N., McLelland, B.R. and Stanford, J.A. (1991) Shrimp stocking, salmon collapse and eagle displacement. Cascading interactions in the food web of a large aquatic system. *BioScience*, **41**, 14–21.

Stewart, T.W., Miner, J.G. and Lowe, R.L. (1999) A field experiment to determine *Dreissena* and predator effects on zoobenthos in a nearshore, rocky habitat of western Lake Erie. *Journal of the North American Benthological Society*, **18**, 488–498.

Stone, J.H., Bahr, L.M. Jr, Day, J.W. and Darnell, R.M. (1982) Ecological effects of urbanization on Lake Pontchartrain, Louisiana, between 1953 and 1978, with implications for management. In Bornkamm, R., Lee, J.A. and Deaward, M.R.D. (eds) *Urban Ecology. The Second European Ecological Symposium*, pp. 243–252. Oxford: Blackwell.

Stoner, J.H., Gee, A.S. and Wade, K.R. (1983) The effects of acidification on the ecology of streams in the Upper Tywi Catchment in West Wales. *Environmental Pollution*, **35**, 125–157.

Strickland, R.M. (1983) *The Fertile Fjord*. Seattle: University of Washington Press.

Strong, D.R. (1992) Are trophic cascades all wet? Differentiation and donor-control in species ecosystems. *Ecology*, **73**, 747–754.

Suarez, G.A., Ramiro, M.E.G. and Tapanes, J.J.D. (1972) Influencia del dimecron en la supervivencia de la langosta *Panulirus argus* en relacion con la circulacion de Agua sobre la Plataforma de Cuba. In Ruivo, M. (ed.) *Marine Pollution and Sea Life*, pp. 238–242. London: FAO and Fishing News Books Ltd.

Swan, S.H., Elkin, E.P. and Fenster, L. (1997) Have sperm densities declined? A reanalysis of global trend data. *Environmental Health Perspectives*, **105**, 1228–1232.

Talling, J.F. and Lemoalle, J. (1998) *Ecological Dynamics of Tropical Inland Waters*. Cambridge: Cambridge University Press.

Thayer, G.W. (ed.) (1992) *Restoring the Nation's Marine Environment*. Maryland Sea Grant Publication, USA.

Thistle, D. (1980) The response of a Harpacticoid copepod community to a small-scale natural disturbance. *Journal of Marine Research*, **38**, 381–395.

Thistle, D. (1981) Natural physical disturbance and communities of marine soft bottoms. *Marine Ecology Progress Series*, **6**, 223–228.

Thorsteinsson, G. (1992) The use of square mesh cod-ends in the Icelandic shrimp (*Pandalus borealis*) fishery. *Fisheries Research*, **13**, 255–266.

Thrush, S.F. (1986) The sublittoral macrobenthic community structure of an Irish Sea lough: effects of decomposing accumulation of seaweed. *Journal of Experimental Marine Biology and Ecology*, **96**, 199–212.

Thrush, S.F. and Roper, D.S. (1988) Merits of macrofaunal colonisation of intertidal mudflats for pollution monitoring: preliminary study. *Journal of Experimental Marine Biology and Ecology*, **116**, 219–233.

Thrush, S.F., Hewitt, J.E., Cummings, V.J. and Dayton, P.K. (1995) The impact of habitat disturbance by scallop dredging on marine benthic communities – what can be predicted from the results of experiments? *Marine Ecology Progress Series*, **129**, 141–150.

Thrush, S.F., Hewitt, J.E., Cummings, V.J., *et al.* (1998) Disturbance of the marine benthic habitat by commercial fishing: impacts at the scale of the fishery. *Ecological Applications*, **8**, 66–879.

Thrush, S.F., Pridmore, R.D., Hewitt, J.E. and Cummings, V.J. (1992) Adult infauna as facilitators of colonization on intertidal sandflats. *Journal of Experimental Marine Biology and Ecology*, **159**, 253–265.

Toppari, J., Larsen, J.C., Christiansen, P., *et al.* (1995) *Male Reproductive Health and Environmental Chemicals with Estrogenic Effects*. Miljøprojekt 290, Copenhagen, Danish Ministry of Energy and Environment.

Townsend, C.R. (1996) Invasion biology and ecological impacts of brown trout *Salmo trutta* in New Zealand. *Biological Conservation*, **78**, 13–22.

Tramier, B., Aston, G.H.R., Durrieu, M., *et al.* (1981) *A Field Guide to Coastal Oil Spill Control and Clean-up Techniques.* Den Haag, *CONCAWE,* **112.**

Trendall, J. (1988) The distribution and dispersal of introduced fish at Thumbi West Island in Lake Malawi, Africa. *Journal of Fish Biology,* **33,** 357–369.

Tuck, I.D., Hall, S.J., Roberston, M.R., Armstrong, E. and Basford, D.J. (1998) Effects of physical trawling disturbance in a previously unfished sheltered Scottish sea loch. *Marine Ecology Progress Series,* **162,** 227–242.

Underwood, A.J. (1993a) Exploitation of species on the rocky coast of New South Wales (Australia) and options for its management. *Ocean and Coastal Management,* **20,** 41–62.

Underwood, A.J. (1993b) The mechanics of spatially replicated sampling programs to detect environmental impacts in a variable world. *Australian Journal of Ecology,* **18,** 99–116.

Unsal, M. (1988) Effects of sewage on the distribution of benthic fauna in Golden Horn (Turkey). *Revue Internationale d'Oceanographie et Medicale,* **91,** 105–124.

Usui, A. (1984) *Eel Culture.* Oxford: Fishing News Books.

Van Beek, F.A., Leeuwen, P.I. and Rijnsdorp, A.D. (1990) On the survival of plaice and sole discards in the otter trawl and beam trawl fisheries in the North Sea. *Netherlands Journal of Sea Research,* **26,** 151–160.

van Breeman, N. (1995) How *Sphagnum* bogs down other plants. *Trends in Ecology and Evolution,* **10,** 270–275.

Van Dessel, J.P. (1991) Marine mammals and seismic research in the Dutch continental shelf, pp. 573–577. 1st SPE Conference on Health, Safety and Environment in Oil and Gas Exploration and Production, 11–14 November 1991, The Hague, Netherlands.

Van Haren, R.J.F. and Kooijman, S. (1993) Application of a dynamic energy budget model to *Mytilus edulis* (L). *Netherlands Journal of Sea Research,* **31,** 119–133.

Van Marlen, B., Redant, F., Polet, H., *et al.* (1998) Research into *Crangon* fisheries unerring effect.

(RESCUE) – EU-Study report 94/044. RIVO Report C054/97.

Van Seters, T.E. and Price, J.S. (2001) The impact of peat harvesting and natural regeneration on the water balance of an abandoned cutover bog, Quebec. *Hydrological Processes,* **15,** 233–248.

Vanderpoorten, A. and Klein, J-P. (1999) A comparative [sic] study of the hydrophyte flora from the Alpine Rhine to the Middle Rhine. Application to the conservation of the upper Rhine aquatic ecosystems. *Biological Conservation,* **87,** 163–172.

Vickery, J. (1995) Access. In Sutherland, W.J. and Hill, D.A. (eds) *Managing Habitats for Conservation,* pp. 42–58. Cambridge: Cambridge University Press.

von Bertalanffy, L. (1957) Quantitative laws in metabolism and growth. *Quarterly Review of Biology,* **32,** 217–221.

von Brandt, A. (1984) *Fish Catching Methods of the World,* 3rd edition. Farnham: Fishing News Books.

Ward, J.V. (1998) Riverine landscape: biodiversity patterns, disturbance regimes, and aquatic conservation. *Biological Conservation,* **83,** 269–278.

Water Research Centre (1990) *Guide to Marine Treatment of Sewage.* Medmanham, Water Research Centre.

Weatherley, N.S., Ormerod, S.J., Thomas, S.P. and Edwards, R.W. (1988) The response of macroinvertebrates to experimental episodes of low pH with different forms of aluminium during a natural spate. *Hydrobiologia,* **169,** 225–232.

Weston, D.P. (1986) The environmental effects of floating mariculture in Puget Sound. Report to the Washington Department of Fisheries and the Washington Department of Ecology, from the School of Oceanography, University of Washington, Seattle.

Wheeler, A. (1977) The origin and distribution of the freshwater fishes of the British Isles. *Journal of Biogeography,* **4,** 1–24.

Wheeler, A. (2000) Status of the crucian carp, *Carassius carasius* (L.), in the UK. *Fisheries Management and Ecology,* **7,** 315–322.

White, H.H. and Champ, M.A. (1983) The great bioassay hoax and alternatives. In Conway, R.A. and Gulledge, W.P. (eds) *Symposium on Hazardous and Industrial Solid Waste Testing,* pp. 200–312. American Society for Testing and Materials.

Widdows, J. and Page, D.S. (1993) Effects of tributyltin and dibutyltin on the physiological energetics of the Mussel, *Mytilus edulis*. *Marine Environmental Research*, **35**, 233–249.

Widdows, J., Nasci, C. and Fossato, V.U. (1997) Effects of pollution on the scope for growth of mussels (*Mytilus galloprovincialis*) from the Venice Lagoon, Italy. *Marine Environmental Research*, **43**, 69–79.

Wigley, T.M.L. and Jones, P.D. (1987) England and Wales precipitation: a discussion of recent changes in variability and an update to 1985. *Journal of Climatology*, **7**, 231–246.

Williams, A.T. and Simmons, S.L. (1999) Sources of riverine litter: the River Taff, South Wales, UK. *Water, Air and Soil Pollution*, **112**, 197–216.

Wilson, W.H. (1986) Importance of predatory infauna in marine soft-bottom communities. *Marine Ecology Progress Series*, **32**, 35–40.

Wisniewski, P. (1993) Martin Mere. *Biologist*, **40**, 194–196.

Worrall, P., Peberdy, K.J. and Millett, M.C. (1997) Constructed wetlands and nature conservation. *Water Science and Technology*, **35**, 205–213.

Wright, K.K. and Li, J.L. (1998) Effects of recreational activities on the distribution of *Dicosmoecus gilvipes* in a mountain stream. *Journal of the North American Benthological Society*, **17**, 535–543.

Wynberg, R.P. and Branch, G.M. (1991) An assessment of bait-collecting for *Callianassa kraussi* Stebbing in Langebaan Laggon, Western Cape, and of associated avian predation. *South African Journal of Marine Science*, **11**, 141–152.

Wynberg, R.P. and Branch, G.M. (1994) Disturbance associated with bait-collection for sand prawns (*Callianassa kraussi*) and mud prawns (*Upogebia africana*): long-term effects on the biota of intertidal mudflats. *Journal of Marine Research*, **52**, 523–558.

Yodzis, P. (1989) *Introduction to Theoretical Ecology*. New York: Harper and Row.

Young, G.J., Dooge, J.C.I. and Rodda, J.C. (1994) *Global Water Resource Issues*. Cambridge: Cambridge University Press.

Zachariassen, K. and Jakupsstovu, S.H. (1998) Grid sorting in a trawl for lemon sole. *ICES Fishing Technology and Fish Behaviour Working Group Meeting* (Working paper). La Coruna, Spain.

Glossary

Abiotic – referring to the non-living component of the environment; e.g. climate, temperature, pH (cf. **Biotic**).

Acidification – increase in the acidity of water. Normally refers to anthropogenic activity.

Acidity – the concentration of alkalinity-buffering ions in water. Normally refers to the concentration of hydrogen ions, expressed as pH.

Advect – to move laterally within a water column.

Agenda 21 – a comprehensive plan of action to be taken globally, nationally and locally by organisations of the United Nations system, governments, and major groups in every area in which humanity impacts on the environment. It derived from the United Nations Conference on Environment and Development, held in Rio de Janeiro, Brazil, in June 1992.

Alga (plural **algae**) – photosynthetic organism in the phylum Protoctista. Includes unicellular forms (e.g. diatoms) and multicellular forms (e.g. seaweeds).

Alkalinity – the concentration of acidity-buffering ions in water.

Allochthonous – originating from outside a given site; e.g. allochthonous detritus (cf. **Autochthonous**).

Alluvium – sediment deposited by a river.

Anadromous – migrating from the sea to fresh water to breed.

Anoxic – deprived of oxygen.

Antagonistic – with respect to pollution: the term describing the process by which one pollutant decreases the detrimental effect of another.

Anthropogenic – caused by human activity.

Antifouling – see **Fouling**.

Aquaculture – raising of aquatic organisms in culture. Includes **Mariculture**.

Artisanal fishing – small-scale fishing using traditional methods.

Assimilation – the process of incorporating ingested items (food) into body tissues.

Attenuation – reduction in intensity of light by absorption or reflection as it passes through water.

Autochthonous – originating from within a given site; e.g. autochthonous primary production (cf. **Allochthonous**).

Autotroph (primary producer) – an organism that obtains energy from inorganic substances through primary production (cf. **Heterotroph**).

Autotrophy – primary production.

Bankfull discharge – the discharge which fills a river to its maximum capacity, without flooding beyond the perimeter of the channel.

Benthic – referring to organisms living on or in the bed of a water body.

Benthos – collective term for benthic organisms.

Bioaccumulation – incorporation of a contaminant into the body tissues of an organism.

Biochemical oxygen demand – the amount of oxygen consumed by a sample of water. Oxygen consumption is a product of microbial respiration (**Biological oxygen demand**) and chemical oxidation of contaminants (**Chemical oxygen demand**).

Biodegradable – referring to materials that break down in the environment as a result of microbial activity.

Biodiversity – see **Biological diversity**.

Biofilm – the layer of bacteria, algae, fine particulate organic matter and extracellular bacterial secretions which develops on solid surfaces in water. Occasionally referred to as 'aufwuchs'.

Biological diversity – the variability among living organisms from all sources, including diversity within species, between species and of ecosystems. See **Section 1.6**.

Biological oxygen demand – see **Biochemical oxygen demand**.

Biomagnification – increase in concentration of a contaminant in the body tissues of organisms as it passes up the food chain.

Biomass – the mass of living organisms.

Biospecies – a group of separate species within a community that are assumed to feed on the same range of prey and in turn are preyed on by the same suite of predators.

Biota – the living organisms within an area.

Biotic – referring to living components of the environment, or to products derived from living components; e.g. detritus (cf. **Abiotic**).

Biotic Index – a water quality scoring system based upon the presence of living organisms.

Bioturbation – disturbance caused by biological activity.

Bloom – a rapid increase in the population of algae in response to a sudden input of nutrients.

BOD – Biochemical oxygen demand (also sometimes used as an abbrevation for **Biological oxygen demand**).

Buffering capacity – the ability of a substance to resist change in pH caused by addition of an acid or alkali.

Bycatch – animals accidentally caught during harvesting. Includes the **discard fraction**.

Calefaction – artificially induced change in water temperature.

Carrying capacity – the maximum population of a species that can be supported by the available resources.

Catadromous (catadromy) – migration from fresh water to the sea to breed.

Catch quota – the maximum total catch of a harvestable species allowed for a given unit exploiting a wild population within a given season. The 'unit' may be an individual person, boat or even country.

Catchment – the land surface that drains into a given river.

Channelisation – engineering a channel to make it straighter and more homogeneous in structure than the natural channel.

Charismatic species – species that attract much human interest and sympathy, normally because of their large size or visual beauty.

Chemical oxygen demand – see **Biochemical oxygen demand**.

Chemoautotrophy – see **Primary production**.

Chemosynthesis – see **Primary production**.

Climax – a term referring to the hypothetical final stable state of a community at the end of a successional process.

Closed area – an area within a water body where fishing activity is totally prohibited or restricted.

COD – Chemical oxygen demand.

Community – the assemblage of interacting living organisms within a location or habitat.

Compensation depth (Compensation point) – the depth at which energy produced by aquatic algae through photosynthesis exactly matches their requirements for respiration.

Conductivity – a measure of salinity of water, through its ability to conduct electricity. Can be used to determine concentration of solutes in water.

Connectance – the actual number of feeding links in a food web as a proportion of the theoretical maximum (see **Section 1.5d**).

Contaminant – the substance added that causes contamination.

Contamination – the introduction, directly or indirectly, of substances or energy into the environment such that levels are altered from those that would have existed without human activity.

Creel – a traditional term describing various designs of shellfish trap, e.g. lobster pot.

Critical pathway – the series of steps that brings a pollutant from its source to its point of biological impact.

CSO – combined sewage overflows: mechanisms that allow sewage pipes to overflow into rivers, estuaries and the sea in the event of major flooding or heavy rain.

Cyanobacteria – 'blue-green algae': photosynthetic bacteria formerly known as Cyanophyta and classed as algae.

DDE – dichloro-diphenyl ethane; a breakdown product of DDT.

DDT – dichloro-diphenyl trichloroethane; a persistent organochlorine compound widely used as an insecticide.

Demersal – living adjacent to, and in association with, the sea floor, e.g. demersal fish.

Density dependent – referring to a process whose influence on a population is determined by the total population size.

Density independent – referring to a process whose influence is not affected by the total population size.

Deposition zone – the zone within which a river deposits its sediment load.

Detritus – particulate organic matter derived from formerly living organisms.

Diffuse source – referring to input of a pollutant over a large area.

Discard fraction – that part of the catch not wanted due to its size or species composition.

Discharge – (a) with reference to river flow, the total volume of water moving past a given point; (b) with reference to pollution and contamination, the input of a contaminant into a water body.

Disturbance – a discrete event which removes, damages or impairs the normal function of organisms.

Diversity – see **Biological diversity**.

DOC – dissolved organic carbon.

DOM – dissolved organic matter.

Drawdown – the reduction in volume of water and consequent exposure of lake bed caused by extraction from a reservoir.

Drift – passive movement with the current of a water body. See **Section 2.3d**.

Ecosystem – the biotic components of the environment along with the abiotic components with which they interact.

Ecotourism – tourism whose aim is to observe wildlife or experience 'natural' environments.

Endocrine disrupter (ED) – a toxin that disrupts hormone-mediated processes.

Epibenthic – attached to or living on the bed of a water body.

Epifauna – organisms living or active on the surface of a water body.

Epilimnion – the surface layer of a stratified water body, above the thermocline. A limnological term equivalent to the mixed layer in marine systems.

Epilithic – attached to stones.

Epiphyte – an organism growing on a plant.

Epiphytic – attached to plants.

Erosion zone – the zone within a river channel in which net movement of sediment is from the bed into the water column.

Euphotic zone – see **Photic zone**.

Euryhaline – tolerant of a wide range of salinities.

Eutrophic – referring to water containing high concentrations of nutrients, relative to an oligotrophic or mesotrophic water body.

Eutrophication – the increase in nutrient concentrations in a water body, normally used to refer to the effects of anthropogenic pollution.

Evaporite – the solid residue remaining when water evaporates. It comprises salts formerly in solution.

Exploitation – the use by humans of water or its products for their own benefit.

FAO – United Nations Food and Agriculture Organisation.

Fecundity – the number of offspring produced per individual within a population.

Fish weir – a traditional name for certain designs of fish trap.

Flood refugium – an area in which river organisms are able to persist during floods, normally because flow rates are reduced relative to the main water column.

Floodplain – the area flooded by a river with a predictable, normally seasonal frequency.

Food chain – a linear representation of feeding interactions, incorporating a single species (or biospecies) at each trophic level.

Food web – a representation of feeding interactions within a community.

Fouling – the attachment of organisms onto solid surfaces, leading to an impairment in function of the structure affected. Treated by physical removal or application of **antifouling** agents – pesticides that kill the biofouling agent.

Ghost fishing – the continued operation of fishing gear lost or discarded by fishers.

Groundwater – water in sediments beneath the Earth's surface.

Guild – a group of species within a community that exploit the same resource in similar ways.

Harvesting – removal of organisms for human use.

Headwaters – stream channels within a river basin close to their source.

Heterotroph – an organism that obtains its energy from eating other living organisms or their by-products (cf. **Autotroph**).

Holdfast – the adhesive pad on a multicellular marine alga ('seaweed') by which it attaches itself to a solid substrate.

Holoplankton – see **Plankton**.

Hydraulic head – difference in potential energy between two bodies of water, e.g. the difference in height between the water bodies above and below a dam.

Hypertrophic – extremely eutrophic.

Hypolimnion – the deep layer of a stratified lake, below the thermocline. A limnological term equivalent to the deep ocean in marine systems (see **Box 1.3**).

Hyporheic – referring to the sediments and interstitial water beneath a water body (**Section 2.3e**).

Hyporheos – organisms inhabiting the hyporheic zone.

IBM – integrated basin management.

ICM – (a) integrated coastal management; (b) integrated catchment management. To avoid confusion with integrated coastal management, IBM is preferred when referring to a river catchment.

ICZM – integrated coastal zone management.

IMO – International Maritime Organisation. An international governmental organisation responsible primarily for issues relating to shipping, including pollution from ships, ballast water and ship safety.

Imposex – the condition whereby females develop male sexual features in addition to their normal female reproductive structures. Most often seen in neogastropods such as dog whelks and whelks.

Infauna – benthos living within the bed sediments, rather than on its surface.

Interbasin transfer (IBT) – artificial transfer of water from one river basin to another.

Interspecific competition – competition occurring between individuals of different species.

Intertidal – referring to the part of the coastal zone that is periodically exposed and inundated by tidal movements.

Intraspecific competition – competition occurring between individuals within the same species.

IRBM – integrated river basin management.

Irrigation – artificial application of water onto agricultural land to enhance growth of crops.

Keystone species – a species whose role is critical to the maintenance of a community.

Lacustrine – referring to lakes.

Liming – the addition of calcium carbonate to a water body to raise its pH.

Linkage density – the mean number of feeding links per species within a food web.

Littoral zone – (a) in lakes, that part of the water body in which the photic zone extends to the bed (also referred to as the **lake littoral**); (b) in the sea, the intertidal zone.

Long line – a fishing device using a hook (often baited) on the end of a line.

Macrofauna (macroinvertebrates) – animals retained on a 0.5 mm (500 μm) sieve.

Macrophyte – a large multicellular photosynthesing organism; generally applied to vascular plants but also refers to multicellular algae.

Management – mechanisms whereby the exploitation of aquatic systems is controlled or regulated.

Mariculture – aquaculture of marine organisms.

MARPOL – Marine Pollution Annex to **UNCLOS**.

Maximum sustainable yield (MSY) – the greatest harvest of a wild living resource (e.g. fish) which can be taken each year while still leaving a viable population to harvest the following year.

Mesotrophic – referring to water containing a concentration of nutrients intermediate between that of a eutrophic and an oligotrophic water body.

Mire – a wetland whose water supply is independent of surface water bodies.

Mixed layer – see **Epilimnion**.

Monitoring – carrying out periodic measurements of parameters.

Nekton – organisms living within the water column which are able to swim or otherwise move independently of the current (see **Section 1.6a**).

Niche – the place of a species within a community; the sum of its interactions with other species and with components of the abiotic environment.

Nursery ground – the area within which a marine species deposits its eggs or young offspring.

Nutrient cycle – the pathway followed by a nutrient as it changes from an inorganic to an organic form and back.

Nutrient spiral – a condition in which the nutrient cycle is completed only when the nutrient has moved to a different geographical location (see **Section 2.7a**).

Oceanography – the study of marine systems.

Oligotrophic – referring to water containing low concentrations of nutrients, relative to a eutrophic or mesotrophic water body.

Omnivory – consumption of food from two or more trophic levels.

Over-harvesting – harvesting at a higher rate than the target species can replenish its numbers.

Oxycline – a region of rapid change in oxygen concentration.

PCB – poly-chlorinated biphenyl; a persistent substance widely used in electrical components.

Peat – partially decayed and compressed plant material that aggregates in wetlands in the absence of aerobic decomposition.

Peatland – an area dominated by peat.

Pelagic – referring to the water column, as opposed to its bed or edges.

Persistent – with respect to a pollutant, remaining in the environment in its polluting state for long periods.

PFA – pulverised fuel ash: the predominantly silty fuel residue that remains following combustion of coal.

pH – see **Acidity**.

Photic zone (Euphotic Zone) – depth of water to which sunlight penetrates. The bottom of the photic zone is defined as the depth at which only 1% of light intensity at the surface remains.

Photosynthesis – see **Primary production**.

Phytoplankton – see **Plankton**.

Plankter – an individual member of the plankton.

Plankton (planktonic) – organisms inhabiting the pelagic zone which have no independent means of propulsion, or are too small and weak to swim in the horizontal plane. **Holoplankton** – organisms which spend their entire lives as members of the plankton. **Meroplankton** – organisms which spend only part of their life cycle as members of the plankton. **Phytoplankton** – the photosynthesising component of the plankton, mainly single-celled algae but including autotrophic bacteria. **Potamoplankton** – phytoplankton species which are endemic to river systems. **Zooplankton** – animal components of the plankton.

POC – particulate organic carbon (see **Section 1.4c**).

Point source – a single location from which a pollutant derives.

Pollutant – the substance that causes pollution.

Pollution – the introduction of substances or energy into the environment, resulting in deleterious effects to humans, human activities or other living components of the environment.

Polyculture – aquaculture of two or more species simultaneously.

Precautionary principle – a strategy that requires a potential discharger to demonstrate no environmental effect before a discharge is licensed.

Precipitation – (a) water that falls from the atmosphere to the surface of land or sea, including rain, snow, etc.; (b) the process by which a substance dissolved in water separates out of solution and becomes a solid.

Primary production (autotrophy) – production of organic compounds from inorganic components, using energy fixed from an external source. **Photosynthesis** – primary production in which energy is obtained from sunlight. **Chemosynthesis (chemoautotrophy)** – primary production in which energy is derived from chemical oxidation of simple inorganic compounds.

Profundal zone – the zone in a lake beneath the photic zone.

Quota – see **Catch quota**.

Ranching – management or enhancement of a population of a harvested species within the natural environment.

Recruitment – addition of individuals to a population through reproduction.

Redox – an abbreviation for oxidation–reduction, a chemical reaction in which an atom or ion loses electrons to another atom or ion.

Redox discontinuity layer (RDL) – the layer at which a sediment changes from an oxygenated to an anoxic state.

Refractory – not readily assimilated and therefore slow to decompose.

Resources – the components of a water body that are exploited. The term is normally applied to components that can diminish through over-use, such as stocks of fish.

Runoff – rainfall which is not absorbed by soil, but passes into surface water bodies.

Salinisation – increase in concentration of salts in water or soil.

Salinity – the concentration of dissolved ions in water.

Scope for growth – see **Box 9.1**.

Selectivity – the degree to which a harvesting device or technique captures only the target species.

Sewage – human and domestic waste dissolved and/or suspended in water.

Shear stress – the effect of a series of forces of different magnitude acting simultaneously at different parts of an object.

Size–frequency relationship – the distribution of individuals by size in a population. This may be represented as a size-frequency graph or by summary statistics, such as mean or modal size.

Socio-economic – pertaining to human society and cultural interactions.

Solute – a substance dissolved in another substance.

Stochastic – referring to events that do not conform to a regular pattern and therefore, although not strictly unpredictable from a human perspective, are effectively so from the perspective of the organisms affected.

Stock – the population of a harvested species.

Stratification – development of discrete vertical layers within a water body.

Stress – a state in which the normal physiological functioning of an organism is impaired.

Succession – directional change through time at a site by means of colonisation and local extinction (NB succession is a controversial process with several, often contradictory, definitions in ecology).

Sustainability – the management of resources in a way that does not deplete them and therefore ensures their continuation.

Synergistic – with respect to pollution: the term describing the process by which one pollutant enhances the detrimental effect of another.

Synoptic – referring to the examination of multiple influences.

Target species – the species which the harvesting effort is aimed at catching.

Taxon (plural **taxa**) – organisms classed together within the same taxonomic group (differs from species in that the taxonomic group may be genus, family, order, etc.).

TBT – tributyl tin; a toxic organotin compound used as an antifouling pesticide.

Thermocline – zone between the epilimnion (or mixed layer) and hypolimnion (or deep ocean) within which the temperature changes abruptly.

Toxin – a substance that has a detrimental influence on physiological processes.

Translocation – movement and release of individuals of a species beyond its native range.

Trawl – a bag of netting, widest at its mouth, that is pulled through the water column.

Trophic level – an identifiable feeding level within a food web.

Turbidity – the concentration of suspended particles in water.

UNCLOS – United Nations Convention on the Law of the Sea.

Upwelling – the appearance at the surface of a water mass previously within the depths of the water body.

Virtual population analysis – a method of estimating the population (stock) of a harvested species using catch data alone.

Water mass – a body of water characterised by certain values of its conservative properties (i.e. those not altered except by mixing), e.g. salinity, temperature.

WFD – European Commission's Water Framework Directive.

Zooplankton – see **Plankton**.

Organism Index

Subject Index